FAIR Next Generation ScientistS

(FAIRNESS 2014)

Journal of Physics: Conference Series Volume 599

Vietri sul Mare, Italy
22 – 27 September 2014

ISBN: 978-1-5108-0293-3
ISSN: 1742-6588

Printed from e-media with permission by:

Curran Associates, Inc.
57 Morehouse Lane
Red Hook, NY 12571

Printed by Curran Associates, Inc. (2015)

For permission requests, please contact the Institute of Physics
at the address below.

Institute of Physics
Dirac House, Temple Back
Bristol BS1 6BE UK

Phone: 44 1 17 929 7481
Fax: 44 1 17 920 0979

techtracking@iop.org

Additional copies of this publication are available from:

Curran Associates, Inc.
57 Morehouse Lane
Red Hook, NY 12571 USA
Phone: 845-758-0400
Fax: 845-758-2634
Email: curran@proceedings.com
Web: www.proceedings.com

TABLE OF CONTENTS

FAIRNESS 2014 was the third edition in a series of workshops designed to bring together excellent international young scientists with research interests focused on physics at FAIR (Facility for Antiproton and Ion Research) and was held on September 22-27 2014 in Vietri sul Mare, Italy. The topics of the workshops cover a wide range of aspects in both theoretical developments and current experimental status, concentrated around the four scientific pillars of FAIR. FAIR is a new accelerator complex with brand new experimental facilities, that is currently being built next to the existing GSI Helmholtzzentrum for Schwerionenforschung close to Darmstadt, Germany.

The spirit of the conference is to bring together young scientists, e.g. advanced PhD students and postdocs and young researchers without permanent position to present their work, to foster active informal discussions and build up of networks. Every participant in the meeting with the exception of the organizers gives an oral presentation, and all sessions are followed by an hour long discussion period. During the talks, questions are anonymously collected in a box to stimulate discussions. The broad physics program at FAIR is reflected in the wide range of topics covered by the workshop:

- Physics of hot and dense nuclear matter, QCD phase transitions and critical point
- Nuclear structure, astrophysics and reactions
- Hadron Spectroscopy, Hadrons in matter and Hypernuclei
- New developments in atomic and plasma physics
- Special emphasis is put on the experiments CBM, HADES, PANDA, NUSTAR, APPA and related experiments

For each of these different areas one invited speaker was selected to give a longer introductory presentation. The write-ups of the talks presented at FAIRNESS 2014 are the content of this issue of Journal of Physics: Conference Series and have been refereed according to the IOP standard for peer review. This issue constitutes therefore a collection of the forefront of research that is dedicated to the physics at FAIR.

February 2015,

Organizers of FAIRNESS 2014: Marco Destefanis, Tetyana Galatyuk, Fernando Montes, Diana Nicmorus, Hannah Petersen, Claudia Ratti, Laura Tolos, and Sascha Vogel.

Support for holding the conference was provided by:

HIC for FAIR
Helmholtz International Center

EMMI

HGS-HIRe for FAIR
Helmholtz Graduate School for Hadron and Ion Research

FAIR

HELMHOLTZ ASSOCIATION

Journal of Physics: Conference Series **599** (2015) 011002 doi:10.1088/1742-6596/599/1/011002

Peer review statement

All papers published in this volume of *Journal of Physics: Conference Series* have been peer reviewed through processes administered by the proceedings Editors. Reviews were conducted by expert referees to the professional and scientific standards expected of a proceedings journal published by IOP Publishing.

FAIRNESS 2014 IOP Publishing
Journal of Physics: Conference Series **599** (2015) 012001 doi:10.1088/1742-6596/599/1/012001

The $\overline{\text{P}}$ANDA Experiment

Ralf Kliemt (for the $\overline{\text{P}}$ANDA Collaboration)
GSI Helmholtzzentrum für Schwerionenforschung GmbH, Planckstraße 1, D-64291 Darmstadt, Germany
E-mail: r.kliemt@gsi.de

Abstract. The PANDA (anti-Proton ANnihiliation at DArmstadt) experiment will be a multi-purpose apparatus at the future Facility for Antiproton and Ion Research (FAIR) at Darmstadt. Anti-proton induced reactions with 1.5 to 15 GeV/c beam momentum at high luminosities of up to $2{\cdot}10^{32}/(\text{s}{\cdot}\text{cm}^2)$ will be investigated. Exclusive detection of whole events with almost 4π acceptance and high precision are needed for the broad physics program. The focus lies on studying the strong interaction in the charm region, by charmonium, open-charm and baryon spectroscopy, and includes the search for glueballs, hybrids and other exotics, hypernuclear physics, nucleon structure studies as well as in-medium modifications of hadrons.

1. PANDA Physics

$\overline{\text{P}}$ANDA is an experiment with precise antiproton beams at the Facility for Antiproton and Ion Research, FAIR [1], which is being built at the Helmholtz-Center for Heavy Ion Research [2] in Darmstadt, Germany. It is focussed on hadron physics in the charm region, addressing a broad physics program [4]. Its key aspects are:

- Hadron Spectroscopy
- Nucleon Structure
- Hadrons in Nuclear Medium
- Hypernuclear Physics

Hadron Spectroscopy The main focus of the hadron spectroscopy program lies on the spectra of charmonium ($c\bar{c}$) states, open charm mesons carrying one charm quark as well as heavy baryons containing charm and strange quarks. Furthermore the search for exotic states, such as gluonic excitations, glueballs, hybrids, etc. is of interest because such states are predicted but not identified, yet. Figure 1 illustrates how the beam energies cover the region of interest, up to the creation of Ω_c pairs. Due to the nature of antiproton-proton reactions the initial state is not limited to the photon quantum numbers as in the case of electron-positron collisions. This opens the opportunity to measure states in the whole spectrum directly.

One special feature at $\overline{\text{P}}$ANDA is the beam energy scan technique. The measured reaction rates at several energies around the resonance reveal the line shape with great precision. This technique has proven its value at other experiments, e.g. for the measurement of the χ_1 resonance at E835/Fermilab [3]. Besides resonance masses and widths, also certain threshold behaviours are of particular interest.

Journal of Physics: Conference Series **599** (2015) 012001 doi:10.1088/1742-6596/599/1/012001

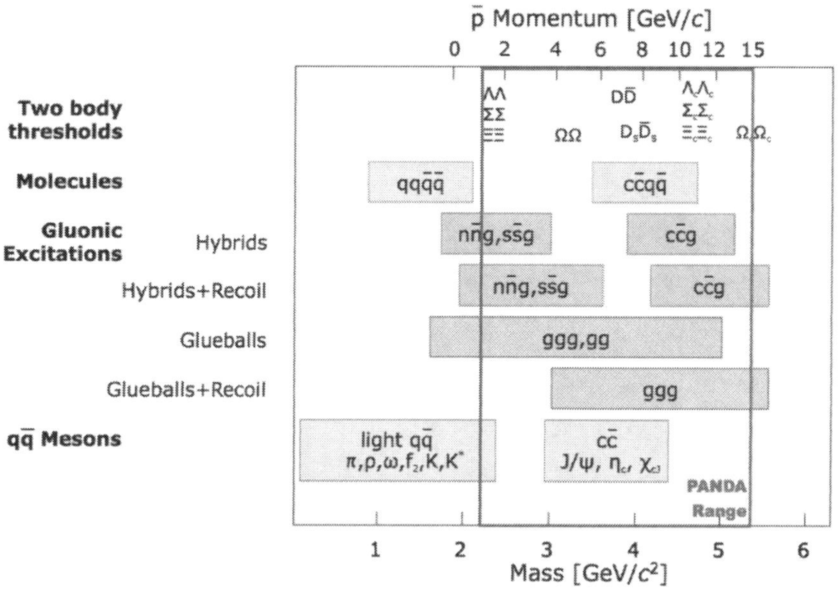

Figure 1: Range of the available energy at the storage ring HESR.

Nucleon Structure Through the electromagnetic annihilation of antiprotons and protons into a lepton pair, $\overline{\text{P}}$ANDA will contribute to the measurement of the time-like electromagnetic form factors with high precision. The nucleon structure can be accessed as well with the means of Transition Distribution Amplitudes, transverse momentum dependent Parton Distribution Functions or Generalised Distribution amplitudes, which can be accessed in electromagnetic channels.

Hadrons in Nuclear Medium By employing a heavy nuclear target, it is possible to study mesons (e.g. charmed mesons) inside the nuclear medium. Mass shifts and width changes are predicted for hadrons in matter. Depending on the applied model not only the mass shift magnitude but sometimes even its direction differs.

Hypernuclear Physics Double strange hyperons, produced by the antiproton beam on a nuclear target are captured in a secondary target, replacing two nucleons with single strange baryons. Gamma spectroscopy will be performed with a dedicated detector upgrade.

2. Experimental Setup

The $\overline{\text{P}}$ANDA apparatus is composed of two parts (see Figure 2), a target spectrometer surrounding the interaction region in a barrel shape and a forward spectrometer up to smaller polar angles. A superconducting solenoid and a dipole magnet provide 2T and 1T, respectively [5] for the momentum measurement of charged particles. Antiprotons of 1.5 to 15 GeV/c interact with a hydrogen or nuclear target in the interaction region. Charged particles are detected in the tracking systems, a Micro Vertex Detector (MVD), a Straw Tube Tracker (STT), Gas Electron Multiplier (GEM) disks as well as Forward Straw Tube Stations (FTS). Electrons and photons are detected by the Electromagnetic Calorimeter (EMC) and particle identification is performed by Detection of Internally Reflected Cherenkov light (DIRC) detectors as well as a time of flight system and muon counters.

Journal of Physics: Conference Series **599** (2015) 012001 doi:10.1088/1742-6596/599/1/012001

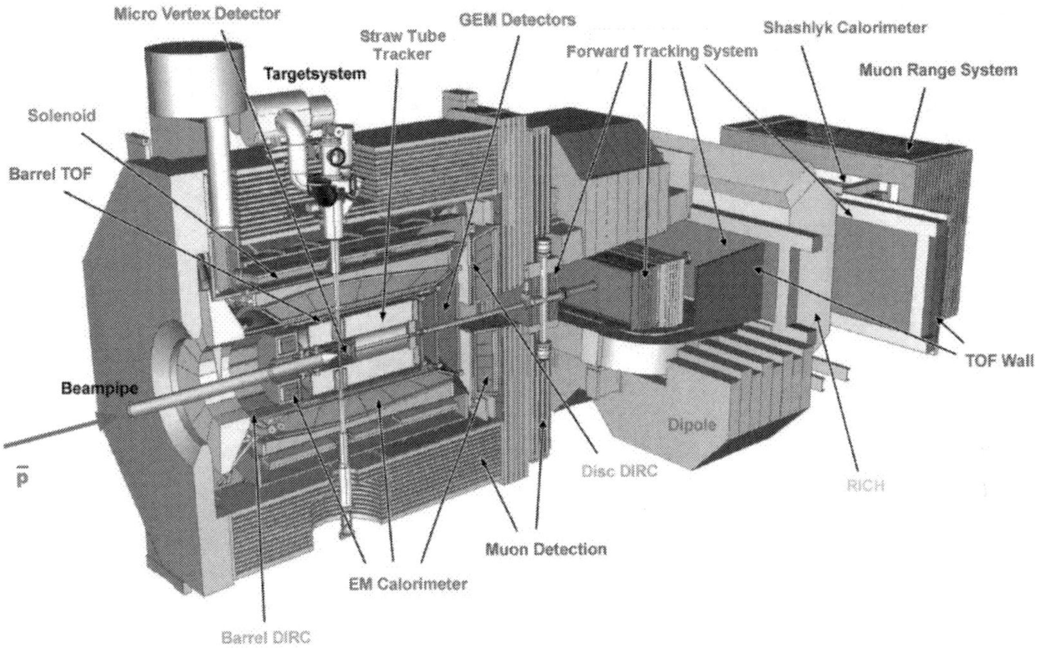

Figure 2: The $\overline{\text{P}}$ANDA detector. The beam enters from the left into the Target Spectrometer, where it reacts with the target and the produced particles with higher transverse momentum are measured.

Target Falling from the top of the detector through a pipe, the target material reacts with the beam of antiprotons. Two options are considered for the hydrogen target, a gaseous cluster jet and a stream of frozen pellets. Both options aim to provide the design luminosity of $10^{32}\text{cm}^{-2}\text{s}^{-1}$. The cluster jet will provide a rather homogeneous density distribution, whereas the advantage of pellets is a strong localisation of the interaction region.

MVD The MVD [6] consists of two barrel and four disk layers with silicon pixel detectors ($100\times100\mu\text{m}^2$ pixel size) as well as two barrel and two disk layers of double sided silicon strip detectors (130µm and 65µm strip pitch). Precise measurements as close as possible to the interaction point provide a high resolution of about 100µm for primary and secondary decay vertices. The front-end electronics will run self-triggered to measure efficiently at the high interaction rates.

STT & FTS There are two straw tube tracking systems, the barrel-shaped STT [7] and the FTS stations in the forward part. The tubes are made from an aluminum-mylar laminate with an argon/carbon dioxide gas mixture inside and a central wire. Glued together, they form self-supporting double layers which are arranged straight along the general magnetic field lines as well as tilted by an stereo angle for polar angle measurements.

EMC Photons and electrons are measured by their electromagnetic showers inside the more than 15000 PWO crystals of the EMC [8], surrounding the interaction region in a barrel shape with end-caps in both, forward and backward directions for a optimal phase space coverage.

Large Area Avalanche Photo Diodes (LAAPD) are foreseen for fast readout inside the strong magnetic field.

PID Charged particles are identified by multiple techniques in the various subsystems. The DIRC detectors, will separate pions, protons and kaons of momenta up to 3.5 GeV/c. Energy loss measurements by the tracking detectors as well as time-of-flight information will be used to improve the PID, especially at the lowest momenta. Resistive plate chambers embedded inside the iron yoke separate muons from pions while the EMC primarily identifies electrons.

DAQ Each subsystem will run in a self-triggered mode, sending data whenever there is a signal. This leads to a large amount of data that will be reconstructed online for filtering. It is expected to deal with a data rate of 200 GB/s. Typical signal signatures contain J/Ψ mesons, $D\overline{D}$ pairs, light hadrons as well as photons which have to be distinguished from the, by a factor of 10^5-10^7 dominant, background channels. Online filtering is pursued by performing a partial physics analysis with the available reconstructed data.

3. Perspectives

Basic Research and development on most components is finished and first component constructions have started. Until the civil construction on site is available, parts of the detector will be assembled at the research center in Jülich. Installation and commissioning at the accelerator is foreseen for 2018 and the start of data taking one year later.

References

[1] http://ww.fair-center.de
[2] http://www.gsi.de
[3] Andreotti M, Bagnasco S, Baldini W, Bettoni D, Borreani G, Buzzo A, Calabrese R and Cester R *et al.*, 2005, "Measurement of the resonance parameters of the $\chi_1(1^3P_1)$ and $\chi_2(1^3P_2)$ states of charmonium formed in $p\bar{p}$ annihilations," *Nucl. Phys.* B **717** 34 [hep-ex/0503022].
[4] The PANDA Collaboration, "Physics Performance Report for PANDA: Strong Interaction Studies with Antiprotons," arXiv:0903.3905 [hep-ex].
[5] The PANDA Collaboration, "Technical Design Report for the: PANDA Solenoid and Dipole Spectrometer Magnets" - arXiv:0907.0169 [physics.ins-det]
[6] The PANDA Collaboration, "Technical Design Report for the: PANDA Micro Vertex Detector" - arXiv:1207.6581 [physics.ins-det]
[7] The PANDA Collaboration, "Technical Design Report for the: PANDA Straw Tube Tracker" - arXiv:1205.5441 [physics.ins-det]
[8] The PANDA Collaboration, "Technical Design Report for PANDA Electromagnetic Calorimeter (EMC)" - arXiv:0810.1216 [physics.ins-det]

Journal of Physics: Conference Series **599** (2015) 012002 doi:10.1088/1742-6596/599/1/012002

Feasibility studies for the Forward Spectrometer

Jacek Biernat[1]

[1]Jagiellonian University

For the $\bar{P}$ANDA collaboration

E-mail: jacek.b.biernat@gmail.com

Abstract. The Forward Spectrometer designed for the $\bar{P}$ANDA detector will consist of many different detector systems allowing for precise track reconstruction and particle identification. Feasibility studies for Forward Spectrometer done by means of specific reactions will be presented. In the first part of the paper, results of simulations focussing on rate estimates of the tracking stations based on straw tubes will be presented. Next, the importance of the Forward Tracker will be demonstrated through the reconstruction of the $\psi(4040) \rightarrow D\bar{D}$ decay. Finally, results from the analysis of the experimental data collected with a straw tube prototype designed and constructed at the Research Center in Juelich will be discussed.

1. Introduction

$\bar{P}$ANDA is one of the major experiments that will be installed at the international FAIR facility in the site of the GSI laboratory (Darmstadt, Germany). It will use the high-intensity phase-space cooled antiproton beams provided by the High Energy Storage Ring (HESR). The $\bar{P}$ANDA experiment will use the antiproton beam from the HESR colliding with an internal proton target to carry out a rich and diversified hadron physics program, which includes the charmonium and open charm spectroscopy, the search for exotic hadrons and the study of in-medium modifications of hadron masses [1]. $\bar{P}$ANDA will be a fixed target experiment consisting of a Target Spectrometer (TS), surrounding the interaction point, and a Forward Spectrometer (FS), covering the acceptance of the spectrometer at low angles. A solenoid and a dipole will provide the magnetic field inside the TS and FS, respectively.The combination of the two spectrometers allows for a full angular coverage and high acceptance for a wide range of energies. The detector has a modular structure and can be easily adopted for the needs of different measurements.

2. The Forward Tracking system (FTS)

The role of the Forward Tracking (FT) is the momentum analysis of charged particles deflected in the field of the $\bar{P}$ANDA dipole magnet. The FT will detect particles emitted within an angular range of (-10°, 10°) and (-5°, 5°) in the horizontal and vertical directions, respectively. The system consists of three pairs of tracking stations (see figure 1): one pair (FT1, FT2) is placed in front, the second (FT5, FT6) behind the dipole magnet and the third pair (FT3, FT4) is placed inside the magnet gap to track low momentum particles.

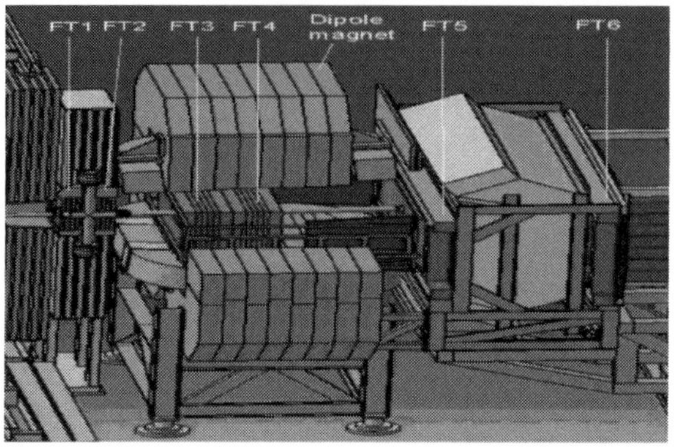

Figure 1. Layout of the three pairs of tracking stations.[2]

Each FT station will consist of four sets of planar double layers of straw tubes oriented at vertical and stereo angles (0°,+5°,-5°,0°) . Each tube will have a 10 mm inner diameter, a 30 μm aluminized Mylar foil as cathode and a 20 μm diameter gold-plated tungsten-rhenium wire as anode. The tubes will be filled with a gas mixture of $ArCO_2$ (90/10) and will be operated at a 2 bar absolute pressure. The tracking stations will be exposed to high local particle fluxes, reaching 10^4 $cm^{-2}s^{-1}$ close to the beam pipe at the maximum interaction rate of $2 \cdot 10^7$ s^{-1} expected in the high luminosity mode of $\bar{P}$ANDA.

3. Simulation of the PANDA detector

Systematic studies of the FT occupancies were performed by means of the PandaROOT software, described in details in [3]. In order to estimate the counting rates in FT, $\bar{p}$-p and $\bar{p}$-N reactions at 15 GeV/c have been simulated using event generators based on Dual Parton Model [4,5] and UrQMD [6], respectively. The results for the three (FT1, FT3, FT5) tracking stations are presented in figure 2. There is a significant increase of the rate in the downstream counters as a function of the distance to the beam pipe (up to 1.4 MHz for the innermost tubes). It is mainly due to secondary particles produced in the interaction with the beam pipe elements. These high counting rates present a challenge for the detector, as for the readout electronics.

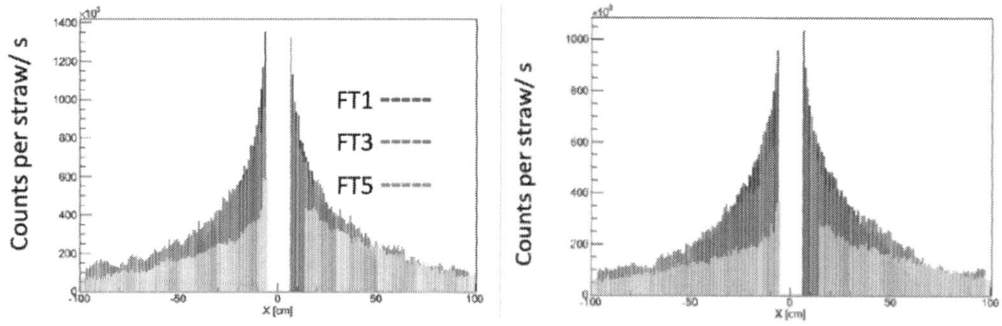

Figure 2. Calculated count rates per straw for $\bar{p}$-p (left), and for $\bar{p}$-A (right) in FT1 (before dipole magnet), FT3 (inside of the dipole magnet) and FT5 (after the dipole magnet)

The impact of the acceptance of the FT for the reconstruction of some physics channels relevant for the $\bar{P}$ANDA physics program has been studied as well. The $\bar{p}$-p $\rightarrow \psi(4040) \rightarrow D^{*+} D^{*-}$ reaction with the anti-proton momentum set to 7.71 GeV/c has been simulated and reconstructed. In particular,

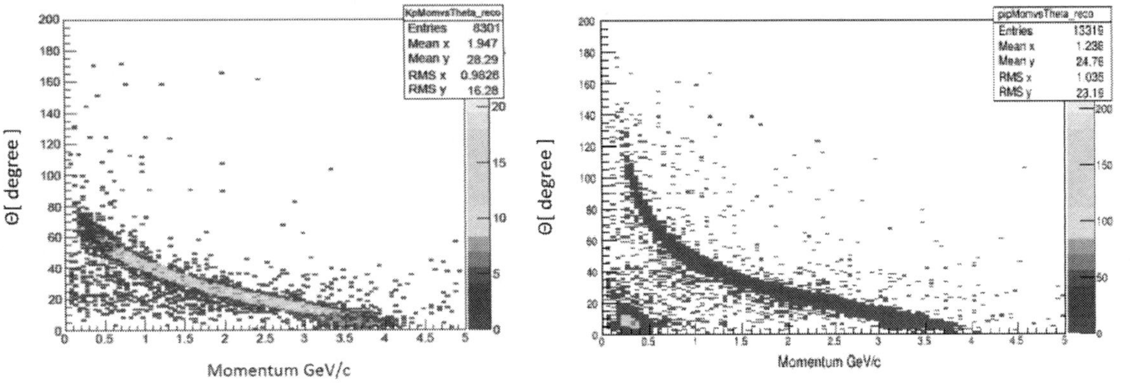

Figure 3. Distribution of theta vs momentum for pions (right) and kaons (left).

the D* mesons decay into a (D^0, π) pair and the subsequent $D^0 \rightarrow K\,\pi$ decay was investigated. A particular feature of this reaction is that, due to the low energy excess above the decay threshold, a significant part of the pions and kaons coming from the D meson decays are emitted below 10 degrees and enter the FTS (see figure 3. left). Thanks to the FTS one can still reconstruct the D^* meson with a good efficiency (see figure 4 -right).

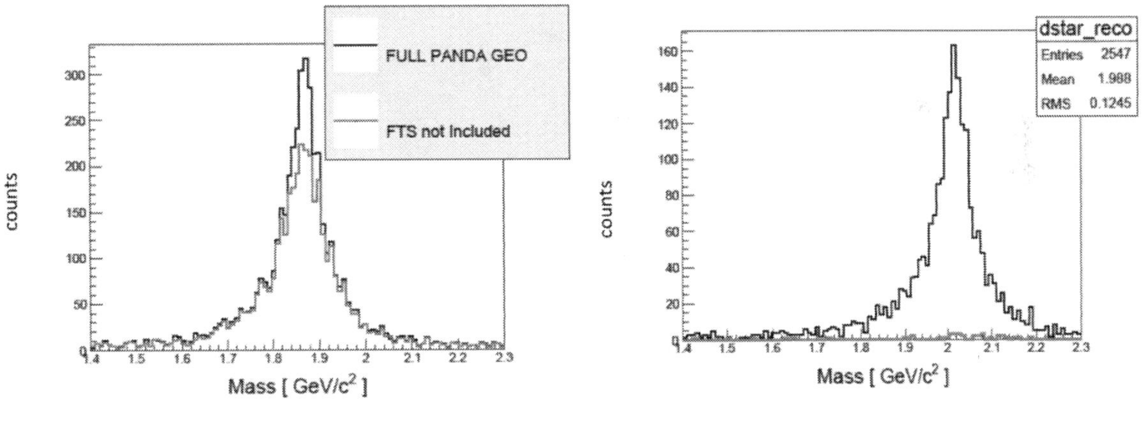

Figure 4. Reconstructed invariant mass of the D^0 (left) and D* meson (right). Black and red lines represent the result of the event reconstruction in spectrometer with and without FTS, respectively.

On the contrary, if the FTS is excluded from the reconstruction there is almost no D* visible in figure 4 right (red line). Although the importance of the FTS has been demonstrated here for this specific reaction, one can in general conclude that such a situation will also occur for other channels with similar kinematics. Figure 5 shows the reconstructed invariant mass of the $\psi(4040)$ meson with a total efficiency of 8.5% .

Journal of Physics: Conference Series **599** (2015) 012002 doi:10.1088/1742-6596/599/1/012002

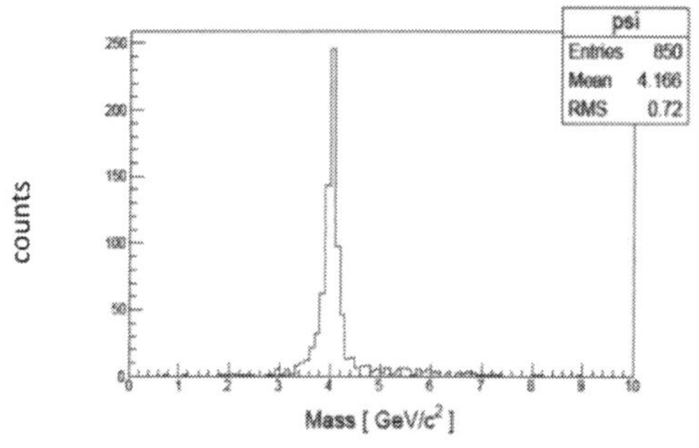

Figure 5. Reconstructed invariant mass of the ψ (4040) meson.

4. Prototype test

A Straw Tube Tracker prototype (referred to as STT) consisting of 32 tubes was tested at the Research Center in Juelich using a collimated proton beam with momentum of 0.9 GeV/c coming from COSY@IKP (Cooler Synchrotron). The beam was crossing the STT prototype, as shown in figure 6, scintillators A and B act as the trigger.

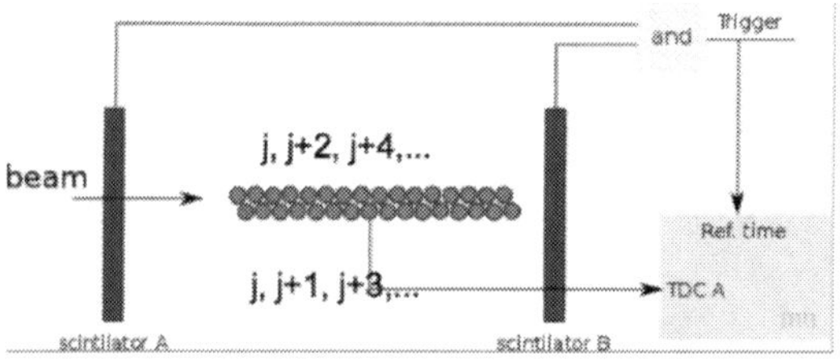

Figure 6. The experimental setup, consisting of 32 straw chamber prototype detector, two scintillators providing a trigger and the readout chain.

To readout the signals coming from the detector specialized front-end (FE) boards were designed and developed. The main part of the FE was a custom ASIC chip, developed by AGH Cracow (figure 7) [7], featuring preamplifier and shaper with a baseline stabilization and tail cancellation circuits. This is especially important for high rate application. Furthermore, the chip provides also a leading edge discriminator allowing for the time and Time over Threshold (ToT) measurement. ToT is related to the signal amplitude and allows for energy loss measurement. Signals from the ASIC board are being processed by Trigger and Readout Board[8], working as a 256 channel TDC.

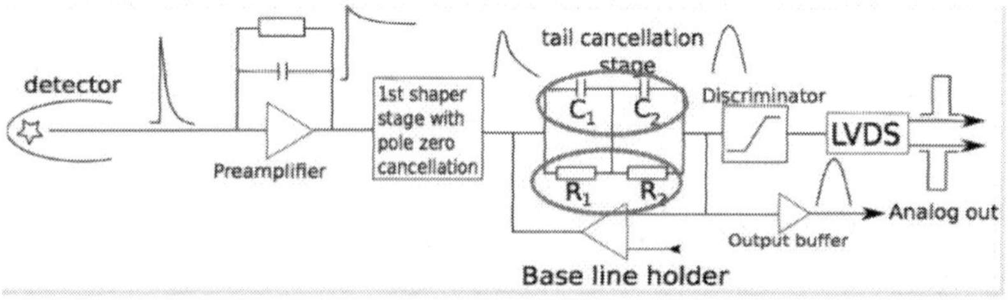

Figure 7. Schematics of the ASIC chip, three stages of signal processing 1st tail cancellation, 2nd pole-zero cancellation and 3rd discriminator .

For each channel an electron drift time spectrum was obtained. Next, the drift radius vs time correlation was obtained using uniform irradiation method. Finally, track reconstruction has been performed on event-by-event basis as described in [9]. In order to extract spatial and energy resolutions of the detector, events with tracks which passed at least 10 straw tubes were considered in the analysis. The obtained residual distribution (distance between the experimental point and the calculated track of the proton) is presented in figure 8 and shows a resolution of 150 μm.

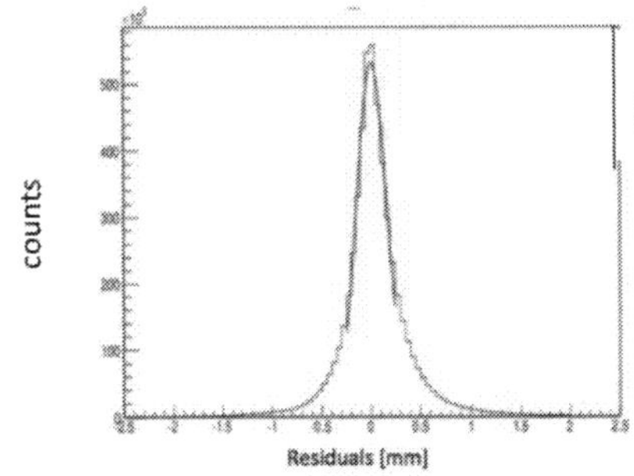

Figure 8. Residual distribution fitted with a Gaussian function.

This resolution meets the requirements of the future $\bar{P}$ANDA experiment, as specified in the respective Technical Design Report for the PANDA Straw Tube Tracker [7].

The energy loss in straw tubes has been calculated by means of the ToT measurement, shown in figure 10 (left) together with a fit using Gaussian function revealing 5.5% resolution.

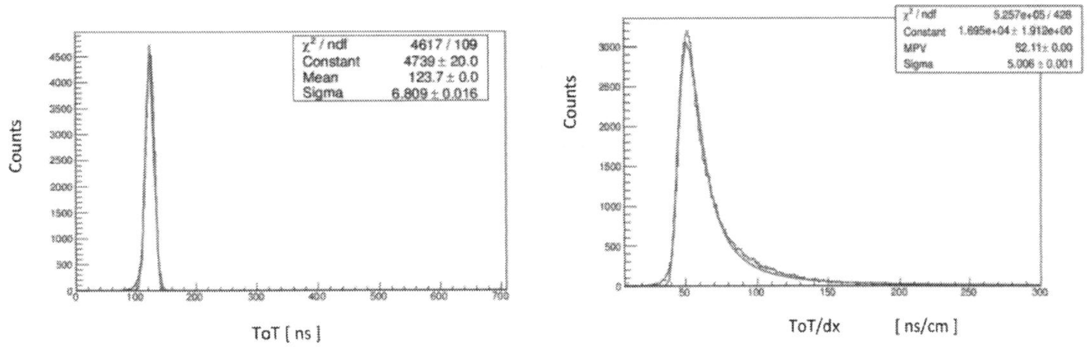

Figure 9. Left: Mean ToT distributions for selected tracks traversing 10 straws together with a fit using Gaussian distribution, Right: a similar distribution but divided by the length of the track (dx)(cm), fitted with Landau function.

Figure 9 right shows the ToT/dx distribution calculated by means of the total track length obtained from the tracking . The distribution follows the expected distribution represented by the Landau function, shown by red line (result of fit). The measurements demonstrate a potential of the ToT method to measure energy loss.

5. Conclusions

Obtained results from simulation and the prototype test show that:

- The FTS will be exposed to a high particle flux with a maximal rate to 1.4 MHz per straw. Such high rates require a dedicated fast front-end electronics with tail cancelation and base line stabilization. Such FEE electronics has been designed and tested using proton beams
- Tests show good performance of the detector and the developed FEE, however further tests are needed to evaluate in details resolution of the energy loss measurement based on ToT technique.
- Selected physics channel simulations show the importance of the FTS in reconstruction of the states above the open charm threshold.

References

[1] PANDA Physics Preformance Report, arXiv:0903.3905

[2] http://www-panda.gsi.de/framework/content/detector/det_mdc.html

[3] http://www-panda.gsi.de/framework/computing.php

[4] S. A. Bass et al., Prog. Part. Nucl. Phys. 41, 225 (1998)

[5] M. Bleicher et al., J. Phys G25, 1859 (1999)

[6] A. Capella, U. Sukhatme, C.-I.Tan, and J. Tran Than Var, Phys. Rept. 236, 225 (1994)

[7] Technical Design Report for the PANDA Straw Tube Tracker, arXiv:1205.5441

[8] Kajetanowicz, M. ; Korcyl, K. ; Krzemien, W. ; Palka, M. ; Salabura, P. et al. "A General Purpose Trigger and Readout Board for HADES and FAIR-Experiments", IEEE Trans. Nucl. Sci 55, 59 (2008)

[9] Costanza S. , Design of the Central Tracker of the PANDA experiment, PhD thesis, Università degli Studi di Pavia, ISBN 978-88-95767-38-3 (2010)

Journal of Physics: Conference Series **599** (2015) 012003 doi:10.1088/1742-6596/599/1/012003

Light Hadron Spectroscopy at BESIII

Jifeng Hu
(on behalf of the BESIII Collaboration)
Physics Department, University of Torino, Via Pietro Giuria 1, Torino
E-mail: hu@to.infn.it

Abstract. With data of 1.3 billion J/ψ events and 106 million $\psi(2S)$ events collected with the BESIII detector operating at the BEPCII, many different analyses were performed. New baryon states were observed in partial wave analyses of charmonium decays. A comparison between the branching fraction of $\mathcal{B}(J/\psi \to \gamma f_0(1710))$ to the recent lattice QCD prediction of J/ψ decaying to glueball ground state $\mathcal{B}(J/\psi \to \gamma G(0^{++}))$ benefits from new results of $J/\psi \to \gamma\eta\eta$. The observation of $\eta' \to \pi^+\pi^-\pi^+(\pi^0)\pi^-(\pi^0)$ via the radiative decay of $J/\psi \to \gamma\eta'$ agrees with the combined model of chiral perturbation theory and vector-meson dominance approach.

1. Introduction

Rich quantum states are allowed within the quantum chromodynamics (QCD) theory. However, glue-balls still remain unknown in experiments due to the difficult separation from light mesons. J/ψ radiative decay providing a very clean laboratory of scalar and tensor glueballs, has long been used for hunting for glueballs. The Crystal Ball Collaboration [1] made the first observation of $f_0(1710)$ via $J/\psi \to \gamma\eta\eta$, but suffered from low statistics. Recent lattice QCD calculations [2, 3] tell that, the mass of glueball ground state with $J^{PC} = 0^{++}$ lies in the region of 1.5 to 1.7 GeV/c^2, and the branching fraction of $J/\psi \to \gamma G(0^{++})$ has a value of 3.8×10^{-3}. An comparison could be made by summing up branching fractions of $J/\psi \to \gamma f_0(1710) \to \gamma K^+K^-(\gamma\omega\omega, \gamma\pi\pi)$ and the new measurement of $J/\psi \to \gamma f_0(1710) \to \gamma\eta\eta$. Another problem known as the "missing baryons" [4, 5] can be further understood by investigating new baryon states beyond fixed target experiments [6, 7, 8], because charmonium decays produced at the BESIII experiment take advantages of 93% acceptance of 4π coverage and high statistics to search for the missing baryons [9] [10] [11, 12]. η' meson interpreted as a singlet state arising due to the axial $U(1)$ anomaly [13, 14], still remains active in theoretical studies aiming at extensions of chiral perturbation theory [17], from its discovery in 1964 [15, 16]. New insight could be made by the four-pion decays of $\eta' \to \pi^+\pi^-\pi^{+(0)}\pi^{-(0)}$, which could be mediated by the pentagon anomaly instead of suppression according to approximate symmetries. In experiment, no observation of the four-pion decays has been made only the best upper limits reported by the CLEO collaboration: $\mathcal{B}_1(\eta' \to \pi^+\pi^-\pi^+\pi^-) < 2.4 \times 10^{-4}$ and $\mathcal{B}_2(\eta' \to \pi^+\pi^-\pi^0\pi^0) < 2.6 \times 10^{-3}$ at the 90% confidence level (C.L.) [18] excluding the branching ratio of 1.0×10^{-3} calculated using the broken-$SU_6 \times O_3$ quark model [19] three decades ago. Recent predictions $\mathcal{B}_1 = (1.0 \pm 0.3) \times 10^{-4}$ and $\mathcal{B}_2 = (2.4 \pm 0.7) \times 10^{-4}$ employed a combined model of chiral perturbation theory (ChPT) and a vector-meson dominance (VMD) approach [20].

In this paper, the results of partial wave analysis (PWA) on $J/\psi \to \gamma\eta\eta$ are presented based on a sample of 225 million J/ψ events [21]. The PWA of the decay $\psi(2S) \to p\bar{p}\pi^0$ is performed using the 106 million $\psi(2S)$ events. The first observation of $\eta' \to \pi^+\pi^-\pi^+\pi^-$ and $\eta' \to \pi^+\pi^-\pi^0\pi^0$ takes advantage of $J/\psi \to \gamma\eta'$ decay with data of 1.3×10^9 J/ψ events (2.25×10^8 events in 2009 and 1.09×10^9 events in 2012) collected at the center of mass energy of 3.097 GeV with the BESIII detector [22].

The BESIII detector is a general-purpose spectrometer located at the Beijing Electron Position Collider (BEPCII) [23], designed with a double-ring e^+e^- collider structure. The designed peak luminosity of 10^{33} cm^{-2}s^{-1} is optimized at the center of mass energy of 3.773 GeV. The cylindrical core of the BESIII detector consists of a helium-based main drift chamber (MDC), a plastic scintillator time-of-flight system (TOF), and a CsI(Tl) electromagnetic calorimeter (EMC), which are all enclosed in a superconducting solenoidal magnet providing a 1.0 T (0.9 T in 2012) magnetic field. The acceptance of charged particles and photons has 93% over 4π coverage. The charged-particle momentum resolution at 1 GeV/c is 0.5%, and the dE/dx resolution is 6%. The photon energy resolution is 2.5% (5%) at 1GeV in the barrel (endcaps). The time resolution of TOF is 80 ps for the barrel and 110 ps for the end caps.

2. Partial wave analysis of $J/\psi \to \gamma\eta\eta$

This analysis[24] was performed using the relativistic covariant tensor amplitude method. The resonance parameters and branching fractions are listed in Table 1. Projections shown in Fig. 1 indicate that the dominant 0^{++} and 2^{++} components are from the $f_0(1710)$, $f_0(2100)$, $f_0(1500)$, $f_2'(1525)$, $f_2(1810)$ and $f_2(2340)$. Among all scalar components, the measured properties of dominant $f_0(1710)$ are consistent with results of $J/\psi \to \gamma K\bar{K}$ [25] and $J/\psi \to \gamma\pi\pi$ [26] at BESII. The production rate for the $f_0(1500)$ is lower than the one for $f_0(1710)$ and $f_0(2100)$ by almost one order. The first experimental evidence for the $f_0(1790)$ was observed in $J/\psi \to \phi\pi\pi$ but no evidence was observed in $J/\psi \to \phi K\bar{K}$ [27]. Tensor components, shown as the histogram in Fig. 1 (i), stand for their total contribution, where the peak component around 1.5 GeV/c^2 is the well-established resonance $f_2'(1525)$ and the components contributing to the bump around 2.1 GeV/c^2 are from $f_2(1810)$ and $f_2(2340)$. A tensor component around 1.8 GeV/c^2 with a statistical significance of 6.4σ exists and can not be distinguished from $f_2(1810)$, $f_2(1910)$ and $f_2(1950)$ with the present statistics, denoted as $f_2(1810)$ in this analysis, and the ambiguous assignment of $f_2(1810)$ or $f_2(1950)$ is considered as a source of systematic error. Other possible tensor resonances, $f_2(2010)$, $f_2(2150)$, $f_J(2220)$, $f_2(2300)$ and $f_2(2340)$, are also considered in alternative combinations to get the optimized solution. The best fit favors the presence of $f_2(2340)$. Since the mass of $f_2(2300)$ is close to $f_2(2340)$, an attempt was made by replacing $f_2(2340)$ with $f_2(2300)$ using the fixed mass and width referenced to PDG [28], and the log likelihood value gets worse by 15. The narrow $f_J(2220)$ (also known as $\xi(2230)$) reported by MarkIII [29] and BES [30] shows no evidence in this analysis. Component $f_J(2220)$ has a significance of 0.4σ. None of the other contributions from scalar mesons, $f_0(1370)$, $f_0(2020)$, $f_0(2200)$ and $f_0(2330)$, have a significance greater than 5.0σ, thus they are excluded.

Table 1. Summary of the PWA results. The first errors are statistical and the second ones are systematic.

Resonance	Mass(MeV/c^2)	Width(MeV/c^2)	$\mathcal{B}(J/\psi \to \gamma X \to \gamma\eta\eta)$	Significance
$f_0(1500)$	1468^{+14+23}_{-15-74}	$136^{+41+28}_{-26-100}$	$(1.65^{+0.26+0.51}_{-0.31-1.40}) \times 10^{-5}$	$8.2\ \sigma$
$f_0(1710)$	$1759\pm6^{+14}_{-25}$	$172\pm10^{+32}_{-16}$	$(2.35^{+0.13+1.24}_{-0.11-0.74}) \times 10^{-4}$	$25.0\ \sigma$
$f_0(2100)$	$2081\pm13^{+24}_{-36}$	273^{+27+70}_{-24-23}	$(1.13^{+0.09+0.64}_{-0.10-0.28}) \times 10^{-4}$	$13.9\ \sigma$
$f_2'(1525)$	$1513\pm5^{+4}_{-10}$	75^{+12+16}_{-10-8}	$(3.42^{+0.43+1.37}_{-0.51-1.30}) \times 10^{-5}$	$11.0\ \sigma$
$f_2(1810)$	1822^{+29+66}_{-24-57}	$229^{+52+88}_{-42-155}$	$(5.40^{+0.60+3.42}_{-0.67-2.35}) \times 10^{-5}$	$6.4\ \sigma$
$f_2(2340)$	$2362^{+31+140}_{-30-63}$	$334^{+62+165}_{-54-100}$	$(5.60^{+0.62+2.37}_{-0.65-2.07}) \times 10^{-5}$	$7.6\ \sigma$

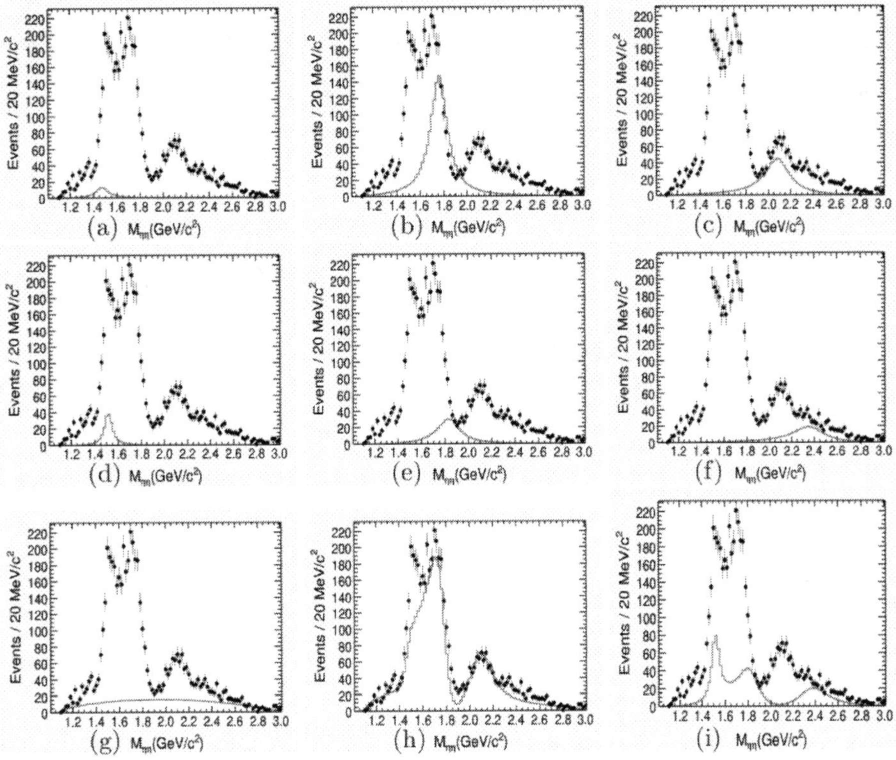

Figure 1. Contribution of the components. (a) $f_0(1500)$, (b) $f_0(1710)$, (c) $f_0(2100)$, (d) $f_2'(1525)$, (e) $f_2(1810)$, (f) $f_2(2340)$, (g) 0^{++} phase space, (h) total 0^{++} component, and (i) total 2^{++} component. The dots with error bars are data with background subtracted, and the solid histograms are the projection of the PWA.

3. Observation of $\eta' \to \pi^+\pi^-\pi^+\pi^-$ and $\eta' \to \pi^+\pi^-\pi^0\pi^0$

This analysis [37] provides clean η' data via J/ψ radiative decay. The $\pi^+\pi^-\pi^+\pi^-$ invariant mass spectrum is shown in Fig. 2(a), where an η' peak is clearly observed in the inset plot. The projections of the fit to $M_{\pi^+\pi^-\pi^{+(0)}\pi^{-(0)}}$ in the η' mass region are shown in Figs. 2(b) and (c), where the shape of the sum of signal and background shapes agree well with data. 199 ± 16 $\eta' \to \pi^+\pi^-\pi^+\pi^-$ events was observed with a statistical significance of 18σ and 84 ± 16 $\eta' \to \pi^+\pi^-\pi^0\pi^0$ events with a statistical significance of 5σ respectively. The $M_{\pi^+\pi^-}$ spectrum of data carrying key information of η' decay mechanism is extracted from fitting the $\pi^+\pi^-\pi^+\pi^-$ mass spectrum and subtracting background. The MC spectrum employs two models, a phase space model and a combined model of ChPT and VMD implemented using decay amplitudes in Ref. [20]. To make comparison, the MC $M_{\pi^+\pi^-}$ spectrum is divided into 38 bins in the region of [0.28, 0.66] GeV$/c^2$ for decay of $\eta' \to \pi^+\pi^-\pi^+\pi^-$, as shown in Fig. 2 (d) (four entries per event), where the errors are statistical only. Clearly, the combined model agrees with data better than the phase space model. Branching fractions of $\mathcal{B}(J/\psi \to \gamma\eta', \eta' \to \pi^+\pi^-\pi^+\pi^-)$ and $\mathcal{B}(J/\psi \to \gamma\eta', \eta' \to \pi^+\pi^-\pi^0\pi^0)$ are determined to be $(4.40 \pm 0.35 \pm 0.30) \times 10^{-7}$ and $(9.38 \pm 1.79 \pm 0.89) \times 10^{-7}$ respectively.

4. Observation of two new N^* resonances in $\psi(3686) \to p\bar{p}\pi^0$

This analysis[31] takes advantage of a data set with larger statistics than that at CLEOc shows more than one N^* state below 1700 MeV$/c^2$, which are easily seen in the $p\pi^0$ and $\bar{p}\pi^0$ mass

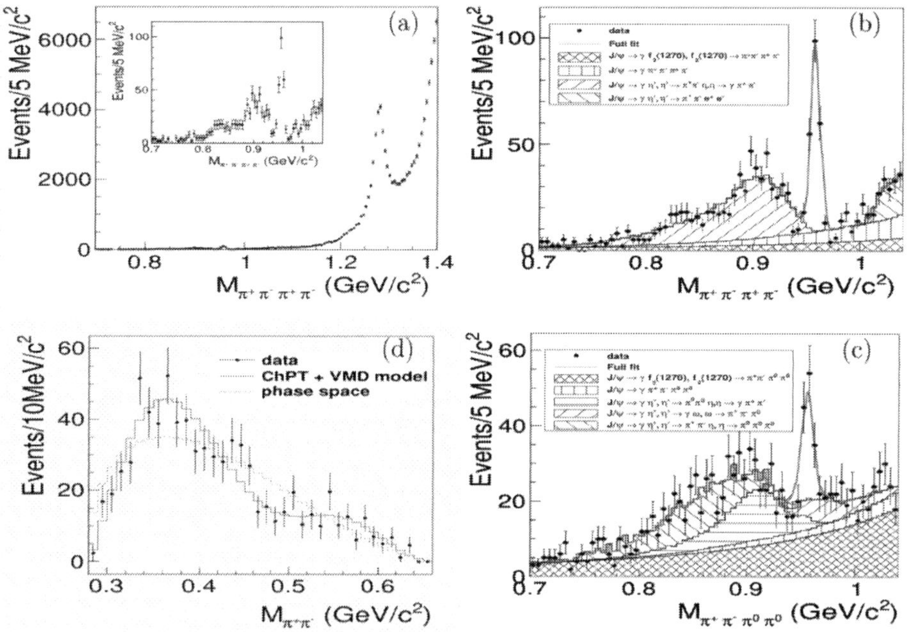

Figure 2. (a) The invariant mass distributions of $\pi^+\pi^-\pi^+\pi^-$. Results of the fits to (b) $M_{\pi^+\pi^-\pi^+\pi^-}$ and (c) $M_{\pi^+\pi^-\pi^0\pi^0}$, where the background contributions are displayed as the hatched histograms. (d) The comparison of $M_{\pi^+\pi^-}$ (four entries per event) between data and two different models, where the dots with error bars are for the background-subtracted data, the solid line is for the ChPT and VMD model, and the dashed line is for the phase space.

spectra, and the threshold enhancement in the $p\bar{p}$ mass spectrum. To better understand the structure of multi-resonances and their interference, a partial wave analysis is performed and components' contributions are shown in Fig. 3. Plot (a) shows the contributions of $N(1440)$, $N(1520)$, $N(1535)$ and $N(1650)$ with clear peaks, the tails at the high mass region come from the interference effects. Plot (b) shows the contributions of $N(940)$, $N(1720)$, $N(2300)$ and $N(2570)$. Two new N^* resonances, $N(2300)$ and $N(2570)$ are observed with number of events 948 ± 68 and 795 ± 45 respectively. No clear evidence for $N(1885)$ or $N(2065)$ were found. More investigations such as $J/\psi \to \eta p\bar{p}$[33], $J/\psi \to \lambda\Sigma^0$[34], $\psi' \to \bar{p}K\Sigma^0$[35], $\chi_{c0} \to p\bar{n}\pi^-$[36] also explored new baryon states.

5. Summary

The PWA results of $J/\psi \to \gamma\eta\eta$ as summarized in Table 1 combining with branching fractions of $J/\psi \to \gamma f_0(1710) \to \gamma XX$, were employed to compare the recent lattice QCD prediction of $J/\psi \to \gamma G(0^{++})$. A series of recently observed baryon states aim to improve further understanding of the quark model. The analyses of $\eta' \to \pi^+\pi^-\pi^+\pi^-$ and $\eta' \to \pi^+\pi^-\pi^0\pi^0$ agree with the combined model of chiral perturbation theory and vector-meson dominance approach.

References

[1] C. Edwards *et al.*, Phys. Rev. Lett. **48**, 458 (1982).
[2] Y. Chen *et al.*, Phys. Rev. D **73**, 014516 (2006).
[3] E. Gregory *et al.*, JHEP **1210**, 170 (2012).
[4] S. Capstick and W. Roberts, Phys. Rev. D **47**, 1994 (1993).

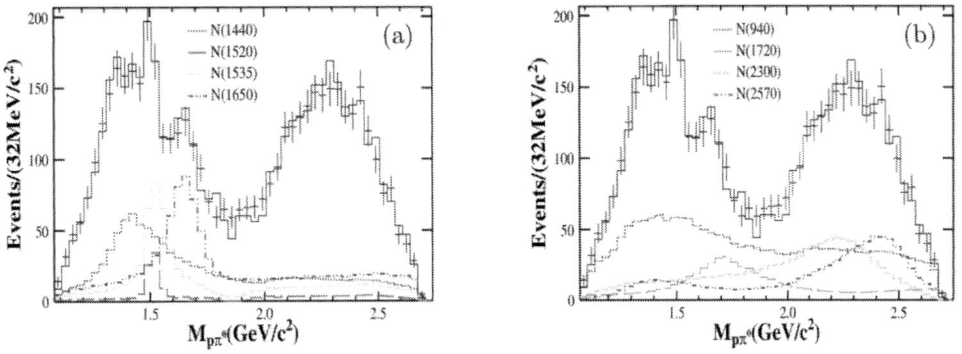

Figure 3. The contribution of each intermediate resonance in the $p\pi^0$ mass spectra. The interferences with other resonances are included.

[5] N. Isgur and G. Karl, Phys. Rev. D **19**, 2653 (1979).
[6] S. Stepanyan *et al.* (CLAS Collaboration) Phys. Rev. Lett. **91**, 252001 (2003)
[7] I. Horn *et al.* (CB-ELSA Collaboration) Phys. Rev. Lett. **101**, 202002 (2008)
[8] E. F. McNicoll *et al.* (Crystal Ball Collaboration at MAMI) Phys. Rev. C **82**, 035208 (2010)
[9] H. B. Li *et al.* (BES Collaboration), Nucl. Phys. A **675**, 189C (2000).
[10] J. Z. Bai *et al.* (BES Collaboration), Phys. Lett. B **510**, 75 (2001).
[11] R. Koniuk and N. Isgur, Phys. Rev. D **21**, 1868(1980).
[12] S. Capstick and W. Roberts, Phys. Rev. D **49**, 4570(1994).
[13] S. Weinberg, Phys. Rev. D **11**, 3583 (1975).
[14] G. 't Hooft, Phys. Rev. D **14** (1976) 3432 [Erratum-ibid. D **18** (1978) 2199].
[15] G. R. Kalbfleisch *et al.*, Phys. Rev. Lett. **12**, 527 (1964).
[16] M. Goldberg *et al.*, Phys. Rev. Lett. **12**, 546 (1964).
[17] J. Gasser, H. Leutwyler, Nucl. Phys. B **250**, 465 (1985).
[18] P. Naik *et al.* [CLEO Collaboration], Phys. Rev. Lett. **102**, 061801 (2009).
[19] D. Parashar, Phys. Rev. D **19**, 268 (1979).
[20] Feng-Kun Guo, Bastian Kubis and Andreas Wirzba, Phys. Rev. D **85**, 014014 (2012).
[21] M. Ablikim *et al.* (BESIII Collaboration), Chin. Phys. C **36**, 915 (2012).
[22] M. Ablikim *et al.* [BESIII Collaboration], Nucl. Instrum. Methods Phys. Res. A **614**, 345 (2010).
[23] J. Z. Bai *et al.* (BES Collaboration), Nucl. Instrum. Meth. A **344**, 319 (1994); Nucl. Instrum. Meth. A **458**, 627 (2001).
[24] M. Ablikim *et al.* [BESIII Collaboration], Phys.Rev.D 87, 092009 (2013).
[25] J. Z. Bai *et al.* (BES Collaboration), Phys. Rev. D **68**, 052003 (2003).
[26] M. Ablikim *et al.* (BES Collaboration), Phys. Lett. B **642**, 441 (2006).
[27] M. Ablikim *et al.* (BES Collaboration), Phys. Lett. B **607**, 243 (2005).
[28] J. Beringer, *et al.* (Particle Data Group), Phys. Rev. D **86**, 010001 (2012).
[29] R. M. Baltrusaitis *et al.* (MARKIII Collaboration), Phys. Rev. Lett. **56**, 107 (1986).
[30] J. Z. Bai *et al.* (BES Collaboration), Phys. Rev. Lett. **81**, 1179 (1998).
[31] M. Ablikim *et al.* (BES Collaboration), Phys. Rev. Lett. **110**, 022001 (2013)
[32] J. C. Chen, G. S. Huang, X. R. Qi, D. H. Zhang and Y. S. Zhu, Phys. Rev. D **62**, 034003 (2000).
[33] M. Ablikim *et al.* [BESIII Collaboration], Phys. Rev. D **88**, 032010 (2013)
[34] M. Ablikim *et al.* [BESIII Collaboration], Phys. Rev. D **87**, 012007 (2013)
[35] M. Ablikim *et al.* [BESIII Collaboration], Phys. Rev. D **86**, 032008 (2012)
[36] M. Ablikim *et al.* [BESIII Collaboration], Phys. Rev. D **86**, 052011 (2012)
[37] M. Ablikim *et al.* [BESIII Collaboration], Phys. Rev. D **112**, 251801 (2014)

Non-conventional mesons at PANDA

Francesco Giacosa

Abstract. Non-conventional mesons, such as glueballs and tetraquarks, will be in the focus of the PANDA experiment at the FAIR facility. In this lecture we recall the basic properties of QCD and describe some features of unconventional states. We focus on the search of the not-yet discovered glueballs and the use of the extended Linear Sigma Model for this purpose, and on the already discovered but not-yet understood X, Y, Z states.

Institute of Physics, Jan Kochanowski University, 25-406 Kielce, Poland and Institute for Theoretical Physics, J. W. Goethe University, Max-von-Laue-Str. 1, 60438 Frankfurt am Main, Germany

1. Introduction

Conventional mesons are bound states made by a quark and an antiquark. Yet, since the very beginning of QCD the search for other possibilities has attracted the attention of both experimentalists and theoreticians [1].

As a prominent example, glueballs were predicted long ago: they are (yet hypothetical) bound states of solely gluons. Computer simulations of QCD on the lattice have found a full spectrum of these states [2, 3], but their firm experimental discovery has not yet taken place. On the other hand, a plenty of mesons, denoted as X, Y, and Z states, has been unambiguously discovered in the last decade in the energy region of charm-anticharm and botton-antibotton states [4, 5]. A clear explanation about the nature of these states is still lacking (tetraquark and molecular hypotheses are a possibility).

The PANDA experiment at the FAIR facility [6] in Darmstadt/Germany is designed to shed light on these questions. It is a proton-antiproton scattering experiment in which the energy of the antiproton can be finely tuned in such a way that a wide energy range in the charmonium region can be scanned. Various resonances can be directly formed in proton-antiproton fusion processes. Glueball, if existent, shall be found by this experiment. On the other hand, many of the X, Y, and Z states can also be formed and/or produced with high statistics. In this lecture a theoretical view concerning the search of non-conventional mesons at PANDA will be presented.

2. Brief recall of the QCD Lagrangian and its symmetries

The Lagrangian of Quantum Chromodynamics (QCD) reads

$$\mathcal{L}_{QCD} = \sum_{i=1}^{N_f} \overline{q}_i(i\gamma^\mu D_\mu - m_i)q_i - \frac{1}{4}G^a_{\mu\nu}G^{a,\mu\nu}\ ,\ D_\mu = \partial_\mu - ig_0 A_\mu \tag{1}$$

where $G^a_{\mu\nu} = \partial_\mu A^a_\nu - \partial_\nu A^a_\mu + g_0 f^{abc} A^b_\mu A^c_\nu$, $A_\mu = A^a_\mu t^a$, $a, = 1, ..., N_c^2 - 1 = 8$, and t^a and f^{abc} are the generators and the structure constants of the group $SU(N_c)$. $\mathcal{L}_{QCD}$ contains N_f quark fields q_i with corresponding bare masses m_i ($i = 1, ..., N_f$, N_f is the number of quark flavors; in Nature $N_f = 6$: u, d, s, c, b, t). Each quark flavor has three colors (red, green, blue). A crucial property of Eq. (1) is that 3-gluon and 4-gluon vertices are present: gluons –contrary to photons– 'shine in their own light'. We list the main symmetries and their meaning.

(i) $\mathcal{L}_{QCD}$ is built under the requirement of invariance under *local* transformations of the $SU(3)$ groups. This means that one can rename at *each space-time point* the color of a quark via a $SU(3)$ matrix $U_C \equiv U_C(x)$ ($U_C U_C^\dagger = 1$, $\det U_C = 1$) and of the gluon field as:

$$q_i \to U_C q_i \ , \ A_\mu \to U_C(A_\mu - i\partial_\mu / g_0)U_C^\dagger \ . \tag{2}$$

One can, for instance, transform a blue quark in a red one in a certain space-time region without changing the properties of the system.

(ii) When the bare quark masses are equal ($m_1 = m_2 = ... = m_{N_f}$), $\mathcal{L}_{QCD}$ is invariant under the interchange of quark flavors via a $N_f \times N_f$ unitary matrix U_V ($U_V U_V^\dagger = 1$) as: $q_i \to U_{V,ij} q_j$. This is flavor symmetry, denoted also as $U(N_f)_V$: in simple terms, it means that gluons are 'democratic' and couple to each quark flavor with the same strength. Thus, one can rename also the quark flavor: it is allowed to interchange a u quark with a d quark, but this can be done only once for all space-time points (the symmetry is global and not local). For $N_f = 2$, $m_u \simeq m_d \simeq 5$ MeV: the symmetry $U(2)_V$ is well realized in Nature (this is isospin symmetry, responsible for instance for the almost equal mass of the three pions and of the proton and the neutron). For $N_f = 3$ the bare strange mass $m_s \simeq 100$ MeV is sizably larger than m_u and m_d: nevertheless, the emergence of flavor multiplets with strange mesons is evident in the PDG, showing that an approximate $U(3)_V$ is also realized [7].

The limit in which all quark masses vanish ($m_1 = ... = m_{N_f} = 0$) is called the *chiral limit* and is important for the understanding of QCD because additional symmetries are present: dilatation invariance and chiral symmetry.

(iii) Dilatation symmetry and its anomalous breaking. In the chiral limit, there is only one parameter in $\mathcal{L}_{QCD}$, the *dimensionless* coupling constant g_0. The theory is classically dilatation invariant. However, upon quantization, a running coupling $g_0 \to g(\mu)$ emerges. An ultraviolet cutoff Λ_{UV} is introduced in the process of regularization in such a way that g_0 is the coupling at this very high energy scale: $g_0 = g(\Lambda_{UV})$. Then, a low-energy scale Λ_{YM} emerges in the theory as

$$g^2(\mu) = \frac{g_0^2}{1 + 2bg_0^2 \log \frac{\mu}{\Lambda_{UV}}} \to \Lambda_{YM} = \Lambda_{UV} e^{-1/(2bg_0^2)} \text{ with } b = \frac{33 - 2N_f}{48\pi^2}. \tag{3}$$

Numerically, $\Lambda_{YM} \simeq 250$ MeV: all quantities in QCD depend crucially on it.

(iv) Chiral symmetry and its spontaneous breaking. One splits the quark field into the right-handed and left-handed components: $q_i = q_{i,L} + q_{i,R} = P_L q_i + P_R q_i$ with $P_{R(L)} = \frac{1}{2}\left(1 \pm \gamma^5\right)$. $\mathcal{L}_{QCD}$ is *separately* invariant under rotations of right-handed quarks and left-handed quarks

$$q_i = q_{i,L} + q_{i,R} \to U_{L,ij} q_{j,L} + U_{R,ij} q_{j,R} \ , \tag{4}$$

where U_R and U_L are two independent unitary matrices. Such a chiral transformation is also denoted as $U(N_f)_R \times U(N_f)_L$. Chiral transformation reduces to a flavor one for $U_V = U_L = U_R$. Conversely, the case $U_A = U_L = U_R^\dagger$ is called axial transformation $U(N_f)_A$ (which is not a group!), which mixes states with different parity, as pseudoscalar and scalar mesons and vector with axial-vector ones. This symmetry is not realized in the hadronic spectrum [7] because it is spontaneously broken by the nonperturbative QCD vacuum. As a consequence the quarks develop -even in the chiral limit- a large constituent (or effective) mass $m \to m^* \simeq \Lambda_{YM}$.

3. Mesons

Quarks and gluons are the basic degrees of freedom of the QCD Lagrangian of Eq. (1). However, these are *not* the asymptotic states that we measure in our detectors. Namely, quarks and gluons are *confined* into hadrons, where each hadron is *white* (i.e., invariant under the local color transformation introduced in Sec. 2).

We use the following *definition*: 'A meson is a strongly interacting particle (a hadron) with integer spin'. This definition is consistent with the PDG [7], in which all mesons are listed together independently of their inner structure.

3.1. Quark-antiquark mesons

A *conventional meson* is a meson constructed out of a quark and an antiquark. Although it represents only one of (actually infinitely many) possibilities to build a meson, the vast majority of mesons of the PDG can be correctly interpreted as belonging to a quark-antiquark multiple [7] (see also the results of the quark model [8]).

Mesons can be classified by their spatial angular momentum L, the spin S, the total angular momentum J and by parity P and charge conjugation C (summarized in J^{PC}). The lightest mesons are pseudoscalar states with $L = S = 0 \rightarrow J^{PC} = 0^{-+}$. Indeed, the pions and the kaons are pseudoscalar (quasi-)Goldstone bosons emerging upon the spontaneous breaking of chiral symmetry. As an example, we write down the wave function for the state K^+ (radial, angular, spin, flavor, color):

$$\left|K^+\right\rangle = |n=1\rangle\,|L=0\rangle\,|S=0(\uparrow\downarrow - \downarrow\uparrow)\rangle\,|u\bar{s}\rangle\,\left|\bar{R}R+\bar{G}G+\bar{B}B\,\right\rangle . \tag{5}$$

For $L = 0$, $S = 1$ one constructs the vector mesons (such as ρ and ω), for $L = S = 1$ one has three multiplets: tensor mesons $J^{PC} = 2^{++}$, axial-vector mesons $J^{PC} = 1^{++}$ and scalar mesons $J^{PC} = 0^{++}$ (scalar states are in the center of a long debate, see e.g. Refs. [9, 10, 11] and refs. therein). By further increasing L one can obtain many more multiplets [7].

It is interesting to notice that the quantum numbers $J^{PC} = 0^{+-}$ *cannot* be obtained in a quark-antiquark system, but is possible for unconventional mesonic states (such as glueballs). The experimental discovery of mesons with such *exotic quantum numbers* naturally points to a non-quarkonium inner structure.

3.2. Glueballs search and the eLSM

According to lattice QCD many glueballs should exist [2, 3], but up to now no glueball state has been unambiguously identified (although for some of them some candidates exist).

A suitable theoretical framework to study the decays of glueballs is the so-called extended linear Sigma Model (eLSM), which is an effective model of QCD built accordingly to the two fundamental symmetries mentioned in Sec. 2: chiral symmetry and dilatation invariance. The former is spontaneously broken by a Mexican-hat potential, the latter explicitly broken in order to mimic the trace anomaly of QCD of Eq. (3), see Ref. [12]. As a consequence, the eLSM Lagrangian contains only a finite number of terms. Moreover, (axial-)vector d.o.f. are included from the very beginning. The eLSM was first developed for $N_f = 2$ in Refs. [13, 14], for $N_f = 3$ in Refs. [15, 16], and for $N_f = 4$ in Ref. [17]. In particular, in Ref. [15] a fit to many experimental data was performed and a good description of low-hadron phenomenology (up to about 1.7 GeV) was obtained. Here we briefly recall the main results concerning glueballs.

The scalar glueball is the lightest gluonic state predicted in QCD and is naturally an element of the eLSM as the excitation of the dilaton field [14, 15, 16]. The result of the recent study of Ref. [16] shows that the scalar glueball is predominantly contained in the resonance $f_0(1710)$, in agreement with the lattice result of Ref. [18]. The eLSM makes predictions for the lightest (and peculiar) glueball state in a chiral framework, completing previous phenomenological works on the subject [9].

The pseudoscalar glueball is related to the chiral anomaly and couples in a chirally invariant way to light mesons [19], where it was shown that it decays predominantly in $\pi\pi K$ (50% of all decays into (pseudo)scalar mesons) and that it does *not* decay in $\pi\pi\pi$: these are simple and testable theoretical predictions which can be helpful in the experimental search at the PANDA experiment, where the pseudoscalar glueball can be directly formed in proton-antiproton fusion process.

A similar program can be carried out for a tensor glueball with a mass of about 2.2 GeV, e.g. Ref. [20], as well as for heavier glueballs, such as the (pseudo)vector ones.

3.3. X, Y, Z states and other non-quarkonium candidates

The discovery in the last years of a plenty of enigmatic resonances -the so called X, Y, Z states- shows that there are now many candidates of resonances beyond the standard quark-antiquark picture, see e.g. [4, 5] ($X(3872)$ was the first to be experimentally found by BELLE in 2003).The interpretation of these states is subject to ongoing debates: tetraquarks and molecular interpretations are at the top of the list, but it is difficult to distinguish among them [5, 21]. Moreover, distortions due to quantum fluctuations of nearby threshold(s) take place and make the understanding of these resonances more complicated [22]. Remarkably, the Z states are *charged* states in the charmonium region: a system made of four quarks is here necessary to understand them since a charmonium is necessarily chargeless (see e.g. Ref. [23]).

There are also other mesonic states which are not yet understood. An example is the strange-charmed scalar state $D_{S0}(2317)$, which is too light to be a $c\bar{s}$ state and could be a four-quark or a dynamically generated state. Historically, the scalar mesons below 1 GeV were among the first to be interpreted as non-quarkonium objects, but as a nonet of tetraquarks [11] or as dynamically generated states [10].

4. PANDA: formation and production of mesons

In the future PANDA experiment at the FAIR facility in Darmstadt [6], antiprotons reach a three-momentum $\mathbf{k}$ in the range $|\mathbf{k}| = 2.2\text{-}10$ GeV and hit protons at rest. We consider here the case in which the proton and the antiproton completely annihilate and generate a particle X (as for instance a glueball) with mass m_X. The four-momentum of the antiproton reads $k_{\bar{p}} = (E_{\bar{p}} = \sqrt{\mathbf{k}^2 + m_p^2}, \mathbf{k})$ while that of the proton is $k_p = (m_p, \mathbf{0})$. By denoting $k_X = (\sqrt{\mathbf{k}^2 + m_X^2}, \mathbf{k})$ as the four-momentum of X, we obtain out of $k_p + k_{\bar{p}} = k_X$ that:

$$m_X = \sqrt{2m_p\left(m_p + E_{\bar{p}}\right)} = 2.25\text{-}4.53 \text{ GeV}. \tag{6}$$

By looking at the lattice spectrum of Ref. [2], we realize that -besides the scalar glueball which is too light- all non-exotic glueball states could be directly formed at the PANDA experiment. This represents a clean environment to study experimentally their decays. Glueballs with exotic quantum numbers (called oddballs) cannot be directly formed because a proton-antiproton system undergoes the same limitations of a quark-antiquark system for what concerns J^{PC} quantum numbers. Nevertheless, oddballs will also be produced together with other resonances and will be studied as well. Besides the search for glueballs, all mesonic states discussed above will be experimentally investigated.

5. Conclusions and outlook

In this work we have given a brief overview of some aspects of the theoretical as well as experimental search for unconventional mesons. The PANDA experiment will play a decisive role in the future of hadron physics, since it will help to clarify many open questions of hadron spectroscopy in general and of exotic mesons in particular. The search for and hopefully the firm discovery of glueballs (and hybrids), as well as the confirmation and measurement with high statistics of the X, Y, Z states will be important milestones toward a better understanding of QCD.

Acknowledgments
The author thanks all the members of the group for chiral field theories of Frankfurt for many valuable cooperations and discussions.

References
[1] C. Amsler and N. A. Tornqvist, Phys. Rept. **389**, 61 (2004); E. Klempt and A. Zaitsev, Phys. Rept. **454**, 1 (2007) [arXiv:0708.4016 [hep-ph]]; F. Giacosa, Phys. Rev. D **80** , 074028 (2009) [arXiv:0903.4481 [hep-ph]].
[2] Y. Chen *et al.*, Phys. Rev. D **73**, 014516 (2006) [arXiv:hep-lat/0510074];
[3] C. J. Morningstar and M. J. Peardon, Phys. Rev. D **60**, 034509 (1999) [arXiv:hep-lat/9901004]; E. Gregory *et al.* JHEP **1210**, 170 (2012) [arXiv:1208.1858 [hep-lat]].
[4] N. Brambilla *et al.*, Eur. Phys. J. C **71** (2011) 1534 [arXiv:1010.5827 [hep-ph]].
[5] E. Braaten, C. Langmack and D. H. Smith, Phys. Rev. D **90** (2014) 014044 [arXiv:1402.0438 [hep-ph]]; G. T. Bodwin *et al.*, arXiv:1307.7425.
[6] M. F. M. Lutz *et al.* [PANDA Collaboration], arXiv:0903.3905 [hep-ex].
[7] K.A. Olive *et al.* (Particle Data Group), Chin. Phys. C **38**, 090001 (2014).
[8] S. Godfrey and N. Isgur, Phys. Rev. D **32** (1985) 189.
[9] C. Amsler and F. E. Close, Phys. Lett. B **353** (1995) 385; W. J. Lee and D. Weingarten, Phys. Rev. D **61** (2000) 014015 [arXiv:hep-lat/9910008]; F. E. Close and A. Kirk, Eur. Phys. J. C textbf21 (2001) 531 [arXiv:hep-ph/0103173]; H. Y. Cheng, C. K. Chua and K. F. Liu, Phys. Rev. D **74** (2006) 094005 [arXiv:hep-ph/0607206]. F. Giacosa, T. Gutsche, V. E. Lyubovitskij and A. Faessler, Phys. Rev. D **72**, 094006 (2005) [arXiv:hep-ph/0509247]; F. Giacosa, T. Gutsche and A. Faessler, Phys. Rev. C **71** (2005) 025202 [hep-ph/0408085].
[10] E. van Beveren*et al.*, Z. Phys. C **30**, 615 (1986) [arXiv:0710.4067 [hep-ph]]; N. A. Tornqvist, Z. Phys. C **68**, 647 (1995) [arXiv:hep-ph/9504372]; M. Boglione and M. R. Pennington, Phys. Rev. D **65**, 114010 (2002) [arXiv:hep-ph/0203149]; ; J. R. Pelaez, Phys. Rev. Lett. **92**, 102001 (2004) [arXiv:hep-ph/0309292]; J. A. Oller and E. Oset, Nucl. Phys. A **620**, 438 (1997) [arXiv:hep-ph/9702314].
[11] R. L. Jaffe, Phys. Rev. D **15**, 267 (1977); L. Maiani, F. Piccinini, A. D. Polosa and V. Riquer, Phys. Rev. Lett. **93**, 212002 (2004) [arXiv:hep-ph/0407017]; F. Giacosa, Phys. Rev. D **74**, 014028 (2006) [arXiv:hep-ph/0605191]; A. H. Fariborz, R. Jora and J. Schechter, Phys. Rev. D **72**, 034001 (2005) [arXiv:hep-ph/0506170]; M. Napsuciale and S. Rodriguez, Phys. Rev. D **70**, 094043 (2004); F. Giacosa, Phys. Rev. D **75** (2007) 054007 [arXiv:hep-ph/0611388].
[12] A. A. Migdal and M. A. Shifman, Phys. Lett. B **114**, 445 (1982); R. Gomm, P. Jain, R. Johnson and J. Schechter, Phys. Rev. D **33**, 801 (1986).
[13] S. Gallas, F. Giacosa and D. H. Rischke, Phys. Rev. D **82**, 014004 (2010) [arXiv:0907.5084 [hep-ph]]; D. Parganlija, F. Giacosa and D. H. Rischke, Phys. Rev. D **82**, 054024 (2010) [arXiv:1003.4934 [hep-ph]].
[14] S. Janowski, D. Parganlija, F. Giacosa and D. H. Rischke, Phys. Rev. D **84**, 054007 (2011) [arXiv:1103.3238 [hep-ph]].
[15] D. Parganlija, P. Kovacs, G. Wolf, F. Giacosa and D. H. Rischke, Phys. Rev. D **87**, 014011 (2013) arXiv:1208.0585 [hep-ph]].
[16] S. Janowski, F. Giacosa and D. H. Rischke, Phys. Rev. D **90** (2014) 114005 [arXiv:1408.4921[hep-ph]].
[17] W. I. Eshraim, F. Giacosa and D. H. Rischke, arXiv:1405.5861 [hep-ph].
[18] L. -C. Gui *et al.*, Phys. Rev. Lett. **110** (2013) 021601 [arXiv:1206.0125 [hep-lat]].
[19] W. I. Eshraim, S. Janowski, F. Giacosa and D. H. Rischke, Phys. Rev. D **87**, 054036 (2013) [arXiv:1208.6474 [hep-ph]]; W. I. Eshraim, S. Janowski, A. Peters, K. Neuschwander and F. Giacosa, Acta Phys. Polon. Supp. **5**, 1101 (2012) [arXiv:1209.3976 [hep-ph]].
[20] F. Giacosa, T. Gutsche, V. E. Lyubovitskij, A. Faessler, Phys. Rev. D **72**, 114021 (2005) [hep-ph/0511171]; L. Burakovsky and J. T. Goldman, Phys. Rev. D **57**, 2879 (1998) [hep-ph/9703271].
[21] L. Maiani, A. D. Polosa and V. Riquer, Phys. Rev. Lett. **99** (2007) 182003 [arXiv:0707.3354 [hep-ph]].
[22] S. Coito, G. Rupp and E. van Beveren, Eur. Phys. J. C **73** (2013) 3, 2351 [arXiv:1212.0648 [hep-ph]].
[23] L. Maiani, F. Piccinini, A. D. Polosa and V. Riquer, Phys. Rev. D **89** (2014) 114010 [arXiv:1405.1551 [hep-ph]].

FAIRNESS 2014 IOP Publishing
Journal of Physics: Conference Series **599** (2015) 012005 doi:10.1088/1742-6596/599/1/012005

Lattice study of hybrid static potentials

Philipp Wolf, Marc Wagner

Goethe-Universität Frankfurt am Main, Institut für Theoretische Physik, Max-von-Laue-Straße 1, D-60438 Frankfurt am Main, Germany

E-mail: pwolf@th.physik.uni-frankfurt.de

Abstract. We report about a recently started project with the aim to compute hybrid static potentials using lattice gauge theory. First preliminary results for pure SU(2) Yang-Mills theory are presented.

1. Introduction

One of the goals of the PANDA experiment at FAIR will be the search for gluonic excitations, e.g. glueballs and hybrid mesons. While glueballs consist only of gluons, hybrid mesons are quark antiquark states with excited gluonic fields, which contribute to the quantum numbers. Hybrid mesons are, therefore, not restricted to quark model quantum numbers: J^{PC} with spin $S = 0, 1$ and orbital angular momentum $L = 0, 1, 2, ...$, where $P = (-1)^{L+1}$ and $C = (-1)^{L+S}$. Thus, exotic states with $J^{PC} = 0^{+-}, 0^{--}, 1^{-+}, \ldots$ can be realized by an excited gluonic field. Examples for $J^{PC} = 1^{-+}$ states are $\pi_1(1400)$ and $\pi_1(1600)$, which are hybrid meson and tetraquark candidates [1].

The ordinary static potential can be extracted from Wilson loop averages, which are straightforward to compute in lattice gauge theory. Similarly, hybrid static potentials require the computation of Wilson loop-like observables with more complicated spatial transporters, which realize the non-trivial gluonic quantum numbers. For existing lattice studies cf. [2, 3, 4, 5, 6, 8, 7, 9, 10, 11, 12, 13, 14, 15].

2. A brief introduction to lattice QCD hadron spectroscopy

To determine the mass of a hadron, one first has to define a suitable hadron creation operator $\mathcal{O}$ composed of quark and gluon field operators ψ, $\bar{\psi}$ and A_μ, which creates the quantum numbers of the hadron of interest, when applied to the vacuum $|\Omega\rangle$. In a second step one computes the temporal correlation function $C(T)$ of this hadron creation operator using lattice QCD,

$$C(T) \equiv \langle\Omega|\mathcal{O}^\dagger(T)\mathcal{O}(0)|\Omega\rangle = \frac{1}{Z}\int \mathcal{D}[A_\mu,\psi,\bar{\psi}]\,\mathcal{O}^\dagger(T)\mathcal{O}(0)e^{-S_{E,\mathrm{QCD}}[A_\mu,\psi,\bar{\psi}]}, \quad (1)$$

where $S_{E,\mathrm{QCD}}[A_\mu,\psi,\bar{\psi}]$ is the QCD action. The hadron mass M can then be obtained from the exponential decay of $C(T)$ at large temporal separations:

$$C(T) = \sum_n \Big|\langle n|\mathcal{O}|\Omega\rangle\Big|^2 e^{-(E_n-E_\Omega)T} = \Big|\langle 0|\mathcal{O}|\Omega\rangle\Big|^2 e^{-MT}\Big(1+\mathcal{O}(e^{-(E_1-E_0)T})\Big), \quad (2)$$

where $|n\rangle$ denote eigenstates of the QCD Hamiltonian with the quantum numbers of the hadron and $M \equiv E_0 - E_\Omega$.

While a straightforward way to determine M is to fit an exponential function Ae^{-MT} to the lattice results for $C(T)$ at large temporal separations T, it is also common to study the so-called effective mass

$$M_{\text{eff}}(T) \equiv \frac{1}{a}\ln\left(\frac{C(T)}{C(T+a)}\right) \tag{3}$$

(a denotes the lattice spacing) and to fit a constant to the plateau at sufficiently large T,

$$M =_{\text{large } T} M_{\text{eff}}(T). \tag{4}$$

For a more detailed introduction to lattice hadron spectroscopy cf. e.g. [16].

3. The static potential

The static potential $V(R)$ is defined as the energy difference of the lowest state containing a static quark Q and a static antiquark $\bar{Q}$ at separation r and the vacuum. In the absence of quarks of finite mass $V(R)$ is linear for large separations, $\sim \sigma R$, where σ denotes the string tension (cf. fig. 2b), and, therefore, displays the confining property of Yang-Mills theory and QCD. Quite often it is used to set the scale for lattice simulations, e.g. by identifying the resulting lattice string tension with its physical value $\sigma \approx 5.5/\text{fm}^2 \ldots 7.5/\text{fm}^2$.

To compute the static potential, one typically uses the creation operator

$$\mathcal{O} \equiv \overline{Q}(\mathbf{x})\mathcal{S}(\mathbf{x},\mathbf{y})Q(\mathbf{y}) \quad , \quad |\mathbf{x}-\mathbf{y}| = R \tag{5}$$

where $\overline{Q}(\mathbf{x})$ and $Q(\mathbf{y})$ are operators creating a static antiquark and a static quark at $\mathbf{x}$ and $\mathbf{y}$, respectively. $\mathcal{S}(\mathbf{x},\mathbf{y})$ is a product of spatial gauge links along a straight line connecting $\mathbf{x}$ and $\mathbf{y}$, i.e. a parallel transporter realizing a gauge invariant static potential creation operator $\mathcal{O}$.

The integration over the static quark fields Q and $\bar{Q}$ in the path integral (1) can be performed analytically. In a computation without dynamical sea quarks the result is

$$C(T) \propto \Big\langle W(R,T)\Big\rangle \equiv \frac{1}{Z}\int \mathcal{D}[A_\mu]\, W(R,T)e^{-S_{E,\text{YM}}[A_\mu]} \tag{6}$$

with the Yang-Mills action $S_{E,\text{YM}}[A_\mu]$. $W(R,T)$ is a product of gauge links along a closed rectangular path of spatial extent R and temporal extent T, the well known Wilson loop. For a precise definition in terms of gauge links it can be decomposed in four parts, two spatial parallel transporters $\mathcal{S}$ and two temporal parallel transporters $\mathcal{T}$,

$$\mathcal{S}(\mathbf{x},\mathbf{y};t) \equiv \prod_{j=0}^{R/a-1} U_k(\mathbf{x}+ja\mathbf{e}_k,t) \quad , \quad \mathcal{T}(\mathbf{x};t_1,t_2) \equiv \prod_{j=0}^{T/a-1} U_0(\mathbf{x},t_1+ja).$$

Then

$$W(R,T) \equiv \text{Tr}\Big(\mathcal{S}(\mathbf{x},\mathbf{y};0)\mathcal{T}(\mathbf{y};0,t)\Big(\mathcal{S}(\mathbf{x},\mathbf{y};t)\Big)^\dagger\Big(\mathcal{T}(\mathbf{x};0,t)\Big)^\dagger\Big) \tag{7}$$

(cf. fig. 1a).

For each quark antiquark separation R an effective mass or effective potential

$$V_{\text{eff}}(R,T) \equiv \frac{1}{a}\ln\left(\frac{\langle W(R,T)\rangle}{\langle W(R,T+a)\rangle}\right) \tag{8}$$

has to be computed (cf. fig. 2a) as explained in the previous section. Fitting a constant to $V_{\text{eff}}(R,T)$ in the plateau-like region at $T/a = 6 \ldots 8$ yields the static potential $V(R)$ (cf. fig. 2b).

Journal of Physics: Conference Series **599** (2015) 012005 doi:10.1088/1742-6596/599/1/012005

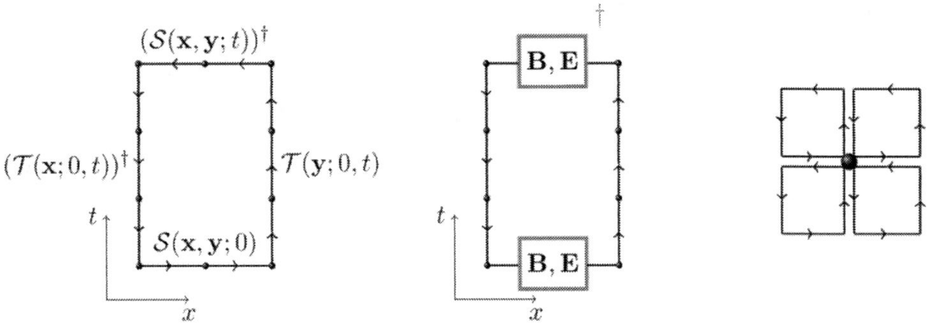

Figure 1: **(a)** Ordinary Wilson loop $W(R = 2, T = 3)$ in the x-t plane. **(b)** Generalized Wilson loop: **B** or **E** fields inserted at the centers of the spatial transporters. **(c)** Average of neighboring plaquettes representing a chromomagnetic or chromoelectric field **B** or **E**.

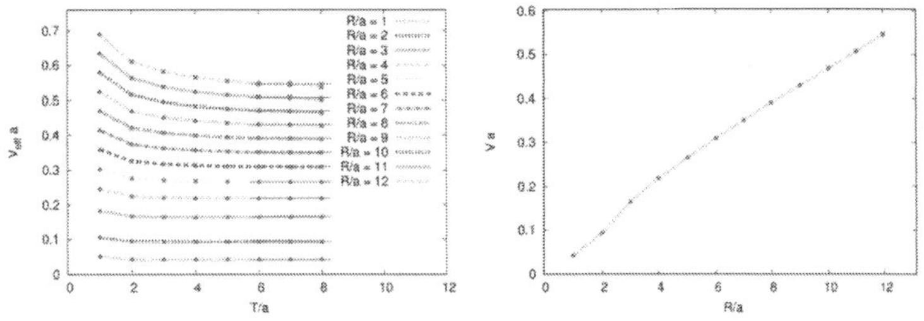

Figure 2: **(a)** Effective potential $V_{\text{eff}}(R, T)$ as a function of T for different quark antiquark separations R in units of the lattice spacing $a \approx 0.073\,\text{fm}$. **(b)** Static potential $V(R)$ in units of the lattice spacing.

4. Hybrid static potentials

States with a static quark and a static antiquark can be classified according to the following three quantum numbers (for a more detailed discussion cf. e.g. [17, 18]):

- Angular momentum J with respect to the axis of separation of the quark antiquark pair. States with $J = 0, \pm 1, \pm 2, \ldots$ are also labeled by $\Sigma, \Pi, \Delta, \ldots$.
- The combination of parity and charge conjugation $P \circ C$. States with $P \circ C = +, -$ are also labeled by g, u.
- The rotational invariant Σ states are either symmetric or antisymmetric with respect to spatial reflections along an axis perpendicular to the axis of separation of the quark antiquark pair denoted by $P_x = +, -$.

For example the ordinary static potential discussed in the previous section has quantum numbers $J^{P_x}_{P \circ C} = \Sigma^+_g$.

Hybrid static potentials correspond to a quark antiquark configuration with an excited gluonic string realizing quantum numbers different from Σ^+_g. One way to implement such quantum numbers in a hybrid static potential creation operator similar to (5) is to insert a suitable gluonic operator at the center of the spatial parallel transporter $\mathcal{S}$. Typically this insertion is a chromomagnetic or chromoelectric field operator **B** or **E** possibly combined with the separation vector $\mathbf{R} \equiv \mathbf{y} - \mathbf{x}$ and/or the covariant derivative **D** (cf. fig. 1b). On a lattice **B** and **E** are expressed in terms of the average of four neighboring plaquettes of gauge links (cf. fig. 1c). The creation operators used in this work are collected in tab. 1. A more detailed discussion of such operators can e.g. be found in [11].

quantum numbers $J^{P_x}_{P\circ C}$	operator insertions
Σ_g^+	1 , $\mathbf{R}\cdot\mathbf{E}$, $\mathbf{R}\cdot(\mathbf{D}\times\mathbf{B})$
Π_g	$\mathbf{R}\times\mathbf{E}$, $\mathbf{R}\times(\mathbf{D}\times\mathbf{B})$
Σ_u^-	$\mathbf{R}\cdot\mathbf{B}$, $\mathbf{R}\cdot(\mathbf{D}\times\mathbf{E})$
Π_u	$\mathbf{R}\times\mathbf{B}$, $\mathbf{R}\times(\mathbf{D}\times\mathbf{E})$
Σ_g^-	$(\mathbf{R}\cdot\mathbf{D})(\mathbf{R}\cdot\mathbf{B})$

Table 1: Hybrid static potential creation operators.

5. Preliminary numerical results

5.1. Lattice setup

We have evaluated Wilson loops and generalized Wilson loops on more than 700 essentially independent gauge link configurations with 24^4 lattice sites. These configurations have been generated with the SU(2) standard Wilson plaquette action using a heatbath algorithm and gauge coupling $\beta = 2.50$. The corresponding lattice spacing is $a \approx 0.073\,\text{fm}$, when identifying the Sommer parameter r_0 with $0.46\,\text{fm}$ [19].

5.2. Hybrid static potentials

In fig. 3 our preliminary results for hybrid static potentials with quantum numbers Σ_g^+, Π_u, Σ_u^- and Σ_g^- are shown. For each of the sectors Σ_g^+, Π_u and Σ_u^- two different creation operators have been used resulting, as expected, in potentials, which are identical within statistical errors. For example the ordinary static potential with $J^{P_x}_{P\circ C} = \Sigma_g^+$ has been extracted from Wilson loops (green curve) and from generalized Wilson loops with insertions $\mathbf{R}\cdot\mathbf{E}$ (yellow curve). Clearly the green curve is of better quality, i.e. exhibits less fluctuations and smaller statistical errors. Consequently, ordinary Wilson loops are much better suited to determine the Σ_g^+ static potential than Wilson loops with insertions $\mathbf{R}\cdot\mathbf{E}$. Similarly, $\mathbf{R}\times\mathbf{B}$ (magenta curve) is superior to $\mathbf{R}\times(\mathbf{D}\times\mathbf{E})$ (orange curve), when computing the Π_u static potential, and $\mathbf{R}\cdot\mathbf{B}$ (blue curve) is superior to $\mathbf{R}\cdot(\mathbf{D}\times\mathbf{E})$ (black curve), when computing the Σ_u^- static potential. For small separations the potentials Π_u and Σ_g^- seem to be repulsive. To exclude that this is merely a cutoff effect, which are expected to be large for $R/a < 2$, we plan to perform similar computations at significantly smaller lattice spacing in the near future.

6. Outlook

Even though we have used a sizable number of more than 700 gauge link configurations, the statistical errors of our hybrid static potential results shown in fig. 3 are quite large, when e.g. compared to results from [11]. This might be due to a possibly different structure of the employed hybrid static potential creation operators. While for our current results we have used local insertions of $\mathbf{B}$ and $\mathbf{E}$ field operators, which might generate a rather poor ground state overlap, in [6, 8] spatially extended creation operators have been proposed, which could result in correlation functions, which are dominated by the ground state already at small temporal separations. We plan to explore this issue in the near future by implementing additional hybrid static potential creation operators.

The ultimate goal is, of course, to arrive at precise results for SU(3) Yang-Mills theory and QCD. This would allow to estimate masses of hybrid mesons by solving a non-relativistic Schrödinger equation with the computed hybrid static potentials, as e.g. proposed in [3, 4, 6, 8, 12, 13, 14]. Moreover, in the context of effective field theories like pNRQCD there is considerable

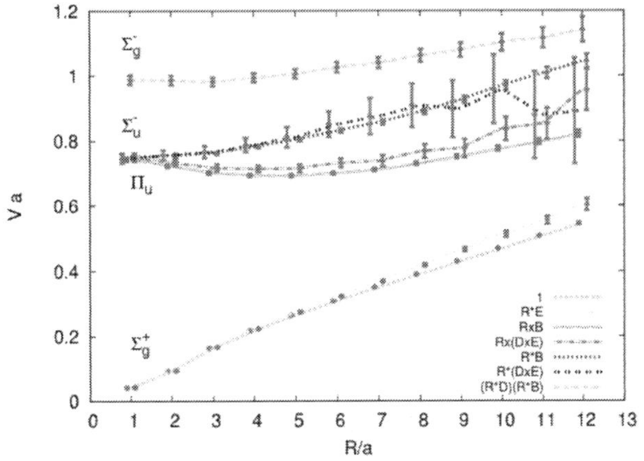

Figure 3: Hybrid static potentials in SU(2) Yang-Mills theory in units of the lattice spacing $a \approx 0.073\,\mathrm{fm}$ (to be able to distinguish different curves, error bars have slightly been shifted horizontally).

interest in the short distance behavior of hybrid static potentials, which is related to gluelump masses [15, 20]. It would also be interesting to compare such lattice results with corresponding model calculations, e.g. [21].

Acknowledgments

We acknowledge helpful discussions with Joshua Berlin, Owe Philipsen and Björn Wagenbach. M.W. acknowledges support by the Emmy Noether Programme of the DFG (German Research Foundation), grant WA 3000/1-1. This work was supported in part by the Helmholtz International Center for FAIR within the framework of the LOEWE program launched by the State of Hesse.

References

[1] W. M. Yao *et al.* [Particle Data Group Collaboration], J. Phys. G **33**, 1 (2006).
[2] K. J. Juge, J. Kuti and C. J. Morningstar, Nucl. Phys. Proc. Suppl. **63**, 326 (1998) [hep-lat/9709131].
[3] M. J. Peardon, Nucl. Phys. Proc. Suppl. **63**, 22 (1998) [hep-lat/9710029].
[4] K. J. Juge, J. Kuti and C. J. Morningstar, AIP Conf. Proc. **432**, 136 (1998) [hep-ph/9711451].
[5] C. Morningstar, K. J. Juge and J. Kuti, hep-lat/9809015.
[6] C. Michael, Nucl. Phys. A **655**, 12 (1999) [hep-ph/9810415].
[7] D. Toussaint, Nucl. Phys. Proc. Suppl. **83**, 151 (2000) [hep-lat/9909088].
[8] C. Michael, PoS **HeavyFlavours8**, 001 (1999) [hep-ph/9911219].
[9] G. S. Bali *et al.* [SESAM and TχL Collaborations], Phys. Rev. D **62**, 054503 (2000) [hep-lat/0003012].
[10] C. Morningstar, AIP Conf. Proc. **619**, 231 (2002) [nucl-th/0110074].
[11] K. J. Juge, J. Kuti and C. Morningstar, Phys. Rev. Lett. **90**, 161601 (2003) [hep-lat/0207004].
[12] C. Michael, hep-lat/0302001.
[13] K. J. Juge, J. Kuti and C. Morningstar, AIP Conf. Proc. **688**, 193 (2004) [nucl-th/0307116].
[14] C. Michael, hep-ph/0308293.
[15] G. S. Bali and A. Pineda, Phys. Rev. D **69**, 094001 (2004) [hep-ph/0310130].
[16] M. Wagner, S. Diehl, T. Kuske and J. Weber, arXiv:1310.1760 [hep-lat].
[17] G. S. Bali *et al.* [SESAM Collaboration], Phys. Rev. D **71**, 114513 (2005) [hep-lat/0505012].
[18] M. Wagner [ETMC Collaboration], PoS LATTICE **2010**, 162 (2010) [arXiv:1008.1538 [hep-lat]].
[19] O. Philipsen and M. Wagner, Phys. Rev. D **89**, 014509 (2014) [arXiv:1305.5957 [hep-lat]].
[20] N. Brambilla, A. Pineda, J. Soto and A. Vairo, Rev. Mod. Phys. **77**, 1423 (2005) [hep-ph/0410047].
[21] O. Andreev, Phys. Rev. D **86**, 065013 (2012) [arXiv:1207.1892 [hep-ph]].

Lattice investigation of heavy meson interactions

Björn Wagenbach[1], **Pedro Bicudo**[2], **Marc Wagner**[1,3]

[1] Goethe-Universität Frankfurt am Main, Institut für Theoretische Physik, Max-von-Laue-Straße 1, D-60438 Frankfurt am Main, Germany

[2] Dep. Física and CFTP, Instituto Superior Técnico, Av. Rovisco Pais, 1049-001 Lisboa, Portugal

[3] European Twisted Mass Collaboration (ETMC)

E-mail: wagenbach@th.physik.uni-frankfurt.de

Abstract. We report on a lattice investigation of heavy meson interactions and of tetraquark candidates with two very heavy quarks. These two quarks are treated in the static limit, while the other two are up, down, strange or charm quarks of finite mass. Various isospin, spin and parity quantum numbers are considered.

1. Introduction

We study the potential of two static quarks in the presence of two quarks of finite mass. While in [1, 2, 3] we have exclusively considered two static antiquarks and two light quarks ($\bar{Q}\bar{Q}ll$), where $l \in \{u, d\}$, here we also use s and c quarks, i.e. investigate $\bar{Q}\bar{Q}ss$ and $\bar{Q}\bar{Q}cc$, to obtain certain insights regarding the quark mass dependence of the static antiquark-antiquark interaction. We also discuss first steps regarding the static quark-antiquark case, i.e. $\bar{Q}Q\bar{l}l$, $\bar{Q}Q\bar{s}s$ and $\bar{Q}Q\bar{c}c$.

$\bar{Q}\bar{Q}qq$ systems as well as $\bar{Q}Q\bar{q}q$ systems have been studied also by other groups (cf. e.g. [4, 5, 6, 7, 8, 9, 10, 11, 12]).

2. Creation operators and trial states

The $\bar{Q}\bar{Q}qq$ and $\bar{Q}Q\bar{q}q$ potentials $V(r)$ are extracted from correlation functions

$$C(t) \equiv \langle\Omega|\mathcal{O}^\dagger(t)\mathcal{O}(0)|\Omega\rangle \tag{1}$$

according to

$$V(r) =_{\text{large } t} V_{\text{eff}}(r,t) \quad , \quad V_{\text{eff}}(r,t) \equiv \frac{1}{a}\ln\left(\frac{C(t)}{C(t+a)}\right), \tag{2}$$

where a is the lattice spacing and $\mathcal{O}$ denote suitable creation operators, which are discussed in detail below. For an introduction to lattice hadron spectroscopy cf. e.g. [13].

2.1. Static-light mesons ("B and $\bar{B}$ mesons")

The starting point are static-light mesons, which either consist of a static quark Q and an antiquark $\bar{q}$ or of a static antiquark $\bar{Q}$ and a quark q with $q \in \{u, d, s, c\}$. These mesons can be labeled by parity $\mathscr{P} = \pm$, by the z-component of the light quark spin $j_z = \pm 1/2$ ($j = 1/2$, because we do not consider gluonic excitations) and in case of $q \in \{u, d\}$ by the z-component of isospin $I_z = \pm 1/2$ ($I = 1/2$). The lightest static-light meson has $\mathscr{P} = -$ and is commonly denoted by S, its heavier parity partner with $\mathscr{P} = +$ by P_-. The static-light meson S is an approximation for B/B^*, B_s/B_s^* and B_c listed in [14].

We use static-light meson trial states

$$\mathcal{O}|\Omega\rangle \equiv \bar{Q}\Gamma q|\Omega\rangle \tag{3}$$

with $\Gamma \in \{\gamma_5, \gamma_0\gamma_5, \gamma_j, \gamma_0\gamma_j\}$ for the S and $\Gamma \in \{1, \gamma_0, \gamma_j\gamma_5, \gamma_0\gamma_j\gamma_5\}$ for the P_- meson. For a more detailed discussion of static-light mesons cf. [15, 16].

2.2. $B\bar{B}$ systems

We are interested in the potential of two static-light mesons, i.e. their energy as a function of their separation r. W.l.o.g. we separate the mesons along the z-axis, i.e. their static antiquark $\bar{Q}$ and quark Q are located at $\vec{r}_1 = (0, 0, +r/2)$ and $\vec{r}_2 = (0, 0, -r/2)$, respectively. The corresponding $B\bar{B}$ trial states are

$$\mathcal{O}|\Omega\rangle \equiv \Gamma_{AB}\tilde{\Gamma}_{CD}\Big(\bar{Q}_C^a(\vec{r}_1)q_A^{(f_1)a}(\vec{r}_1)\Big)\Big(\bar{q}_B^{(f_2)b}(\vec{r}_2)Q_D^b(\vec{r}_2)\Big)|\Omega\rangle \tag{4}$$

($A, B, \ldots$ are spin indices, a, b color indices and $(f_1), (f_2)$ flavor indices). Since there are no interactions involving the static quark spins, one should not couple static spins and spins of finite mass, but contract the static spin indices with $\tilde{\Gamma} \in \{\gamma_5, \gamma_0\gamma_5, \gamma_3, \gamma_0\gamma_3, \gamma_1, \gamma_2, \gamma_0\gamma_1, \gamma_0\gamma_2\}$. This results in a non-vanishing correlation function independent of $\tilde{\Gamma}$.

The separation of the static quark and the static antiquark restricts rotational symmetry to rotations around the axis of separation, i.e. the z-axis. Therefore, and since there are no interactions involving the static quark spins, we can label states by the z-component of the light quark spin $j_z = -1, 0, +1$. For $j_z = 0$, i.e. for rotationally invariant states, spatial reflections along an axis perpendicular to the axis of separation are also a symmetry operation (w.l.o.g. we choose the x-axis). The corresponding quantum number is $\mathscr{P}_x = \pm$. $\mathscr{P}_x$ can be used as a quantum number also for $j_z \neq 0$ states, if we use $|j_z|$ instead of j_z. Parity $\mathscr{P}$ is not a symmetry, since it exchanges the positions of the static quark and the static antiquark. However, parity combined with charge conjugation, $\mathscr{P} \circ C$ is a symmetry and, therefore, a quantum number. When $q, \bar{q} \in \{u, d\}$, isospin $I \in \{0, 1\}$ and its z-component $I_z \in \{-1, 0, +1\}$ are also quantum numbers. In summary, there are up to five quantum numbers, which label $B\bar{B}$ states, $(I, I_z, |j_z|, \mathscr{P} \circ C, \mathscr{P}_x)$.

2.3. BB systems (and $\bar{B}\bar{B}$ systems)

We use BB trial states

$$\mathcal{O}|\Omega\rangle \equiv (\mathcal{C}\Gamma)_{AB}\tilde{\Gamma}_{CD}\Big(\bar{Q}_C^a(\vec{r}_1)\psi_A^{(f_1)a}(\vec{r}_1)\Big)\Big(\bar{Q}_D^b(\vec{r}_2)\psi_B^{(f_2)b}(\vec{r}_2)\Big)|\Omega\rangle \tag{5}$$

with $\tilde{\Gamma} \in \{1, \gamma_0, \gamma_3\gamma_5, \gamma_1\gamma_2, \gamma_1\gamma_5, \gamma_2\gamma_5, \gamma_2\gamma_3, \gamma_1\gamma_3\}$ ($\mathcal{C} \equiv \gamma_0\gamma_2$ denotes the charge conjugation matrix). Arguments similar to those of the previous subsection lead to quantum numbers $(I, I_z, |j_z|, \mathscr{P}, \mathscr{P}_x)$. For a more detailed discussion cf. [1, 2].

3. Lattice setup

We use three ensembles of gauge link configurations generated by the European Twisted Mass Collaboration (ETMC) (cf. Table 1). For the $\bar{Q}\bar{Q}qq$ potentials we use $N_f = 2$ ensembles with lattice spacing $a \approx 0.079\,\text{fm}$ for $q \in \{u, d\}$ and an even finer lattice spacing $a \approx 0.042\,\text{fm}$ for $q \in \{s, c\}$, because in the latter case the potentials are quite narrow. Existing $\bar{Q}Q\bar{q}q$ results are

Ensemble	N_f	β	$(L/a)^3 \times (T/a)$	$a\mu_l$	$a\mu_\sigma$	$a\mu_\delta$	a	m_π
A40.24	2	3.90	$24^3 \times 48$	0.00400	-	-	0.079 fm	340 MeV
E17.32	2	4.35	$32^3 \times 64$	0.00175	-	-	0.042 fm	352 MeV
A40.24	2+1+1	1.90	$24^3 \times 48$	0.00400	0.15	0.19	0.086 fm	332 MeV

Table 1. ETMC gauge link ensembles used in this work.

rather preliminary and have been obtained exclusively with $q = c$ and the $N_f = 2{+}1{+}1$ ensemble with $a \approx 0.086\,\text{fm}$. For details regarding these ETMC gauge link ensembles cf. [17, 18, 19, 20, 21].

Correlation functions have been computed using around 100 gauge link configurations from each of the three ensembles. We have checked that these correlation functions transform appropriately with respect to the symmetry transformations (1) twisted mass time reversal, (2) twisted mass parity, (3) twisted mass γ_5-hermiticity, (4) charge conjugation and (5) cubic rotations. In a second step we have averaged correlation functions related by those symmetries to reduce statistical errors.

4. Numerical results

4.1. $\bar{Q}\bar{Q}qq$ potentials

In the following we focus on the attractive channels between ground state static-light mesons (S mesons). For $q \in \{u, d\}$ there is a more attractive scalar isosinglet ($qq = (ud - du)/\sqrt{2}$, $\Gamma = \gamma_5 + \gamma_0\gamma_5$ corresponding to quantum numbers $(I, |j_z|, \mathscr{P}, \mathscr{P}_x) = (0, 0, -, +)$) and a less attractive vector isotriplet ($qq \in \{uu, (ud{+}du)/\sqrt{2}, dd\}$, $\Gamma = \gamma_j{+}\gamma_0\gamma_j$ corresponding to quantum numbers $(I, |j_z|, \mathscr{P}, \mathscr{P}_x) = (1, \{0, 1\}, -, \pm)$). For $qq = ss$ there is only a single attractive channel, the equivalent of the vector isotriplet. To study also the scalar isosinglet with s quarks, we consider two quark flavors with the mass of the s quark, i.e. $qq = (s_1 s_2 - s_2 s_1)/\sqrt{2}$. Similarly we consider $qq = (c_1 c_2 - c_2 c_1)/\sqrt{2}$ to study a charm scalar isosinglet.

Proceeding as in [3] we perform χ^2 minimizing fits of

$$V(r) \quad = \quad -\frac{\alpha}{r} \exp\left(-\left(\frac{r}{d}\right)^p\right) \tag{6}$$

with respect to the parameters d (light isotriplet), (d, α) ($q = s$ or $q = c$) or (d, α, p) (light isosinglet) to the lattice results for the $\bar{Q}\bar{Q}qq$ potentials. The resulting functions $V(r)$ are shown in Figure 1.

To determine, whether the investigated mesons may form a bound state, i.e. a tetraquark, we insert the potentials shown in Figure 1 into Schrödinger's equation with reduced mass $\mu \equiv m(S)/2$ and solve it numerically (cf. [3] for details). While there is strong indication for a bound state in the light scalar isosinglet channel, there seems to be no binding for the light vector isotriplet, or when $q = s$ or $q = c$. To quantify these statements, we list in Table 2 the factor by which the reduced mass μ has to be multiplied to obtain a bound state with confidence level $1\,\sigma$ and $2\,\sigma$, respectively (the factors ≤ 1.0 in the light scalar isosinglet indicate binding). These results clearly show that meson-meson bound states are more likely to exist for B mesons than for B_s or B_c mesons. In other words it seems to be essential for a tetraquark to have both heavy quarks (leading a large reduced mass μ) and light quarks (resulting in a deep and wide potential).

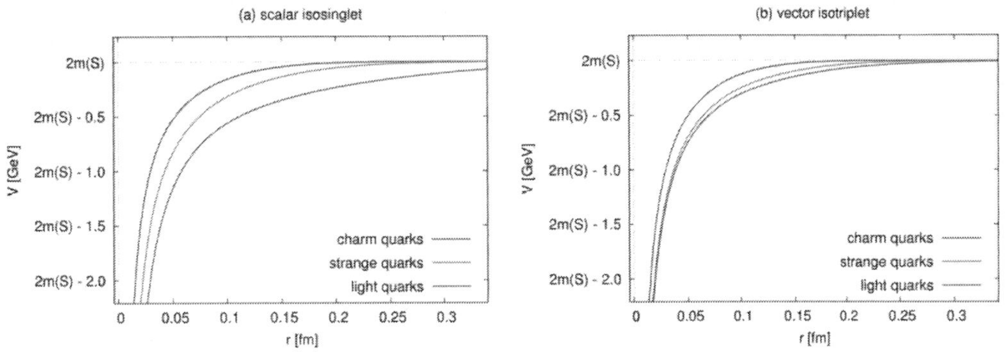

Figure 1. $\bar{Q}\bar{Q}qq$ potentials (6) for $q = u/d$, $q = s$ and $q = c$ (error bands are not shown). **(a)** Scalar isosinglet. **(b)** Vector isotriplet.

flavor	light		strange		charm	
confidence level for binding	$1\,\sigma$	$2\,\sigma$	$1\,\sigma$	$2\,\sigma$	$1\,\sigma$	$2\,\sigma$
scalar isosinglet	0.8	1.0	1.9	2.2	3.1	3.2
vector isotriplet	1.9	2.1	2.5	2.7	3.4	3.5

Table 2. Factors, by which the reduced mass $\mu = m(S)/2$ in Schrödinger's equation has to be multiplied to obtain a four-quark bound state with confidence level $1\,\sigma$ and $2\,\sigma$, respectively.

4.2. $\bar{Q}Q\bar{q}q$ potentials

At the moment there are only preliminary results for $\bar{Q}Q\bar{q}q$ potentials corresponding to isospin $I = 1$ and $q = c$, i.e. $\bar{q}q = (\bar{c}_1 c_2 - \bar{c}_2 c_1)/\sqrt{2}$. Interestingly we observed that all these potentials are attractive, while in the $\bar{Q}\bar{Q}qq$ case only half of them are attractive and the other half is repulsive. This can be understood in a qualitative way by comparing the potential of $\bar{Q}Q$ and of $\bar{Q}\bar{Q}$ generated by one-gluon exchange. For $\bar{Q}\bar{Q}$ the Pauli principle applied to qq implies either a symmetric (sextet) or an antisymmetric (triplet) color orientation of the static quarks corresponding to a repulsive or attractive interaction, respectively. For $\bar{Q}Q$ no such restriction is present, i.e. all channels contain contributions of the attractive color singlet, which dominates the repulsive color octet.

$I = 0$ requires the computation of an additional diagram and u/d and s quarks are more demanding with respect to HPC resources than c quarks. We expect corresponding results to be available soon.

5. Conclusions

We have obtained insights regarding the quark mass dependence of $\bar{Q}\bar{Q}qq$ potentials, which suggest that tetraquark states with two heavy $\bar{b}$ antiquarks seem to be more likely to exist, when there are also two light u/d quarks involved but not s or c quarks.

Preliminary results for $\bar{Q}Q\bar{q}q$ potentials indicate that there are only attractive channels, which is in contrast to the $\bar{Q}\bar{Q}qq$ case.

Acknowledgments
We thank Joshua Berlin, Owe Philipsen, Annabelle Uenver-Thiele and Philipp Wolf for helpful discussions. M.W. acknowledges support by the Emmy Noether Programme of the DFG (German Research Foundation), grant WA 3000/1-1. This work was supported in part by the Helmholtz International Center for FAIR within the framework of the LOEWE program launched by the State of Hesse.

References

[1] M. Wagner [ETM Collaboration], PoS LATTICE **2010**, 162 (2010) [arXiv:1008.1538 [hep-lat]].
[2] M. Wagner [ETM Collaboration], Acta Phys. Polon. Supp. **4**, 747 (2011) [arXiv:1103.5147 [hep-lat]].
[3] P. Bicudo, M. Wagner, Phys. Rev. D **87**, no.11, 114511 (2013) [arXiv:1209.6274 [hep-ph]].
[4] C. Stewart and R. Koniuk, Phys. Rev. D **57**, 5581 (1998) [hep-lat/9803003].
[5] C. Michael *et al.* [UKQCD Collaboration], Phys. Rev. D **60**, 054012 (1999) [hep-lat/9901007].
[6] M. S. Cook and H. R. Fiebig, [hep-lat/0210054].
[7] G. Bali *et al.* [SESAM Collaboration], Phys. Rev. D **71**, 114513 (2005) [hep-lat/0505012].
[8] T. Doi, T. T. Takahashi and H. Suganuma, AIP Conf. Proc. **842**, 246 (2006) [hep-lat/0601008].
[9] W. Detmold, K. Orginos and M. J. Savage, Phys. Rev. D **76**, 114503 (2007) [hep-lat/0703009].
[10] G. Bali *et al.* [QCDSF Collaboration], PoS LATTICE **2010**, 142 (2010) [arXiv:1011.0571 [hep-lat]].
[11] G. Bali *et al.* [QCDSF Collaboration], PoS LATTICE **2011**, 123 (2011) [arXiv:1111.2222 [hep-lat]].
[12] Z. S. Brown and K. Orginos, Phys. Rev. D **86**, 114506 (2012) [arXiv:1210.1953 [hep-lat]].
[13] M. Wagner, S. Diehl, T. Kuske and J. Weber, arXiv:1310.1760 [hep-lat].
[14] K. A. Olive *et al.* [Particle Data Group Collaboration], Chin. Phys. C, 38, 090001 (2014).
[15] K. Jansen *et al.* [ETM Collaboration], JHEP **0812**, 058 (2008) [arXiv:0810.1843 [hep-lat]].
[16] C. Michael *et al.* [ETM Collaboration], JHEP **1008**, 009 (2010) [arXiv:1004.4235 [hep-lat]].
[17] P. Boucaud *et al.* [ETM Collaboration], Comput. Phys. Commun. **179**, 695 (2008) [arXiv:0803.0224 [hep-lat]].
[18] R. Baron *et al.* [ETM Collaboration], JHEP **1008**, 097 (2010) [arXiv:0911.5061 [hep-lat]].
[19] R. Baron *et al.* [ETM Collaboration], JHEP **1006**, 111 (2010) [arXiv:1004.5284 [hep-lat]].
[20] K. Jansen *et al.* [ETM Collaboration], JHEP **1201**, 025 (2012) [arXiv:1110.6859 [hep-ph]].
[21] K. Cichy, K. Jansen and P. Korcyl, Nucl. Phys. B **865**, 268 (2012) [arXiv:1207.0628 [hep-lat]].

Antibaryon–nucleus bound states

J. Hrtánková[1,2], J. Mareš[1]

[1] Nuclear Physics Institute, 25068 Řež, Czech Republic
[2] Czech Technical University in Prague, Faculty of Nuclear Sciences and Physical Engineering, Břehová 7, 115 19 Prague 1, Czech Republic

E-mail: hrtankova@ujf.cas.cz

Abstract. We calculated antibaryon ($\bar{B} = \bar{p}$, $\bar{\Lambda}$, $\bar{\Sigma}$, $\bar{\Xi}$) bound states in selected nuclei within the relativistic mean-field (RMF) model. The G-parity motivated $\bar{B}$–meson coupling constants were scaled to yield corresponding potentials consistent with available experimental data. Large polarization of the nuclear core caused by $\bar{B}$ was confirmed. The $\bar{p}$ annihilation in the nuclear medium was incorporated by including a phenomenological imaginary part of the optical potential. The calculations using a complex $\bar{p}$–nucleus potential were performed fully self-consistently. The $\bar{p}$ widths significantly decrease when the phase space reduction is considered for $\bar{p}$ annihilation products, but they still remain sizeable for potentials consistent with $\bar{p}$–atom data.

1. Introduction

The study of antibaryon–nucleus interactions has attracted increasing interest in recent years at the prospect of future experiments at the FAIR facility [1]. In particular, much attention has been devoted to the antiproton–nucleus interaction and the possibility of formation of $\bar{p}$–nucleus bound states [2, 3, 4]. Exploring the $\bar{p}$–nucleus interaction could provide valuable information about the behavior of the antiproton in the nuclear medium as well as nuclear dynamics. One of the motivations for our study of $\bar{p}$–nucleus bound states is the conjecture that the considerable suppression of the phase space for the $\bar{p}$ annihilation products in the nuclear medium could lead to relatively long living $\bar{p}$ inside the nucleus [2].

In this contribution, we report on our recent fully self-consistent calculations of antibaryon–nucleus bound states within the relativistic mean-field model [5]. The behavior of an antibaryon in the nuclear medium and the dynamical effects caused by the presence of the antibaryon in the nucleus were studied for several selected nuclei. Special attention was devoted to the $\bar{p}$–nucleus interaction including $\bar{p}$ absorption in the nucleus.

In Section 2, a brief description of the underlying model is given. Few representative results of our calculations are presented and discussed in Section 3.

2. Model

In the present work, antibaryon–nucleus bound states are studied within the framework of the RMF approach applied to a system of nucleons and one antibaryon ($\bar{B} = \bar{p}$, $\bar{\Lambda}$, $\bar{\Sigma}$, $\bar{\Xi}$). The interaction among (anti)baryons is mediated by the exchange of the scalar (σ) and vector (ω_μ, $\vec{\rho}_\mu$) meson fields, and the massless photon field A_μ. The standard Lagrangian density $\mathcal{L}_N$ for nucleonic sector is extended by the Lagrangian density $\mathcal{L}_{\bar{B}}$ describing the antibaryon interaction with the nuclear medium (see ref. [6] for details). The variational principle yields the equations of motion for the hadron fields involved. The Dirac equations for nucleons and antibaryon read:

$$[-i\vec{\alpha}\vec{\nabla} + \beta(m_j + S_j) + V_j]\psi_j^\alpha = \epsilon_j^\alpha \psi_j^\alpha, \quad j = N, \bar{B} \;, \tag{1}$$

where

$$S_j = g_{\sigma j}\sigma, \quad V_j = g_{\omega j}\omega_0 + g_{\rho j}\rho_0\tau_3 + e_j\frac{1+\tau_3}{2}A_0 \tag{2}$$

are the scalar and vector potentials. Here, α denotes single particle states, m_j stands for (anti)baryon masses and $g_{\sigma j}, g_{\omega j}, g_{\rho j}$, and e_j are (anti)baryon coupling constants to corresponding fields. The presence of $\bar{B}$ induces additional source terms in the Klein–Gordon equations for the meson fields:

$$\begin{aligned}
(-\triangle + m_\sigma^2 + g_2\sigma + g_3\sigma^2)\sigma &= -g_{\sigma N}\rho_{SN} - g_{\sigma\bar{B}}\rho_{S\bar{B}} \\
(-\triangle + m_\omega^2 + d\omega_0^2)\omega_0 &= g_{\omega N}\rho_{VN} + g_{\omega\bar{B}}\rho_{V\bar{B}} \\
(-\triangle + m_\rho^2)\rho_0 &= g_{\rho N}\rho_{IN} + g_{\rho\bar{B}}\rho_{I\bar{B}} \\
-\triangle A_0 &= e_N\rho_{QN} + e_{\bar{B}}\rho_{Q\bar{B}} \;,
\end{aligned} \tag{3}$$

where $\rho_{Sj}, \rho_{Vj}, \rho_{Ij}$ and ρ_{Qj} are the scalar, vector, isovector and charge densities, respectively, and $m_\sigma, m_\omega, m_\rho$ are the masses of considered mesons. The system of coupled Dirac (1) and Klein–Gordon (3) equations represents a self-consistent problem which is to be solved by iterative procedure.

The values of the nucleon–meson coupling constants and meson masses were adopted from the nonlinear RMF model TM1(2) [7] for heavy (light) nuclei. We used also the density–dependent model TW99 [8, 9] in which the couplings are a function of the baryon density. The hyperon–meson coupling constants for the ω and ρ fields are obtained using SU(6) symmetry relations. The coupling constants for the σ field are constrained by available experimental data — Λ hypernuclei [10], Σ atoms [11], and Ξ production in (K^+, K^-) reaction [12].

In the RMF approach, the nuclear ground state is well described by the attractive scalar potential $S \simeq -350$ MeV and by the repulsive vector potential $V \simeq 300$ MeV. The resulting central potential acting on a nucleon in a nucleus is then approximately $S + V \simeq -50$ MeV deep. The $\bar{B}$–nucleus interaction is constructed from the B–nucleus interaction with the help of the G-parity transformation: the vector potential generated by the ω meson exchange thus changes its sign and becomes attractive. As a consequence, the total potential acting on a $\bar{B}$ will be strongly attractive. In particular, the $\bar{p}$–nuclear potential would be $S - V \simeq -650$ MeV deep. However, the G-parity transformation should be regarded as a mere starting point to determine the $\bar{B}$–meson coupling constants. Various many-body effects, as well as the presence of strong annihilation channels could cause significant deviations from the G-parity values in the nuclear medium. Indeed, the available experimental data from $\bar{p}$ atoms [13] and $\bar{p}$ scattering off nuclei [14] suggest that the depth of the real part of the $\bar{p}$–nucleus potential is in the range $-(100 - 300)$ MeV in the nuclear interior. Therefore, we introduce a scaling factor $\xi \in \langle 0, 1 \rangle$ for the antibaryon–meson coupling constants which are in the following relation to the baryon–meson couplings:

$$g_{\sigma\bar{B}} = \xi\, g_{\sigma N}, \quad g_{\omega\bar{B}} = -\xi\, g_{\omega N}, \quad g_{\rho\bar{B}} = \xi\, g_{\rho N} \;. \tag{4}$$

The annihilation of an antibaryon inside the nuclear medium is an inseparable part of any realistic description of the $\bar{B}$–nucleus interaction. In our calculations, only the $\bar{p}$ absorption in the nucleus has been considered so far. Since the RMF model does not address directly the

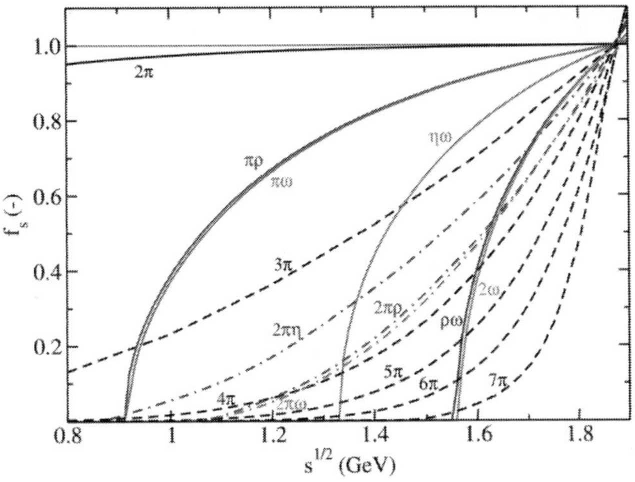

Figure 1. The phase space suppression factors f_s as a function of the c.m. energy $\sqrt{s}$.

absorption of the $\bar{p}$ in the nucleus we adopted the imaginary part of the optical potential in a '$t\rho$' form from optical model phenomenology [13]:

$$2\mu \mathrm{Im} V_{\mathrm{opt}}(r) = -4\pi \left(1 + \frac{\mu}{m_N}\frac{A-1}{A}\right) \mathrm{Im} b_0 \rho(r) \ , \tag{5}$$

where μ is the $\bar{p}$-nucleus reduced mass. While the density ρ was treated as a dynamical quantity determined within the RMF model, the parameter b_0 was constrained by fits to $\bar{p}$-atomic data [13]. The global fits to the $\bar{p}$-atomic data give a single value for the imaginary part of b_0, $\mathrm{Im}b_0 = 1.9$ fm for all nuclei considered.

The energy available for the $\bar{p}$ annihilation in the nuclear medium is usually expressed as $\sqrt{s} = m_{\bar{p}} + m_N - B_{\bar{p}} - B_N$, where $B_{\bar{p}}$ and B_N is the $\bar{p}$ and nucleon binding energy, respectively. Therefore, the phase space available for the annihilation products is considerably suppressed for the deeply bound antiproton.

The phase space suppression factors (f_s) for two body decay are given by [15]

$$f_s = \frac{M^2}{s}\sqrt{\frac{[s-(m_1+m_2)^2][s-(m_1-m_2)^2]}{[M^2-(m_1+m_2)][M^2-(m_1-m_2)^2]}}\Theta(\sqrt{s}-m_1-m_2) \ , \tag{6}$$

where m_1, m_2 are the masses of the annihilation products and $M = m_{\bar{p}} + m_N$.

For channels containing more than 2 particles in the final state the suppression factors f_s were evaluated with the help of Monte Carlo simulation tool PLUTO [16]. In Figure 1, we present the phase space suppression factors as a function of the center-of-mass energy $\sqrt{s}$ for considered annihilation channels. As the energy $\sqrt{s}$ decreases many channels become significantly suppressed or even closed which could lead to much longer lifetime of $\bar{p}$ in a nucleus.

3. Results

We applied the formalism introduced in the previous section to self-consistent calculations of $\bar{p}$, $\bar{\Lambda}$, $\bar{\Sigma}$, $\bar{\Xi}$ bound states in selected nuclei across the periodic table. First, we did not consider absorption of an antibaryon in a nucleus. Our calculations within the TM model confirmed substantial polarization of the nuclear core caused by the antibaryon embedded in the nucleus [2].

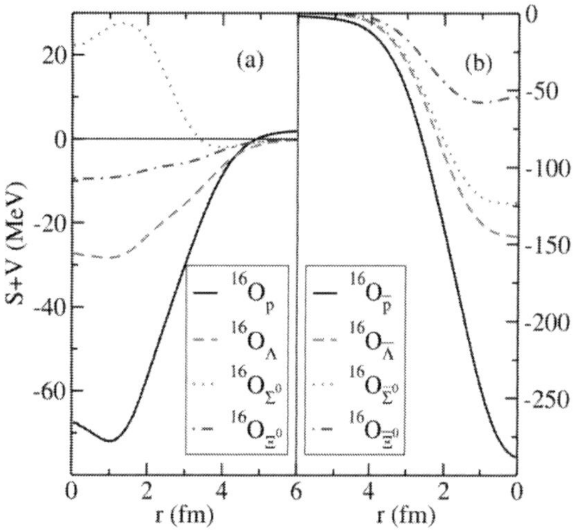

Figure 2. The B–nucleus (a) and $\bar{B}$–nucleus (b) potentials in ^{16}O, calculated dynamically for $\xi = 0.2$ in the TM2 model.

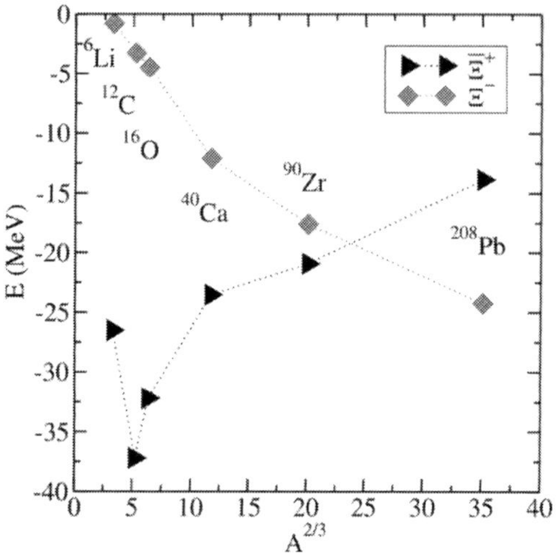

Figure 3. Single particle energies of Ξ^- and $\bar{\Xi}^+$ for $\xi = 0.2$ in various nuclei, calculated dynamically in the TM model.

The nucleon single particle energies are significantly affected by the presence of $\bar{B}$ and the total binding energies increase considerably, as well. The nucleon densities in $\bar{p}$ nuclei reach $2-3$ times the nuclear matter density. The RMF models with constant couplings do not have to describe correctly the properties of nuclear matter when extrapolated to such high densities. Therefore, we performed calculations of $\bar{p}$ nuclei within the density–dependent model TW99, as well. We obtained very similar results as for the TM model and thus confirmed only small model dependence of our calculations.

Figure 2 shows the total potential acting on an extra baryon (a) and on an extra antibaryon (b) in $1s$ state in ^{16}O, calculated dynamically in the TM2 model. The scaling parameter is chosen to be $\xi = 0.2$ as this value gives the $\bar{p}$ potential comparable with the available experimental data. We assume the same scaling parameter also for antihyperons, since there is no reliable experimental information on the in-medium antihyperon potentials. The potentials acting on antibaryons are fairly deep in the central region of the nucleus in contrast to the baryon potentials (notice that the potential for Σ^0 is even repulsive while the potential for $\bar{\Sigma}^0$ is strongly attractive). Such strongly attractive potentials yield deeply bound states of antibaryons in atomic nuclei.

Figure 3 presents a comparison between the Ξ^- and $\bar{\Xi}^+$ $1s$ single particle energies in various nuclei across the periodic table, calculated dynamically in the TM model. The $\bar{\Xi}^+$ coupling constants are scaled by factor $\xi = 0.2$. The binding energy of Ξ^- is increasing with the number of nucleons in the nucleus. The $\bar{\Xi}^+$ binding energy follows the opposite trend and in Pb it is even less bound than Ξ^-. This can be explained by enhanced Coulomb repulsion felt by $\bar{\Xi}^+$ in heavier nuclei.

We performed calculations of $\bar{p}$ nuclei using a complex potential describing the $\bar{p}$ annihilation in the nuclear medium. The results of static as well as dynamical calculations with the real potential, complex potential, and complex potential with the suppression factors f_s for $\bar{p}$ bound in ^{16}O are presented in Table 1. The scaling of the $\bar{p}$–meson coupling constants is chosen to be $\xi = 0.2$. The static calculations, which do not account for the core polarization effects, give approximately the same values of the $\bar{p}$ single particle energy for all three cases. The single particle energies calculated dynamically are larger, which indicates that the polarization of the

Table 1. The $1s$ single particle energies $E_{\bar{p}}$ and widths $\Gamma_{\bar{p}}$ (in MeV) in $^{16}\mathrm{O}_{\bar{p}}$, calculated dynamically (Dyn) and statically (Stat) with the real, complex and complex with f_s potentials (TM2 model), consistent with $\bar{p}$-atom data.

	Real		Complex		Complex + f_s	
	Dyn	Stat	Dyn	Stat	Dyn	Stat
$E_{\bar{p}}$	193.7	137.1	175.6	134.6	190.2	136.1
$\Gamma_{\bar{p}}$	-	-	552.3	293.3	232.5	165.0

core nucleus is significant (even if the $\bar{p}$ absorption is taken into account). When the effect of the phase space suppression is considered the $\bar{p}$ annihilation width is substantially suppressed (compare 552.3 MeV vs. 232.5 MeV in the last row of Table 1). However, the $\bar{p}$ width is still considerable for relevant $\bar{p}$ potentials consistent with the $\bar{p}$ data.

The annihilation of the $\bar{p}$ with a nucleon takes place in a nucleus. Therefore, the momentum dependent term in Mandelstam variable $s = (E_N + E_{\bar{p}})^2 - (\vec{p}_N + \vec{p}_{\bar{p}})^2$ is non-negligible in contrast to two body frame [17]. Our self-consistent evaluation of $\sqrt{s}$ by considering the momenta of annihilating partners leads to an additional downward energy shift. As a consequence, the $\bar{p}$ width in $^{16}\mathrm{O}_{\bar{p}}$ is reduced by additional ≈ 50 MeV. We conclude that even after taking into account the phase space suppression corresponding to self-consistent treatment of $\sqrt{s}$ including the $\bar{p}$ and N momenta, the $\bar{p}$ annihilation widths in nuclei remain sizeable. The corresponding lifetime of the $\bar{p}$ in the nuclear medium is $\simeq 1$ fm/c.

Acknowledgements
This work was supported by GACR Grant No. P203/12/2126. We thank Pavel Tlustý for his assistance during Monte Carlo simulations using PLUTO. J. Hrtánková acknowledges financial support from CTU-SGS Grant No. SGS13/216/OHK4/3T/14.

References
[1] The FAIR facility [online] URL:<http://www.fair-center.eu/index.php?id=1>
[2] Bürvenich T J, Greiner W, Mishustin I N, Satarov L M and Stöcker H 2005 *Phys. Rev.* C **71** 035201
[3] Larionov A B, Mishustin I N, Satarov L M and Greiner W 2010 *Phys. Rev.* C **82** 024602
[4] Gaitanos T, Kaskulov M, Lenske H 2011 *Phys. Lett.* B **703** 193–198
[5] Serot B D and Walecka J D 1986 *Adv. Nucl. Phys.* **16** 1
[6] Hrtánková J 2013, Master's Thesis, Czech Technical University in Prague, URL:<http://physics.fjfi.cvut.cz/publications/ejcf/DP_Jaroslava_Hrtankova.pdf>
[7] Sugahara Y and Toki H 1994 *Nucl. Phys.* A **579** 557–572
[8] Typel S and Wolter H H 1999 *Nucl. Phys.* A **656** 331–364
[9] Long W, Meng J, Van Giai N and Zhou S 2004 *Phys. Rev.* C **69** 034319
[10] Mareš J and Jennings B K 1994 *Phys. Rev.* C **49** 2472
[11] Mareš J, Friedman E, Gal A and Jennings B K 1995 *Nucl. Phys.* A **594** 311–324
[12] Khaustov P et al. 2000 *Phys. Rev.* **61** 054603
[13] Friedman E, Gal A and Mareš J 2005 *Nucl. Phys.* A **761** 283–295
[14] Walker G E, Goodman Ch D and Olmer C (Eds.) 1985 *Antinucleon– and Nucleon–Nucleus Interaction*, (New York: Plenum Press)
[15] *Particle data group* [online] URL:<http://pdg.lbl.gov/2013/reviews/rpp2013-rev-kinematics.pdf>
[16] *A Monte Carlo Simulation tool PLUTO* [online] URL:<http://www-hades.gsi.de/?q=pluto>
[17] Cieplý A, Friedman E, Gal A, Gazda D and Mareš J 2011 *Phys. Lett.* B **702** 402–407

Journal of Physics: Conference Series **599** (2015) 012008 doi:10.1088/1742-6596/599/1/012008

Partons in the chiral periphery of the nucleon

Carlos G. Granados

Uppsala University, Box 516, Uppsala 751 20, Sweden

E-mail: `carlos.granados@physics.uu.se`

Abstract. We introduce transverse densities in the study of parton dynamics in the proton's peripheral region. We calculate these densities using chiral perturbation theory (χPT) and parametrize the long distance structure of the nucleon in a model independent framework in which we identify chiral ($b \sim O(1/M_\pi)$) and molecular ($b \sim O(M_N^2/(M_\pi^3))$) parametrical regions. Through the light cone formulation of the nucleon's electromagnetic current in χPT, one calculates transverse densities from local products of light cone wave functions of a pion-nucleon system. These products are understood as 2-dimensional parton distributions and as such are universal in processes that probe the nucleon's periphery. This universality and also that of the corresponding transverse density can be tested and used in experimental and phenomenological studies of reactions such as high energy proton-nucleon/nucleus collisions and in electron proton scattering as well as in low Q^2 extraction of baryonic form factors.

1. Introduction

The internal structure of the nucleon has been investigated for decades mainly by extracting structure functions from deep inelastic scattering experiments at dedicated facilities. What these structure functions can tell us about the nucleon relies on how they are interpreted through the parton model as superposition of parton distributions in which what the electromagnetic or electro-weak probe sees is the structure of the nucleon in the infinite momentum frame as a collection of quasi-free point-like particles. The parton model predicted the scale invariance of these distributions at high momentum transfer, away from the resonance region, and quantum chromodynamics (QCD) provided corrections to this approximation by way of evolution equations through which parton distributions take into account the fact that the internal dynamics of the nucleon is ultimately dictated by a relativistic quantum field theory. However, despite their universality, and the acknowledgment of QCD, a first principle calculation of these distributions has been prevented by the nonperturbative nature of hadrons as relativistic multiparticle systems governed by QCD. These limitations significantly reduce the quantitative insight into the origin of the global properties of the nucleon that can be gained through QCD alone, and model dependent descriptions supported by phenomenology have to be invoked.

At low energies however, the dynamical description of the strong interaction can be approached by means of effective theories. Chiral effective field theory in particular provides a general framework which arises from the chiral invariance of the QCD Lagrangian with massless quarks. The spontaneous breaking of this symmetry permits a quantitative description of hadron interactions through perturbative expansions dictated by power counting schemes on the pion mass or momentum. One can also make use of χPT for quantifying certain features of the structure of the nucleon that are relevant at the momentum or spatial scale characteristic of chiral dynamics. As opposed to what happens at high momentum, the structure of the nucleon at large distance scales can be addressed in a model independent manner through χPT. We define the chiral periphery as the region in impact parameter space in which the internal dynamics of the nucleon is governed by leading terms in χPT.

We identify the chiral periphery through transverse densities, which are shown to profile the nucleon's spatial structure in characteristic parametric regions. In the nucleon's periphery,these transverse densities are computed from form factors in turn obtained from invariant χPT, or as it will be explained below, from pion-nucleon wave-functions derived in the light front formulation. The latter allows a mechanical interpretation of transverse densities, giving them their quantum classical meaning in impact parameter space which in turns directly connects the associated experimentally observable form factor to a space-time picture of the nucleon as a relativistic multiparticle system, a feature that form factors by themselves cannot provide.

2. Transverse densities and parametric regions

For the electromagnetic form factors F_1 and F_2 that covariantly expand the electromagnetic current, the corresponding transverse densities are defined through [1],

$$\rho_{1,2}(b) \equiv \int \frac{d^2\Delta}{(2\pi)^2} e^{-i\mathbf{\Delta}_T\cdot\mathbf{b}} F_{1,2}(-\Delta^2). \quad (1)$$

The above equation is modified by using dispersion relations that use the analytic properties of the form factors. Then, to calculate the corresponding transverse densities, it suffices with computing the imaginary part of the form factors or spectral functions along the 2-pion cut in the positive real axis of the complex $t = \Delta^2$ plane [2, 3],

$$\rho_{1,2}(b) \equiv \int_{4M_\pi^2}^{\infty} \frac{dt}{2\pi} K_0(\sqrt{t}b) \frac{\mathrm{Im}F_{1,2}(t)}{\pi}, \quad (2)$$

in which $K_0(\sqrt{t}b)$ is a modified Bessel function which filters out contributions from large momentum modes ($\sqrt{t} > 1/b$) to the transverse density parameterized by b. Transverse densities then become a tool for identifying the kinematic structure and potentially the relevant internal dynamics of the nucleon at a particular region in impact parameter space.

Pion current contribution to the nucleon's electromagnetic form factors F_1 and F_2 can be calculated in invariant perturbation theory using a general chiral Lagrangian (see, e.g., Ref.[4]),

$$\mathcal{L}_{\pi N} = -\frac{g_A}{2F_\pi} \bar{N}\gamma^\mu\gamma_5\tau N \cdot \partial_\mu\pi - \frac{1}{4F_\pi^2} \bar{N}\gamma^\mu\tau N \cdot \pi \times \partial_\mu\pi. \quad (3)$$

The leading terms are generated by a pion nucleon axial vector coupling and by a contact term as illustrated in Fig.1.

Figure 1. Leading pion current diagrams in computing the nucleon electromagnetic form factors in χPT. (a) Pion loop with axial vector $N\pi N$ coupling. Equivalent to pseudoscalar coupling plus contact term with strength $-g_A^2$. (b)Pion loop diagram with contact coupling with strength 1.

Journal of Physics: Conference Series **599** (2015) 012008 doi:10.1088/1742-6596/599/1/012008

In Ref.[2], by computing transverse densities through χPT and Eq.(2), the internal transverse profile of the nucleon is segmented in three parametric regions (see Fig.2): Non chiral ($b < O(M_\pi^{-1})$, chiral ($b \sim O(M_\pi^{-1})$ and molecular($b \sim O(M_N^2/M_\pi^3)$. The non-chiral region is characterized by high momentum kinematics and thus cannot be studied through χEFT. Expansions in the pion's momentum converge in the chiral region while the molecular region arises from a sub threshold singularity ($t_{sub} = 4M_\pi^2 - M_\pi^4/M_N^2$) in the corresponding spectral function which limits the convergence of kinematic expansions around the two pion threshold.

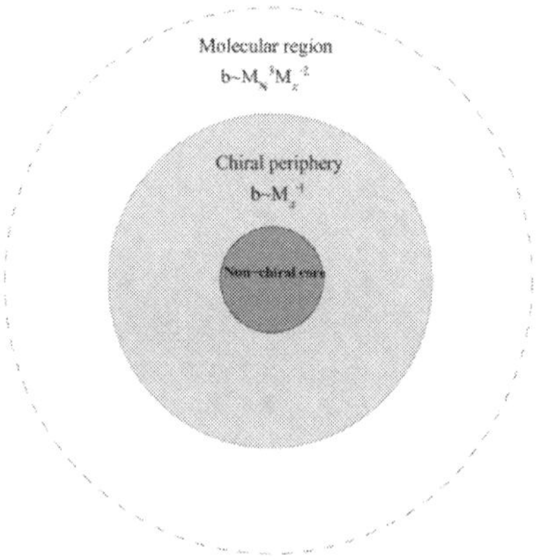

Figure 2. Parametric regions of the nucleon in transverse space.Transverse densities define regions in impact parameter space with characteristic dynamics which controls the relevant kinematics. The structure of the non-chiral core is fitted by phenomenology for instance through vector meson dominance models of the $\gamma^* N$ interaction. The chiral region is described through model independent χPT. In the molecular region χ PT cannot be truncated. It is dominated by the dynamics near the two pion threshold, and because of the large scale distances involved it is mainly only of theoretical interest.

The perturbative nature of the chiral region supports a picture of quasi-free relativistic pions populating the chiral periphery of the nucleon. Furthermore, it is found also that in the chiral region the spin-dependent ($\widetilde{\rho}_2$, see Eq.5) and spin independent (ρ_1) components of the light-cone transverse current density are of the same parametric order and that in general they obey an inequality that guarantees the positive definiteness of the light-cone pion current. These features reinforce a mechanical picture in which the nucleon is composed by a nucleon-like core weakly interacting with a relativistic pion [5, 6].

3. Partonic pions

Form factors can be written as overlap of light-cone wave functions in internal transverse momentum space [7]. In light-front dynamics, one studies the evolution of multiparticle systems in light-cone time ($x^+ = ct + z$) [8]. It makes possible the study of such dynamical systems in a relativistic framework. In the chiral region, the light-cone wavefunctions can be written as pion-nucleon components in a Fock expansion of the nucleon wave function with pseudoscalar coupling ($\bar{N}\gamma_5\tau N\cdot\pi$) (see e.g., Ref.[9]). If the axial-vector coupling provided by the general chiral Lagrangian is used instead in light-cone perturbation theory, one needs to include a contact term correction which contributes of the order of 10% in the chiral region and that is said to account for states with higher masses than the nucleon-like spectator considered here [11].

Using πN light-cone wavefunctions in relative transverse position space computed with a pseudoscalar coupling, the transverse densities are found to take the following form [6],

$$\begin{aligned} \rho_1 &= \frac{1}{2\pi}\int \frac{dy}{2y\bar{y}^3}\left(\psi_1^2\left(y,\frac{b}{\bar{y}}\right)+\psi_2^2\left(y,\frac{b}{\bar{y}}\right)\right) \qquad (4) \\ \widetilde{\rho}_2 &= \frac{\partial}{\partial b}\frac{\rho_2(b)}{2M_N} \end{aligned}$$

$$= \frac{1}{2\pi} \int \frac{dy}{2y\bar{y}^3} 2\psi_1\left(y, \frac{b}{\bar{y}}\right)\psi_2\left(y, \frac{b}{\bar{y}}\right) \quad (5)$$

In which $y = k^+/p^+$ is the fraction of the total light-cone momentum carried by the pion in the πN-system, $\bar{y} = 1 - y$, and $\psi_{1,2}(y, b/\bar{y})$ are the spin non-flip and spin-flip amplitudes of the πN component of the nucleon light-cone wave function. Their explicit form in the invariant formalism as well as various properties that hint to a first quantized approach to this multiparticle system are highlighted at length in Ref.[2]. In these forms (Eq.5), transverse densities are thus recognized in the classical quantum mechanical sense as probability densities in impact parameter space. This association is in close connection with a partonic interpretation of pions in the nucleon's chiral periphery. This is made explicit by identifying the integrand of the above representations with general parton distributions of pions in impact parameter representation [10]. The explicit form agrees with that obtained through the pion light-ray operator in Ref. [11], in which it is argued that gluon GPDs of the nucleon at low x can be computed from gluon GPDs in pions in convolution with pion GPDs in nucleons, and that using the universality of these pion GPDs, gluon GPDs in pions can be extracted from experiments on diffractive hard pion emission in photon or vector meson production off the nucleon since these processes can be factorized (see Fig.3). Further applications and studies of the universal properties of these distributions can also be found in experimental studies of high energy NN exclusive reactions where the periphery can be selected from the kinematics of the final states for instance by selecting high transverse momentum pion-pairs produced in high energy, low $t \sim O(M_\pi^2)$, NN collisions [2].

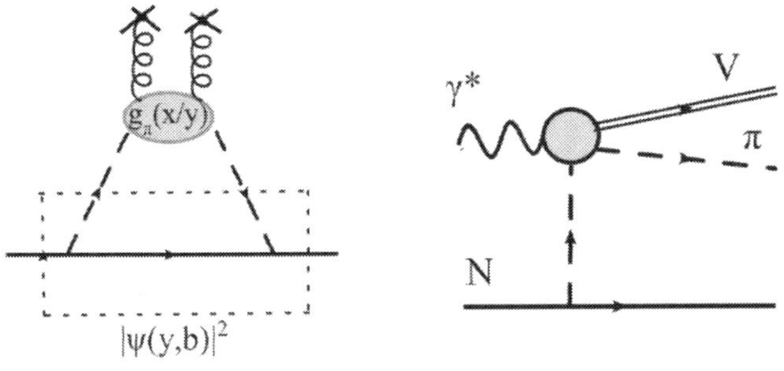

Figure 3. Transverse parton densities in processes at hard kinematics. (Left) Factorization of gluon parton distributions in the nucleon's peripheral region. (Right) Leading chiral contribution to diffractive meson production with hard pion emission.

4. Outlook

We presented a partonic picture of the nucleon's periphery which arises from the onset of chiral dynamics. Parton distributions, or more precisely general peripheral parton distributions of pions can be computed from the pion-nucleon light cone wave functions obtained starting from χPT. Additionally, the moments of these distributions are directly related to distributions of kinematic contributions from parton species to the nucleon's intrinsic physical properties, which opens the possibility of quantifying the peripheral contributions to these properties and of constraining QCD inspired model predictions of these properties as they approach to the chiral limit. Of particular interest is the distribution of orbital angular momentum, motivated by Ji's sum rule [12] connection with GPDs which are experimentally measurable in for instance deeply virtual Compton scattering or time like Compton scattering reactions which are part of the nucleon structure program at PANDA [13].

Journal of Physics: Conference Series **599** (2015) 012008 doi:10.1088/1742-6596/599/1/012008

References

[1] G. A. Miller 2010 Ann. Rev. Nucl. Part. Sci. **60** 1
[2] C. Granados and C. Weiss 2014 JHEP **1401** 092
[3] M. Strikman and C. Weiss 2010 Phys. Rev. C **82** 042201
[4] T. Becher and H. Leutwyler 1999 Eur. Phys. J. C **9** 643
[5] C. Granados and C. Weiss 2014 Int. J. Mod. Phys. Conf. Ser. **25** 1460036
[6] C. Granados and C. Weiss (In preparation)
[7] S. J. Brodsky, D. S. Hwang, B. Q. Ma and I. Schmidt 2001 Nucl. Phys. B **593** 311
[8] H. Leutwyler and J. Stern 1978 Annals Phys. **112** 94
[9] Z. Dziembowski, H. Holtmann, A. Szczurek and J. Speth 1997 Annals Phys. **258** 1
[10] M. Burkardt 2000 Phys. Rev. D 62 071503 [Erratum-ibid. D **66** 119903 (2002)]
[11] M. Strikman and C. Weiss 2004 Phys. Rev. D **69** 054012
[12] X. D. Ji 1997 Phys. Rev. Lett. **78** 610
[13] M. Kotulla *et al.* [PANDA Collaboration] 2004 "Strong interaction studies with antiprotons. Letter of intent for PANDA (Antiproton Annihilations at Darmstadt)"

FAIRNESS 2014 IOP Publishing
Journal of Physics: Conference Series **599** (2015) 012009 doi:10.1088/1742-6596/599/1/012009

Vacuum properties of open charmed mesons in a chiral symmetric model

Walaa I. Eshraim

Institute for Theoretical Physics, Goethe-University, Max-von-Laue-Str. 1, 60438 Frankfurt am Main, Germany

E-mail: weshraim@th.physik.uni-frankfurt.de

Abstract. We present a $U(4)_R \times U(4)_L$ chirally symmetric model, which in addition to scalar and pseudoscalar mesons also includes vector and axial-vector mesons. A part from the three new parameters pertaining to the charm degree of freedom, the parameters of the model are fixed from the $N_f = 3$ flavor sector. We calculate open charmed meson masses and the weak decay constants of nonstrange open charm D and strange open charm D_S. We also evaluate the (OZI-dominant) strong decays of open charmed mesons. The results are turn out to be in quantitative agreement with experimental data.

1. Introduction

Open charmed mesons, composite states of charm quark (c) and up (u), down (d), or strange (s) antiquark, were observed two years later than the discovery of the J/ψ particle in 1974. Since that time, the study of charmed meson spectroscopy and decays has made significant experimental [1, 2, 3] and theoretical process [4, 5, 6, 7]. We show in the present work how the original $SU(3)$ flavor symmetry of hadrons can be extended to $SU(4)$ in the framework of a chirally symmetric model with charm as an extra quantum number. Note that, chiral symmetry is strongly explicitly broken by the current charm quark mass.

The development of an effective hadronic Lagrangian plays an important role in the description of the masses and the interactions of low-lying hadron resonances [8]. To this end, we developed the so-called extended Linear Sigma Model (eLSM) in which (pseudo)scalar and (axial-)vector $q\bar{q}$ mesons and additional scalar and pseudoscalar glueball fields are the basic degrees of freedom. The eLSM has already shown success in describing the vacuum phenomenology of the nonstrange-strange mesons [9, 10, 11, 12, 13]. The eLSM emulates the global symmetries of the QCD Lagrangian; the global chiral symmetry (which is exact in the chiral limit), the discrete C, P, and T symmetries, and the classical dilatation (scale) symmetry. When working with colorless hadronic degrees of freedom, the local color symmetry of QCD is automatically preserved. In eLSM the global chiral symmetry is explicitly broken by non-vanishing quark masses and quantum effects [14], and spontaneously by a non-vanishing expectation value of the quark condensate in the QCD vacuum [15]. The dilatation symmetry is broken explicitly by the logarithmic term of the dilaton potential, by the mass terms, and by the $U(1)_A$ anomaly.

In these proceedings, we present the outline of the extension of the eLSM from the three-flavor case to the four-flavor case including the charm quark [16, 17, 18]. Most parameters of our eLSM are taken directly from Ref.[12] where the nonstrange-strange mesons were considered. There are only new three parameters pertaining to the charm degree of freedom. We compute open charmed meson masses, the weak decay constants of the pseudoscalar D and D_s mesons, and the (OZI-dominant) strong decays of open charmed mesons.

2. The $U(4)_R \times U(4)_L$ Linear Sigma Model

In Refs.[16, 17, 18], we presented the outline of the extension of the eLSM from the three-flavor case to the four-flavor case including charm quark. In this extension, we introduced the (pseudo)scalar and (axial-)vector meson fields in terms of 4×4 (instead of 3×3) matrices, which the charmed mesons appear in the fourth row and fourth column as follows: The matrix of pseudoscalar fields P (with quantum numbers $J^{PC} = 0^{-+}$) reads

$$P = \frac{1}{\sqrt{2}} \begin{pmatrix} \frac{1}{\sqrt{2}}(\eta_N + \pi^0) & \pi^+ & K^+ & D^0 \\ \pi^- & \frac{1}{\sqrt{2}}(\eta_N - \pi^0) & K^0 & D^- \\ K^- & \overline{K}^0 & \eta_S & D_S^- \\ \overline{D}^0 & D^+ & D_S^+ & \eta_c \end{pmatrix}, \tag{1}$$

and the matrix of scalar fields S (with quantum numbers $J^{PC} = 0^{++}$) reads

$$S = \frac{1}{\sqrt{2}} \begin{pmatrix} \frac{1}{\sqrt{2}}(\sigma_N + a_0^0) & a_0^+ & K_0^{*+} & D_0^{*0} \\ a_0^- & \frac{1}{\sqrt{2}}(\sigma_N - a_0^0) & K_0^{*0} & D_0^{*-} \\ K_0^{*-} & \overline{K}_0^{*0} & \sigma_S & D_{S0}^{*-} \\ \overline{D}_0^{*0} & D_0^{*+} & D_{S0}^{*+} & \chi_{c0} \end{pmatrix}, \tag{2}$$

which are used to construct the matrix $\Phi = S + iP$. In the pseudoscalar sector there are: an open charmed state $D^{0,\pm}$, open strange-charmed states $D_S^{\pm}$, and a hidden charmed ground state $\eta_C(1S)$. In the scalar sector there are open charmed $D_0^{*0,\pm}$ and strange charmed meson $D_{S0}^{*\pm}$ which are assigned to $D_0^*(2400)^{0,\pm}$ and $D_{S0}^*(2317)^{\pm}$, respectively.

We now turn to the vector sector. The matrix V^μ which includes the vector degrees of freedom is:

$$V^\mu = \frac{1}{\sqrt{2}} \begin{pmatrix} \frac{1}{\sqrt{2}}(\omega_N + \rho^0) & \rho^+ & K^*(892)^+ & D^{*0} \\ \rho^- & \frac{1}{\sqrt{2}}(\omega_N - \rho^0) & K^*(892)^0 & D^{*-} \\ K^*(892)^- & \overline{K}^*(892)^0 & \omega_S & D_S^{*-} \\ \overline{D}^{*0} & D^{*+} & D_S^{*+} & J/\psi \end{pmatrix}^\mu, \tag{3}$$

where the nonstrange-charmed fields D^{*0}, $D^{*\pm}$ correspond to $\bar{q}q$ resonances $D^*(2007)^0$ and $D^*(2010)^{\pm}$, respectively, while the strange-charmed $D_0^{*\pm}$ is assigned to the resonance $D_0^{*\pm}$ (with mass $m_{D_0^{*\pm}}$), and there is the lowest vector charmonium state $J/\psi(1S)$.

The matrix A^μ describing the axial-vector degrees of freedom is given by:

$$A^\mu = \frac{1}{\sqrt{2}} \begin{pmatrix} \frac{1}{\sqrt{2}}(f_{1,N} + a_1^0) & a_1^+ & K_1^+ & D_1^0 \\ a_1^- & \frac{1}{\sqrt{2}}(f_{1,N} - a_1^0) & K_1^0 & D_1^- \\ K_1^- & \bar{K}_1^0 & f_{1,S} & D_{S1}^- \\ \bar{D}_1^0 & D_1^+ & D_{S1}^+ & \chi_{c,1} \end{pmatrix}^\mu, \tag{4}$$

where the open charmed mesons D_1 and D_{S1} are assigned to $D_1(2420)$ and $D_{S1}(2536)$, respectively, whereas the charm-anticharm state χ_{c1} corresponds to the $c\bar{c}$ resonance $\chi_{c1}(1P)$. From the matrices V^μ and A^μ we construct the left-handed and right-handed vector fields

$L^\mu = V^\mu + A^\mu$ and $R^\mu = V^\mu - A^\mu$, respectively.

This study is a straightforward extension of Ref. [12]. Therefore, the Lagrangian of the $N_f = 4$ model with global chiral invariance has an analogous structure as the corresponding eLSM Lagrangian for $N_f = 3$. However, for a better fit to the masses, we add an additional term $-2\,\mathrm{Tr}[E\Phi^\dagger\Phi]$:

$$\begin{aligned}
\mathcal{L} &= \frac{1}{2}(\partial_\mu G)^2 - \frac{1}{4}\frac{m_G^2}{\Lambda_G^2}\left[G^4 \ln\left(\frac{G}{\Lambda_G}\right) - \frac{G^4}{4}\right] + \mathrm{Tr}[(D^\mu\Phi)^\dagger(D^\mu\Phi)] - m_0^2\left(\frac{G}{G_0}\right)^2 \mathrm{Tr}(\Phi^\dagger\Phi) \\
&- \lambda_1[\mathrm{Tr}(\Phi^\dagger\Phi)]^2 - \lambda_2 \mathrm{Tr}(\Phi^\dagger\Phi)^2 + \mathrm{Tr}[H(\Phi+\Phi^\dagger)] - 2\,\mathrm{Tr}[E\,\Phi^\dagger\Phi] + c(\det\Phi - \det\Phi^\dagger)^2 \\
&+ \mathrm{Tr}\left\{\left[\left(\frac{G}{G_0}\right)^2\frac{m_1^2}{2} + \Delta\right][(L^\mu)^2 + (R^\mu)^2]\right\} - \frac{1}{4}\mathrm{Tr}[(L^{\mu\nu})^2 + (R^{\mu\nu})^2] \\
&+ ic_{\tilde{G}\Phi}\tilde{G}(\det\Phi - \det\Phi^\dagger) + i\frac{g_2}{2}\{\mathrm{Tr}(L_{\mu\nu}[L^\mu, L^\nu]) + \mathrm{Tr}(R_{\mu\nu}[R^\mu, R^\nu])\} \\
&+ \frac{h_1}{2}\mathrm{Tr}(\Phi^\dagger\Phi)\mathrm{Tr}[(L^\mu)^2 + (R^\mu)^2] + h_2 \mathrm{Tr}[(\Phi R^\mu)^2 + (L^\mu\Phi)^2] + 2h_3 \mathrm{Tr}(\Phi R_\mu \Phi^\dagger L^\mu) + \ldots, \qquad (5)
\end{aligned}$$

where the field G denotes the dilaton field and the parameter $\Lambda_G \sim N_C \Lambda_{QCD}$ sets the energy scale of the gauge theory. The dilaton potential breaks the dilatation symmetry explicitly. $D^\mu\Phi \equiv \partial^\mu\Phi - ig_1(L^\mu\Phi - \Phi R^\mu)$ is the covariant derivative; $L^{\mu\nu} \equiv \partial^\mu L^\nu - \partial^\nu L^\mu$, and $R^{\mu\nu} \equiv \partial^\mu R^\nu - \partial^\nu R^\mu$ are the left-handed and right-handed field strength tensors. In eLSM Lagrangian (5) the dots refer to further chirally invariant terms listed in Ref. [12]: these terms do not affect the masses and decay widths studied in the present work and we therefore omitted them. The term $ic_{\tilde{G}\Phi}\tilde{G}\left(\det\Phi - \det\Phi^\dagger\right)$ describes the interaction between the pseudoscalar glueball $\tilde{G} \equiv |gg>$ and (pseudo-)scalar mesons, which is used to study the phenomenology of the pseudoscalar glueball in the case of $N_f = 3$ [19]. The terms $\mathrm{Tr}[H(\Phi + \Phi^\dagger)]$ with $H = 1/2\,\mathrm{diag}\{h_{0N}, h_{0N}, \sqrt{2}h_{0S}, \sqrt{2}h_{0C}\}$, $-2\,\mathrm{Tr}[E\Phi^\dagger\Phi]$ with $E = \mathrm{diag}\{\varepsilon_N, \varepsilon_N, \varepsilon_S, \varepsilon_C\}$, $\varepsilon_i \propto m_i^2$, $\varepsilon_N = \varepsilon_S = 0$, and $\mathrm{Tr}\left[\Delta(L^{\mu 2} + R^{\mu 2})\right]$ with $\delta = \mathrm{diag}\{\delta_N, \delta_N, \delta_S, \delta_C\}$, $\delta_i \sim m_i^2$, $\delta_N = \delta_S = 0$, break chiral symmetry due to nonzero quark masses and are especially important for mesons containing the charm quark. When $m_0^2 < 0$ spontaneous symmetry breaking occurs and the scalar-isoscalar fields condense as well as the glueball field $G = G_0$. To implement this breaking we shift σ_N, σ_S, G, and χ_{C0} by their respective vacuum expectation values ϕ_N, ϕ_S, G_0, and ϕ_C [16, 17, 18] as

$$\sigma_N \to \sigma_N + \phi_N,\ \sigma_S \to \sigma_S + \phi_S\ , G \to G + G_0,\ \text{and}\ \chi_{C0} \to \chi_{C0} + \phi_C\,. \qquad (6)$$

Most of the parameters of the model were already fixed in the three-flavor study of Ref. [12] by requiring a minimum error of 5% for experimental quantities entering the fit which is obtained a reduced χ^2 of about 1.23. There are only three new parameters appear and all of them are related to the bare mass of the charm quark. We then slightly change our fit strategy by enlarging the experimental errors to 7% Ref. [16] of the respective 12 (open and hidden) charmed mesons masses listed by the PDG [20] to obtain the χ^2 of a bout one from

$$\chi^2 \equiv \sum_i^{12}\left(\frac{M_i^{th} - M_i^{exp}}{\xi M_i^{exp}}\right)^2, \qquad (7)$$

where ξ is a constant. Note that, we do not use the experimental errors for the masses to reach the same precision with our effective model. As an outcome, the charm-anticharm condensate is sizable, $\phi_C = 178 \pm 28$ MeV.

3. Results

The weak-decay constants of the pseudoscalar open charmed mesons D and D_S [16, 17, 18] are

$$f_D = \frac{\phi_N + \sqrt{2}\phi_C}{\sqrt{2}Z_D} = (254 \pm 17)\ \text{MeV}\ , \qquad f_{D_S} = \frac{\phi_S + \phi_C}{Z_{D_S}} = (261 \pm 17)\ \text{MeV}\ .$$

The experimental values $f_D = (206.7 \pm 8.9)\,\text{MeV}$ and $f_{D_s} = (260.5 \pm 5.4)\,\text{MeV}$ show how a slightly too large theoretical result for f_D and a good agreement for f_{D_S} [20].

The results for the open charmed meson masses are reported in Table **1** [16, 17]. They have been obtained through a fit to experimental data

Table 1: Masses of open charmed meson.

Resonance	Our Value [MeV]	Experimental Value[MeV]
D^0	1981 ± 73	1864.86 ± 0.13
$D_S^{\pm}$	2004 ± 74	1968.50 ± 0.32
$D_0^*(2400)^0$	2414 ± 77	2318 ± 29
$D_{S0}^*(2317)^{\pm}$	2467 ± 76	2317.8 ± 0.6
$D^*(2007)^0$	2168 ± 70	2006.99 ± 0.15
D_s^*	2203 ± 69	2112.3 ± 0.5
$D_1(2420)^0$	2429 ± 63	2421.4 ± 0.6
$D_{S1}(2536)^{\pm}$	2480 ± 63	2535.12 ± 0.13

The results of (OZI-dominant) strong decay widths of the open charmed mesons described by the resonances D_0^*, D^*, and D_1 are summarized in Table **2** [16, 18].

Table 2: Decay widths of charmed mesons

Decay Channel	Theoretical result [MeV]	Experimental result [MeV]
$D_0^*(2400)^0 \to D\pi$	139^{+243}_{-114}	full width $\Gamma = 267 \pm 40$
$D_0^*(2400)^+ \to D\pi$	51^{+182}_{-51}	full width: $\Gamma = 283 \pm 24 \pm 34$
$D^*(2007)^0 \to D^0\pi^0$	0.025 ± 0.003	< 1.3
$D^*(2007)^0 \to D^+\pi^-$	0	not seen
$D^*(2010)^+ \to D^+\pi^0$	$0.018^{+0.002}_{-0.003}$	0.029 ± 0.008
$D^*(2010)^+ \to D^0\pi^+$	$0.038^{+0.005}_{-0.004}$	0.065 ± 0.017
$D_1(2420)^0 \to D^*\pi$	65^{+51}_{-37}	full width: $\Gamma = 27.4 \pm 2.5$
$D_1(2420)^0 \to D^0\pi\pi$	0.59 ± 0.02	seen
$D_1(2420)^0 \to D^+\pi^-\pi^0$	$0.21^{+0.01}_{-0.015}$	seen
$D_1(2420)^0 \to D^+\pi^-$	0	not seen; $\Gamma(D^+\pi^-)/\Gamma(D^{*+}\pi^-) < 0.24$
$D_1(2420)^+ \to D^*\pi$	65^{+51}_{-36}	full width: $\Gamma = 25 \pm 6$
$D_1(2420)^+ \to D^+\pi\pi$	0.56 ± 0.02	seen
$D_1(2420)^+ \to D^0\pi^0\pi^+$	0.22 ± 0.01	seen
$D_1(2420)^+ \to D^0\pi^+$	0	not seen

4. Conclusion

In this work we have presented the outline of the extension of the eLSM from the three-flavor case to the four-flavor case including the charm quark has been presented. Most parameters are determined in the low-energy study for the nonstrange-strange sector [12]. Three new unknown parameters have been fixed in a fit to the experimental values (details are presented in Ref.[16]). The weak decay constants of nonstrange charm D and strange charm D_S have been calculated. The open charmed meson masses in the eLSM (5) have been computed, which being in reasonably good agreement with experimental data [20]. We have evaluated the (OZI-dominant) decays of open charmed mesons. The results are compatible with the results and the upper bounds listed by the PDG [20]. Moreover, the decay of the vector and axial-vector chiral partners $D^*(2010)$ and $D_1(2420)$ are well described. That leads to the fact that an qualitative description is obtained by using a chiral model and by using the parameters determined by a study of $N_f = 3$ mesons, means that a remnant of chiral symmetry is present also in the sector of charmed mesons. The parameters of the eLSM do not vary too much as a function of the energy at which they are probed. All of this shows that chiral symmetry is still valid for charmed mesons.

Acknowledgments

The author thanks F. Giacosa and D. H. Rischke for cooperation. Financial support from the Deutscher Akademischer Austausch Dienst (DAAD) is acknowledged.

References

[1] Li H B 2012 *Nucl. Phys. Proc. Suppl.* **233** 185
[2] Albrecht H *et al.* [ARGUS Collaboration] 1993 *Phys. Lett. B* **308** 435; Muramatsu H *et al.* [CLEO Collaboration] 2002 *Phys. Rev. Lett.* **89** 251802
[3] Bartelt J E 1995 *Annu. Rev. Nucl. Part. Sci.***45** 133-61; Anastassov A *et al.* [CLEO Collaboration] 2002 *Phys. Rev. D* **65** 032003
[4] Godfrey S and Isgur N 1985 *Phys. Rev. D* **32** 189; Godfrey S and Kokoski R 1991 *Phys. Rev. D* **43** 1679
[5] Capstick S and Isgur N 1986 *Phys. Rev. D* **34** 2809; Isgur N and Wise M B 1991 *Phys. Rev. Lett.* **66** 1130; Eichten E J, Hill C T and Quigg C 1993 *Phys. Rev. Lett.* **71** 4116; Maki Z and Umemura I 1978 *Prog. Theor. Phys.* **59** 507
[6] Bali G S 2006 *Int. J. Mod. Phys. A* **21** 5610; Kalinowski M and Wagner M 2012 *PoS ConfinementX* 303
[7] Neubert M 1994 *Phys. Rept.* **245** 259; Casalbuoni R, Deandrea A, Bartolomeo N D, Gatto R, Feruglio F and Nardulli G 1997 *Phys. Rept.* **281** 145; Georgi H 1990 *Phys. Lett. B* **240** 447; Wise M B 1992 *Phys. Rev. D* **45** 2188
[8] Amsler C and Tornqvist N A 2004 *Phys. Rept.* **389** 61; Klempt E and Zaitsev A 2007 *Phys. Rept.* **454** 1
[9] Lenaghan J T, Rischke D H and Schaffner-Bielich J 2000 *Phys. Rev. D* **62** 085008
[10] Bando M, Kugo T and Yamawaki K 1988 *Phys. Rept.* **164** 217; Ecker G, Gasser J, Pich A and Rafael E de 1989 *Nucl. Phys. B* **321** 311; Jenkins E E, Manohar A V and Wise M B 1995 *Phys. Rev. Lett.* **75** 2272
[11] Parganlija D, Giacosa F and Rischke D H 2010 *Phys. Rev. D* **82** 054024; Gallas S, Giacosa F and Rischke D H 2010 *Phys. Rev. D* **82** 014004
[12] Parganlija D, Kovacs P, Wolf G, Giacosa F and Rischke D H 2013 *Phys. Rev. D* **87** 014011
[13] Giacosa F, Parganlija D, Kovacs P and Wolf G 2012 *EPJ Web Conf.* **37** 08006; D. Parganlija, Kovacs P, Wolf G, Giacosa F and Rischke D H 2012 *Acta Phys. Polon. Supp.* **5** 1109
[14] Hooft G t 1976 *Phys. Rev. D* **14** 3432; Hooft G t 1976 *Phys. Rev. Lett.* **37** 8; Hooft G t 1986 *Phys. Rept.* **142** 357
[15] Vafa C and Witten E 1984 *Nucl. Phys. B* **234** 173; Giusti L and Necco S 2007 JHEP **0704** 090
[16] Eshraim W I, Giacosa F and Rischke D H arXiv:1405.5861 [hep-ph]
[17] Eshraim W I 2013 *PoS* **QCD-TNT-III**, 049
[18] Eshraim W I and Giacosa F 2014 *EPJ Web Conf.* **81** 05009; Eshraim W I arXiv:1411.2218 [hep-ph].
[19] Eshraim W I, Janowski S, Giacosa F and Rischke D H 2013 *Phys. Rev. D* **87** 054036; Eshraim W I, Janowski S, Peters A, Neuschwander K and Giacosa F 2012 *Acta Phys. Polon. Supp.* **5** 1101; Eshraim W I and Janowski S 2012 *PoS* **ConfinementX** 118; Eshraim W I and Janowski S 2013 *J. Phys. Conf. Ser.* **426** 012018
[20] Beringer J *et al.* [Particle Data Group Collaboration] 2012 *Phys. Rev. D* **86** 010001

FAIRNESS 2014 IOP Publishing
Journal of Physics: Conference Series **599** (2015) 012010 doi:10.1088/1742-6596/599/1/012010

Chiral phase transition in the vector meson extended linear sigma model

Péter Kovács[1], Zsolt Szép[2] and György Wolf[1]

[1] Institute for Particle and Nuclear Physics, Wigner Research Centre for Physics, Hungarian Academy of Sciences, H-1525 Budapest, Hungary

[2] MTA-ELTE Statistical and Biological Physics Research Group, H-1117 Budapest, Hungary

E-mail: kovacs.peter@wigner.mta.hu

Abstract. In the framework of an SU(3) (axial)vector meson extended linear sigma model with additional constituent quarks and Polyakov loops, we investigate the effects of (axial)vector mesons on the chiral phase transition. The parameters of the Lagrangian are set at zero temperature and we use a hybrid approach where in the effective potential the constituent quarks are treated at one-loop level and all the mesons at tree-level. We have four order parameters, two scalar condensates and two Polyakov loop variables and their temperature and baryochemical potential dependence are determined from the corresponding field equations. We also investigate the changes of the tree-level scalar meson masses in the hot and dense medium.

1. Introduction

Forthcoming heavy ion experiments, such as the planned CBM experiment at FAIR will explore the QCD phase diagram in the high density and moderately high temperature region. This region is very interesting since it is believed that the critical endpoint (CEP) – which separates the crossover and first order phase transition regions along the phase boundary – if it exists, should be found there.

Earlier experiments such as RHIC, SPS, or LHC investigated the low density, high temperature region. No sign of the chiral phase transition has been observed experimentally, which is not so surprising considering that the phase transition is of crossover type there, which is very hard to observe experimentally. Theoretically, if the transition is of first order – and the system is infinite – certain quantities have discontinuities crossing the phase boundary, which gives a much higher chance to observe it. Although, since in real physical experiments the system is always finite, there will not be any observable discontinuities, but just peaks.

Due to the difficulties of the observation it is very important to gather as much theoretical information as possible from the phase boundary, in order to assist future experiments. According to that, we would like to investigate the chiral phase transition in the framework of a (axial)vector meson extended linear sigma model with additional constituent quarks and Polyakov loops.

The paper is organized as follows. The model is briefly presented in Sec. 2 and its parametrization is discussed in Sec. 3. The four field equations which provide the temperature and chemical potential dependence of the order parameters are given in Sec. 4. Finally, the results are discussed in Sec. 5, where we also conclude.

2. The Model

In terms of matrix valued (pseudo)scalar and (axial)vector meson fields the Lagrangian is,

$$\begin{aligned}
\mathcal{L} &= \mathrm{Tr}[(D_\mu\Phi)^\dagger(D_\mu\Phi)] - m_0^2\,\mathrm{Tr}(\Phi^\dagger\Phi) - \lambda_1[\mathrm{Tr}(\Phi^\dagger\Phi)]^2 - \lambda_2\,\mathrm{Tr}(\Phi^\dagger\Phi)^2 \\
&- \frac{1}{4}\mathrm{Tr}(L_{\mu\nu}^2 + R_{\mu\nu}^2) + \mathrm{Tr}\left[\left(\frac{m_1^2}{2} + \Delta\right)(L_\mu^2 + R_\mu^2)\right] + \mathrm{Tr}[H(\Phi + \Phi^\dagger)] \\
&+ c_1(\det\Phi + \det\Phi^\dagger) + i\frac{g_2}{2}(\mathrm{Tr}\{L_{\mu\nu}[L^\mu, L^\nu]\} + \mathrm{Tr}\{R_{\mu\nu}[R^\mu, R^\nu]\}) \\
&+ \frac{h_1}{2}\mathrm{Tr}(\Phi^\dagger\Phi)\,\mathrm{Tr}(L_\mu^2 + R_\mu^2) + h_2\,\mathrm{Tr}[(L_\mu\Phi)^2 + (\Phi R_\mu)^2] + 2h_3\,\mathrm{Tr}(L_\mu\Phi R^\mu\Phi^\dagger) \\
&+ g_3[\mathrm{Tr}(L_\mu L_\nu L^\mu L^\nu) + \mathrm{Tr}(R_\mu R_\nu R^\mu R^\nu)] + g_4[\mathrm{Tr}\,(L_\mu L^\mu L_\nu L^\nu) \\
&+ \mathrm{Tr}\,(R_\mu R^\mu R_\nu R^\nu)] + g_5\,\mathrm{Tr}\,(L_\mu L^\mu)\,\mathrm{Tr}\,(R_\nu R^\nu) + g_6[\mathrm{Tr}(L_\mu L^\mu)\,\mathrm{Tr}(L_\nu L^\nu) \\
&+ \mathrm{Tr}(R_\mu R^\mu)\,\mathrm{Tr}(R_\nu R^\nu)] + \bar\Psi i\gamma_\mu D^\mu\Psi - g_F\bar\Psi\,(\Phi_S + i\gamma_5\Phi_{PS})\,\Psi,
\end{aligned} \tag{1}$$

where $D^\mu\Phi = \partial^\mu\Phi - ig_1(L^\mu\Phi - \Phi R^\mu) - ieA_e^\mu[T_3, \Phi]$, $L^{\mu\nu} = \partial^\mu L^\nu - ieA_e^\mu[T_3, L^\nu] - \{\partial^\nu L^\mu - ieA_e^\nu[T_3, L^\mu]\}$, $R^{\mu\nu} = \partial^\mu R^\nu - ieA_e^\mu[T_3, R^\nu] - \{\partial^\nu R^\mu - ieA_e^\nu[T_3, R^\mu]\}$, and $D^\mu = \partial^\mu - iG^\mu$. The field content of the Lagrangian is as follows, Φ is the scalar/pseudoscalar field, L^μ and R^μ are the left and right handed vector fields, $\Psi = (u, d, s)^{\mathrm{T}}$ stands for the constituent quark fields, G^μ is the gluon field, while H is the external field. As usual, the nonstrange and strange scalar fields are shifted by their expectation values ϕ_N and ϕ_S (scalar condensates). A previous version of this model, without the constituent quarks and Polyakov loops and with a different anomaly term was soundly analyzed at zero temperature in [1].

3. Determination of the parameters of the Lagrangian

In Eq. (1), not considering the unused g_3, g_4, g_5, and g_6, there are 14 unknown parameters, namely m_0 – the bare (pseudo)scalar mass, λ_1, and λ_2 – the (pseudo)scalar self-couplings, c_1 – the $U_A(1)$ anomaly coupling, m_1 – the bare (axial)vector mass, h_1, h_2 and h_3 – the (axial)vector–(pseudo)scalar couplings, δ_S – the (axial)vector explicit break coupling, ϕ_N and ϕ_S – the scalar condensates, g_F – the Yukawa coupling, and g_1 and g_2 – two (axial)vector couplings. Compared to the parametrization done in [1], in the present case we have two modifications: i) there are two additional equations for the constituent quark masses, $m_{u/d} = g_F\phi_N/2$ and $m_s = g_F\phi_s/\sqrt{2}$, for which we use the values $m_{u/d} = 330$ MeV and $m_s = 500$ MeV; ii) the inclusion of the fermion vacuum fluctuations modifies at $T = \mu = 0$ the masses and decay widths of the model from which the parameters were calculated in [1]. This is because in the present approach the masses used in the parametrization are the curvature masses which are the second derivatives of the grand canonical potential with respect to the meson fields at the minimum. The curvature masses consist of two parts: a usual tree-level part, which can be found in [1], and the vacuum and thermal contributions of the one-loop fermion potential (bosonic fluctuations are neglected). The later one can be found in [2, 3] with Polyakov loops, and in [4] without Polyakov loops. Scanning through the parameter space, curvature masses and decay widths are calculated and compared using a χ^2 minimization method [5] to the experimental data taken from the PDG [6] and the above values of $m_{u/d}$ and m_s, as described in [1].

It is important to note, that in the scalar sector there are more physical particles than we can describe with one $q\bar q$ nonet, as in nature there are two a_0, two $K_0^\star$ and five f_0 particles (see [6]). Since in one scalar nonet there are one a_0, one $K_0^\star$ and two f_0's, we have $2\cdot 2\cdot\binom{5}{2} = 40$ different possibilities for matching the scalar sector with physical particles. However, we consider here only cases where a_0 and $K_0^\star$ correspond to the $a_0(980)$ and $K_0^\star(800)$ physical particles.

We apply two different parametrization scenarios here. In the first one we do not fit the very uncertain isoscalars (f_0^L, f_0^H), consequently m_0 and λ_1 always appear in the same combination $C_1 = m_0^2 + \lambda_1\left(\phi_N^2 + \phi_S^2\right)$ in all the expressions, thus we can not determine them separately. A

Table 1. Parameters determined by χ^2 minimization in the two cases

Parameter	Value(1*b*)	Value(2)	Parameter	Value(1*b*)	Value(2)
ϕ_N [GeV]	0.1359	0.1333	h_2	4.8765	2.7065
ϕ_S [GeV]	0.1400	0.13823	h_3	4.6523	3.7935
m_0^2 [GeV^2]	−0.0103	0.0394	δ_S [GeV^2]	0.1114	0.1178
m_1^2 [GeV^2]	0.5600	0.5508	c_1 [GeV]	1.5293	1.600
λ_1	0 (undetermined)	−1.2200	g_1	5.5737	5.5761
λ_2	23.0638	23.7957	g_2	2.1263	1.3889
h_1	0 (undetermined)	2.6007	g_F	4.3650	4.4217

similar combination $C_2 = m_1^2 + \frac{h_1}{2}\left(\phi_N^2 + \phi_S^2\right)$ appears in the vector sector for m_1 and h_1 [1]. Practically, we choose $\lambda_1 = h_1 = 0$ in this scenario. Even if we do not fit the isoscalars, we still have the possibility to get parametrizations with different values of $m_{f_0^L}$, which has a huge effect, as will be seen immediately, on the finite T/μ_B behavior. We consider two subcases, labeled by '1*a*' and '1*b*': one with a high and one with a low value of $m_{f_0^L}$, that is 1326 MeV and 402 MeV, respectively. In the second scenario, labeled as case '2', we fit the isoscalars to $f_0(500)$ and $f_0(1370)$. The parameter values for case '1*b*' and case '2' are given in Table 1.

4. T/μ_B dependence of the order parameters and curvature masses

In medium, the values of the order parameters – which are the two scalar condensates ϕ_N and ϕ_S and the two Polyakov loop variables Φ and $\bar{\Phi}$ – change with the temperature/chemical potential. The two scalar condensates encodes the effect of both spontaneous and explicit symmetry breakings, while the Polyakov loop variables mimic some properties of the quark confinement, which naturally emerge in mean field approximation, if one calculates free fermion grand canonical potential on a constant gluon background (for more details see [7]).

In order to determine the T/μ_B dependence of the order parameters and curvature masses, we use four coupled stationarity equations (field equations), which require the vanishing of the first derivatives of the grand canonical potential with respect to the order parameters. We apply here a hybrid approach, where we only consider vacuum and thermal fluctuations for the fermions and not for the bosons. In this case the equations are given by

$$-\frac{d}{d\Phi}\left(\frac{U(\Phi,\bar{\Phi})}{T^4}\right) + \frac{2N_c}{T^3}\sum_{q=u,d,s}\int\frac{d^3\mathbf{p}}{(2\pi)^3}\left(\frac{e^{-\beta E_q^-(p)}}{g_q^-(p)} + \frac{e^{-2\beta E_q^+(p)}}{g_q^+(p)}\right) = 0, \tag{2}$$

$$-\frac{d}{d\bar{\Phi}}\left(\frac{U(\Phi,\bar{\Phi})}{T^4}\right) + \frac{2N_c}{T^3}\sum_{q=u,d,s}\int\frac{d^3\mathbf{p}}{(2\pi)^3}\left(\frac{e^{-\beta E_q^+(p)}}{g_q^+(p)} + \frac{e^{-2\beta E_q^-(p)}}{g_q^-(p)}\right) = 0, \tag{3}$$

$$m_0^2\phi_N + \left(\lambda_1 + \frac{1}{2}\lambda_2\right)\phi_N^3 + \lambda_1\phi_N\phi_S^2 - h_N + \frac{g_F}{2}N_c\left(\langle u\bar{u}\rangle_T + \langle d\bar{d}\rangle_T\right) = 0, \tag{4}$$

$$m_0^2\phi_S + (\lambda_1+\lambda_2)\phi_S^3 + \lambda_1\phi_N^2\phi_S - h_S + \frac{g_F}{\sqrt{2}}N_c\langle s\bar{s}\rangle_T = 0, \tag{5}$$

where $U(\Phi,\bar{\Phi})$ is the Polyakov loop potential, for which we used a polynomial form with coefficient taken from [8], and

$$g_q^+(p) = 1 + 3\left(\bar{\Phi} + \Phi e^{-\beta E_q^+(p)}\right)e^{-\beta E_q^+(p)} + e^{-3\beta E_q^+(p)},$$

$$g_q^-(p) = 1 + 3\left(\Phi + \bar{\Phi} e^{-\beta E_q^-(p)}\right)e^{-\beta E_q^-(p)} + e^{-3\beta E_q^-(p)},$$

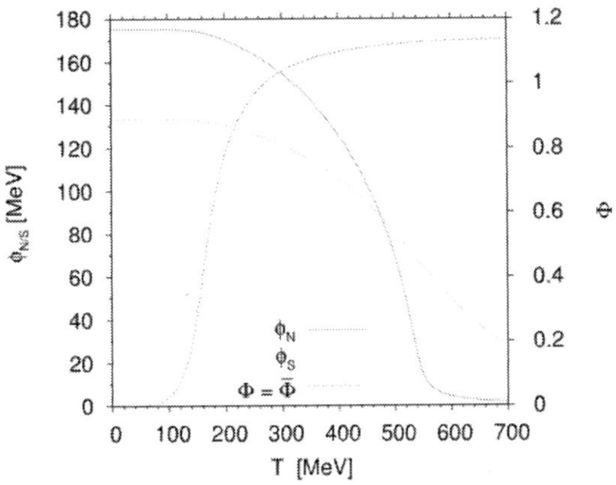

Figure 1. Temperature dependence of the order parameters with the parametrization '$1a$' ($m_{f_0^L} = 1326$ MeV).

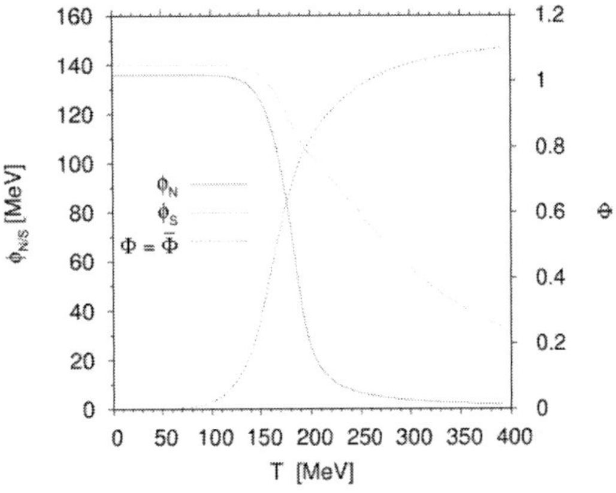

Figure 2. Temperature dependence of the order parameters with the parametrization '$1b$' ($m_{f_0^L} = 402$ MeV).

$$E_q^{\pm}(p) = E_q(p) \mp \mu_B/3,\ E_{u/d}(p) = \sqrt{p^2 + m_{u/d}^2},\ E_s(p) = \sqrt{p^2 + m_s^2},$$

$$\langle q\bar{q}\rangle_T = -4m_q \int \frac{d^3\mathbf{p}}{(2\pi)^3} \frac{1}{2E_q(p)} \left(1 - f_{\Phi}^-(E_q(p)) - f_{\Phi}^+(E_q(p))\right),$$

with the modified Fermi−Dirac distribution functions

$$f_{\Phi}^+(E_p) = \frac{\left(\bar{\Phi} + 2\Phi e^{-\beta(E_p-\mu_q)}\right) e^{-\beta(E_p-\mu_q)} + e^{-3\beta(E_p-\mu_q)}}{1 + 3\left(\bar{\Phi} + \Phi e^{-\beta(E_p-\mu_q)}\right) e^{-\beta(E_p-\mu_q)} + e^{-3\beta(E_p-\mu_q)}},$$

$$f_{\Phi}^-(E_p) = \frac{\left(\Phi + 2\bar{\Phi} e^{-\beta(E_p+\mu_q)}\right) e^{-\beta(E_p+\mu_q)} + e^{-3\beta(E_p+\mu_q)}}{1 + 3\left(\Phi + \bar{\Phi} e^{-\beta(E_p+\mu_q)}\right) e^{-\beta(E_p+\mu_q)} + e^{-3\beta(E_p+\mu_q)}}.$$

5. Results and Conclusion

By solving the system of Eqs. (2)-(5), the temperature and baryochemical potential dependence of the order parameters and the curvature masses can be determined. In Fig. 1 the order parameters can be seen with the parametrization '$1a$' at $\mu_B = 0$. Here the pseudocritical temperature (T_c) is around 550 MeV, which is much higher than the continuum lattice result [9], which is $T_c = 151$ MeV. In Fig. 2 the order parameters are shown with the parametrization '$1b$', where $m_{f_0^L} = 402$ MeV. In this case T_c is between $150 - 200$ MeV, which is in the range of the lattice results.

It is a common belief that there is a critical endpoint (CEP) along the phase boundary in the $T - \mu_B$ plane. For this to happen, the phase transition should be of first order as a function of μ_B along the $T = 0$ axis. However, as can be seen in Fig. 3 (case '$1b$') the phase transition at $T = 0$ is of crossover type, while in Fig. 4 with parametrization '2' the transition is of first order. In this case the value of the pseudocritical temperature T_c at $\mu_B = 0$ is close to that of case '$1b$'. The temperature dependence of the pseudo(scalar) curvature masses can be seen in Fig. 5 and Fig. 6 for the parametrization of case '2'.

In conclusion we can say that the parametrization which fulfills the two physical requirements of having a pseudocritical temperature $T_c \approx 150$ MeV at $\mu_B = 0$ and a first order transition in μ_B at $T = 0$ is realized in case '2', that is when $f_0(500)$ and $f_0(1370)$ are used for parametrization. Similar results were found in [2, 3] without (axial)vector mesons.

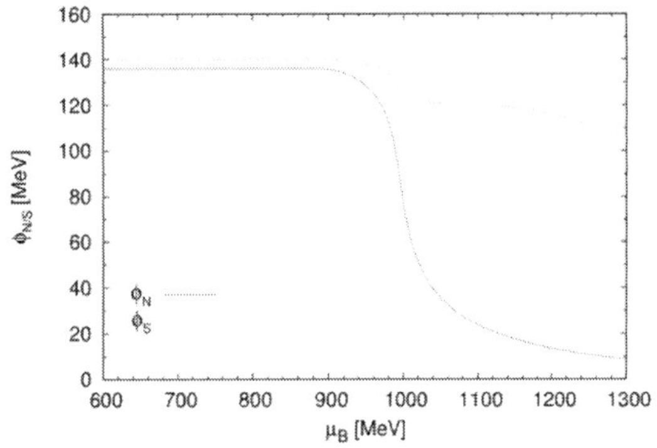

Figure 3. μ_B dependence of the condensates with the parametrization '1*b*'.

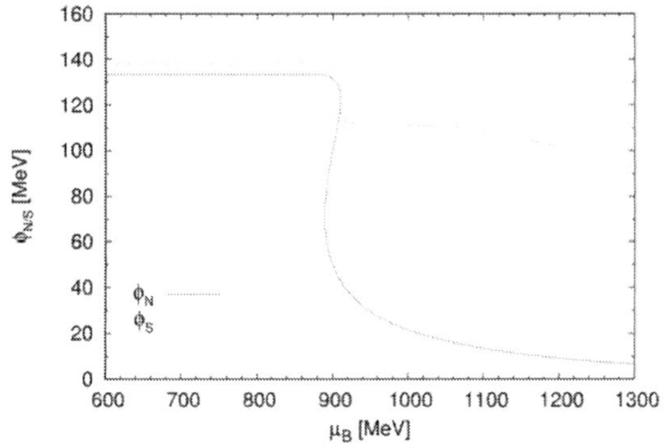

Figure 4. μ_B dependence of the condensates with the parametrization '2'.

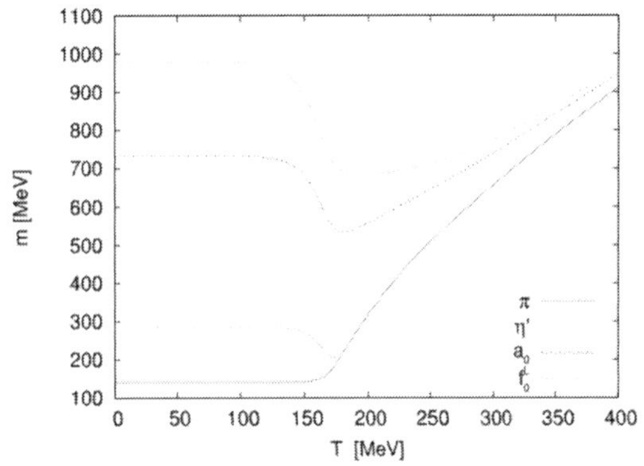

Figure 5. Temperature dependence of the π, η', a_0 and $f_0^L(i.e.\ \sigma)$ particle masses.

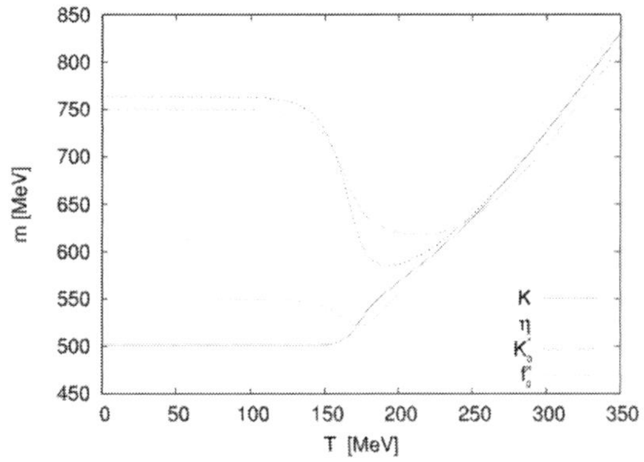

Figure 6. Temperature dependence of the K, η, $K_0^\star$ and f_0^H particle masses.

Acknowledgments
The authors were supported by the Hungarian OTKA fund K109462 and by the HIC for FAIR Guest Funds of the Goethe University Frankfurt.

References

[1] Parganlija D, Kovacs P, Wolf G, Giacosa F and Rischke D H 2013 *Phys. Rev.* D **87** 014011
[2] Tiwari V K 2013 *Phys. Rev.* D **88** 074017
[3] Chatterjee S and Mohan K A 2012 *Phys. Rev.* D **85** 074018
[4] Schaefer B J and Wagner M 2009 *Phys. Rev.* D **79** 014018
[5] James F and Roos M 1975 *Comput. Phys. Commun.* **10** 343
[6] Beringer J *et al* (Particle Data Group) 2012 *Phys. Rev.* D **86** 010001
[7] Fukushima K 2004 Phys. Lett. B **591** 277
[8] Ratti C, Thaler M A and Weise W 2006 *Phys. Rev.* D **73** 014019
[9] Aoki Y, Fodor Z, Katz S D and Szabo K K 2006 *Phys. Lett.* B **643** 46

Journal of Physics: Conference Series **599** (2015) 012011 doi:10.1088/1742-6596/599/1/012011

τ vector and axial vector spectral functions in the extended linear sigma model

A. Habersetzer[1], **F. Giacosa**[1,2]

[1]Institut für Theoretische Physik, Goethe-Universität, Max-von-Laue-Str. 1, 60438 Frankfurt, Germany

[2] Institute of Physics, Jan Kochanowski University, 25-406 Kielce, Poland

E-mail: habersetzer@th.physik.uni-frankfurt.de, fgiacosa@ujk.edu.pl

Abstract. The extended linear sigma model describes the vacuum phenomenology of scalar, pseudoscalar, vector and axial-vector mesons at energies $\simeq$ 1 GeV. We combine the chiral $U(2)_L \times U(2)_R$ symmetry of this model with a local $SU(2)_L \times U(1)_Y$ symmetry and obtain a gauge invariant effective description for electroweak interaction of hadrons in the vacuum. Vector and axial-vector spectral functions can be described well by two intermediate resonances ρ and a_1. They are implemented into this model as chiral partners and yield the predominant contributions to both spectral functions. However, the contributions that arise from the non-resonant decay channels of the weak charged W bosons are essential for reproducing the lineshapes of the spectral functions.

1. Introduction

The weak decay of the τ lepton as e.g. measured by the ALEPH collaboration [1] provides an excellent ground for studying electroweak interactions at low energies. As a phenomenological approach vector meson dominance describes the interaction between the photon and hadrons by $\bar{q}q$ vacuum fluctuations of the photon that form a neutral ρ meson which then interacts with other hadrons (e.g. pions). We can also understand weak interaction with hadrons in terms of such vacuum fluctuations that manifest themselves as mixing of the charged weak currents with vector and axial-vector mesons. The question whether the axial vector resonance $a_1(1260)$ can indeed be described as a $\bar{q}q$ state within the quark model, where it is the $J^{PC} = 1^{++}$ member of the $N_F = 2$ axial-vector multiplet, or whether it might not be a dynamically generated ρ-π molecule-like state has not been unambiguously answered, yet [2, 3]. Related to this is also the question whether chiral models can describe the a_1 and ρ resonance as chiral partners. Including weak interaction is thus not only useful to describe electroweak interactions of hadrons, it also provides a basis for understanding more about the nature of the $a_1(1260)$ resonance and the applicability of an effective chiral model to the vacuum phenomenology.

Effective chiral models (such as e.g. [4]) that apply the principles of QCD to hadronic observables are widely used to study the low-energy phenomenology in heavy-ion collisions. In the extended Linear Sigma Model (eLσM, see Refs. [5, 6, 7, 8]) that we present here scalar, pseudoscalar, vector, and axial-vector fields are considered genuine degrees of freedom that can be constructed from a $\bar{q}q$ picture. We show that with the eLσM we can indeed describe the τ vector and axial-vector resonances as emerging from the mixing of the chiral partners a_1 and ρ with the charged weak W boson.

Journal of Physics: Conference Series **599** (2015) 012011 doi:10.1088/1742-6596/599/1/012011

2. Lagrangian

The hadronic degrees of freedom are described by the matrix-valued fields

$$\Phi = (S_a + iP_a)t_a\,,\ L^\mu = (V_a^\mu + A_a^\mu)t_a\,,\ R_a^\mu = (V_a^\mu - A_a^\mu)t_a \tag{1}$$

on the basis of the strong isospin algebra with generators $t_a = (1/2)\sigma_a$. Their masses, strong and weak interactions are obtained from the Lagrangian

$$\begin{aligned}\mathscr{L} = &\ \mathrm{Tr}\big[(D^\mu\Phi)^\dagger(D^\mu\Phi)\big] - m_0^2\mathrm{Tr}[\Phi^\dagger\Phi] - \lambda_1\mathrm{Tr}[(\Phi^\dagger\Phi)]^2 - \lambda_2\mathrm{Tr}[(\Phi^\dagger\Phi)^2] - \frac{1}{4}\mathrm{Tr}[(L^{\mu\nu})^2 + (R^{\mu\nu})^2] \\ &+ \frac{m_1}{2}\mathrm{Tr}\Big[\Big(L^{\mu 2} + i\frac{g_2}{2}\big(\mathrm{Tr}\big[L_{\mu\nu}[L_\mu, L_\nu]\big] + \mathrm{Tr}\big[R_{\mu\nu}[R_\mu, R_\nu]\big]\big) + R^{\mu 2}\Big)\Big] + \mathrm{Tr}[H(\Phi + \Phi^\dagger)] \\ &+ c_1(\det\Phi - \det\Phi^\dagger)^2 + 2h_3\mathrm{Tr}[\Phi R_\mu \Phi^\dagger L^\mu] + \frac{g\delta_w}{2}\mathrm{Tr}[W_{\mu\nu}L^{\mu\nu}] + \frac{g'\delta_{\mathrm{em}}}{2}\mathrm{Tr}[R_{\mu\nu}B^{\mu\nu}] \\ &+ \frac{1}{4}\mathrm{Tr}[(W^{\mu\nu})^2 + (B^{\mu\nu})^2] + \frac{g}{2\sqrt{2}}\left(W_\mu^- \bar{u}_{\nu_\tau}\gamma_\mu(1-\gamma_5)u_\tau + \mathrm{h.c.}\right) + \mathscr{L}_{\Phi LR} + \mathscr{L}_{LR}\,.\end{aligned} \tag{2}$$

The Lagrangian is invariant under a **global** linear chiral $U(2)_L \times U(2)_R$ transformation

$$\Phi \to U_L \Phi U_R^\dagger\,,\ L^\mu \to U_L L^\mu U_L^\dagger\,,\ R^\mu \to U_R R^\mu U_R^\dagger\,, \tag{3}$$

and it is also invariant under a **local** $SU(2)_L \times U(1)_Y$ transformation by

$$\Phi \to U_L \Phi U_Y^\dagger\,,\ L^\mu \to U_L L^\mu U_L^\dagger\,,\ R^\mu \to U_Y R^\mu U_Y^\dagger. \tag{4}$$

The bare gauge fields $W^\mu = W_i^\mu t_i$ and $B^\mu = B^\mu t_3$ transform in the adjoint representation of $SU(2)_L \times U(1)_Y$ as

$$W^\mu \xrightarrow{SU(2)_L} U_L W^\mu U_L^\dagger + \frac{i}{g}U_L\partial^\mu U_L^\dagger\,,\ B^\mu \xrightarrow{U(1)_Y} U_Y B^\mu U_Y^\dagger + \frac{i}{g'}U_Y\partial^\mu U_Y^\dagger\,. \tag{5}$$

After rotating the bare fields into the physical fields A_μ and Z_μ by the Weinberg mixing angle θ_W, we obtain the covariant derivative

$$\begin{aligned}D^\mu\Phi = &\ \partial^\mu\Phi - ig_1(L^\mu\Phi - \Phi R^\mu) - ig\cos\theta_C(W_1^\mu t_1 + W_2^\mu t_2)\Phi - ie[A^\mu t_3, \Phi] \\ &- ig\cos\theta_W(Z^\mu t_3\Phi + \tan\theta_W \Phi Z^\mu t_3)\,.\end{aligned} \tag{6}$$

The Cabibbo angle θ_C in (6) rotates the flavor eigenstates into the weak eigenstates. The global chiral symmetry is broken spontaneously by the scalar condensate and explicitly by the term $\sim H$ which modulates non-vanishing but equal current quark masses and also by the 't Hooft determinant that corresponds to the explicitly broken $U(1)_A$ symmetry. In (2) we included two terms $\sim \delta_w$ and $\sim \delta_{em}$. These terms describe a mixing between the electroweak interaction fields and the (axial-)vector fields. This mixing arises from the hadronic loop contributions of the electroweak bosons and is also described by vector meson dominance in its *first representation* as explained in [9]. The W-ρ mixing is only given by $\sim \delta_w s$ while the W-a_1 mixing contains an additional contribution $\sim (g_1\phi^2 - \delta_w s)$ by the chiral condensate.

3. τ-decay and spectral functions

In the vector channel τ^- decays predominantly into $\pi^-\pi^0\nu_\tau$ and in the axial-vector channel predominantly by two isospin channels into $2\pi^0\pi^-$ and $2\pi^-\pi^+$. The Feynman diagrams we use are shown in Fig. 1. Because of the identical particles in the final state of the axial-vector channel we have to integrate over m_{12}^2 and m_{23}^2 and include the corresponding symmetry factor $1/N! = 1/2$.

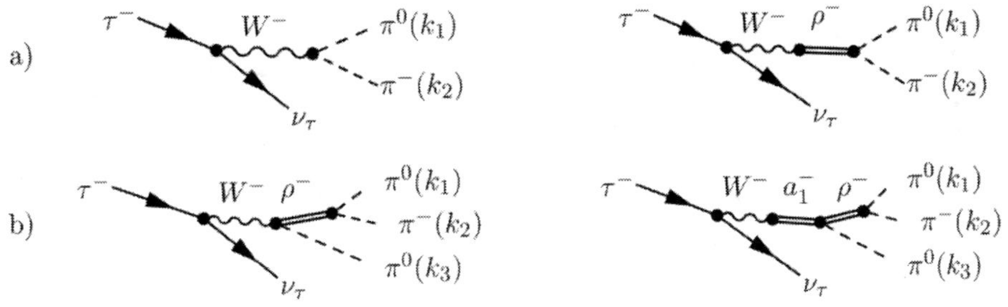

Figure 1: Feynman diagrams of a) the 2π decays in the vector channel and b) the 3π decays of the axial-vector channel.

In principle there are also the contributions from the direct $W \to 3\pi$ and $W \to a_1 \to 3\pi$ decays, but they are expected to be small and not included, yet. Decay rates and spectral functions are calculated on tree-level. We use the optical theorem [10] that relates the tree level amplitude to the imaginary part of the self-energy $\mathrm{Im}[\Sigma(s)] = \sqrt{s}\Gamma(s)$ and define vector and axial-vector propagator as

$$D_\rho^{\mu\nu}(s) = \frac{-g_{\mu\nu} + \frac{P_\mu P_\nu}{P.P}}{s - m_\rho^2 + i\sqrt{s}\,\Gamma_\rho(s)}, \quad D_{a_1}^{\mu\nu}(s) = \frac{-g_{\mu\nu} + \frac{P_\mu P_\nu}{P.P}}{s - m_{a_1}^2 + i\sqrt{s}\,\Gamma_{a_1}(s)}. \tag{7}$$

The bare masses are given by

$$m_\rho^2 = m_1^2 + \phi^2(h_1 + h_2 + h_3)/2\,, \ m_{a_1}^2 = m_1^2 + g_1^2\phi^2 + \phi^2(h_1 + h_2 - h_3)/2\,. \tag{8}$$

They are fixed to the physical masses m_ρ and m_{a_1}. We require that the real part of the respective self-energies $\mathrm{Re}\Sigma(s = m^2) = 0$. Then the contributions that arise from the real part of the self-energies will not contribute significantly to the spectral functions in the region around the peak. The sum rules

$$\int_0^\infty ds\, d_\rho(s) = 1\,, \quad \int_0^\infty ds\, d_{a_1}(s) = 1 \tag{9}$$

describe a probability conservation where $d(s)$ is the probability that ρ and/or a_1 meson have mass $m = \sqrt{s}$. The sum rule in (9) holds only at all orders in resummed perturbation theory. Care is needed when non-renormalizable interaction terms are considered and if moreover, $\mathrm{Im}\Sigma(s)$ is only calculated at one loop with its real part being neglected. Therefore we need to introduce a 'cutoff', which is usually of the order of $1.5 - 2$ GeV. Here we choose m_τ and obtain the normalized spectral functions as

$$\frac{1}{N_\rho}\int_0^{m_\tau^2} ds\, d_{\rho^-\to\pi^-\pi^0}(s) = 1\,, \quad \frac{1}{N_{a_1}}\int_0^{m_\tau^2} ds\, d_{a1}(s) = 1\,. \tag{10}$$

We can then obtain vector and axial-vector spectral functions from

$$d_V(s) = \frac{1}{\pi}\frac{\sqrt{s}\,\Gamma_{W^-\to\pi^-\pi^0}(s)}{M_w^4}\,, \quad d_A(s) = \frac{1}{\pi}\frac{\sqrt{s}\,\Gamma_{W^-\to 3\pi}(s)}{M_w^4}\,, \tag{11}$$

where $\Gamma_W(s)$ contains the coherent amplitude squared of the processes depicted in Fig. 1 and where, for $s \ll M_w^2$, the W propagator reduces to a point-like interaction vertex. Notice that, if the direct contributions $W^- \to \pi^0\pi^-$ and $W^- \to \rho^{0/-}\pi^{-/0}$ were neglected we would obtain $d_V(s) = d_{W\to\rho\to 2\pi}(s)$ and $d_A(s) = d_{W\to a_1\to 3\pi}(s)$. But, as we will see in the next section, this limit yields only an approximate description of the spectral functions.

Journal of Physics: Conference Series **599** (2015) 012011 doi:10.1088/1742-6596/599/1/012011

4. Results

All relevant parameters, that is g_1, g_2, h_3, w, and Z, except the mixing parameter δ_w, have been determined in a global fit in [5]. We use their result for masses and decay widths

$$m_\rho = (0.7831 \pm 0.0070)\ \text{GeV}, \qquad \Gamma_\rho = (0.1609 \pm 0.0044)\ \text{GeV}, \tag{12}$$

$$m_{a_1} = (1.186 \pm 0.006)\text{GeV}, \qquad \Gamma_{a_1} = (0.549 \pm 0.043)\text{GeV}, \tag{13}$$

such that δ_w is the only free paramter. The result for the coherent vector-channel spectral function compared to the ALEPH data is seen in Fig. 2.

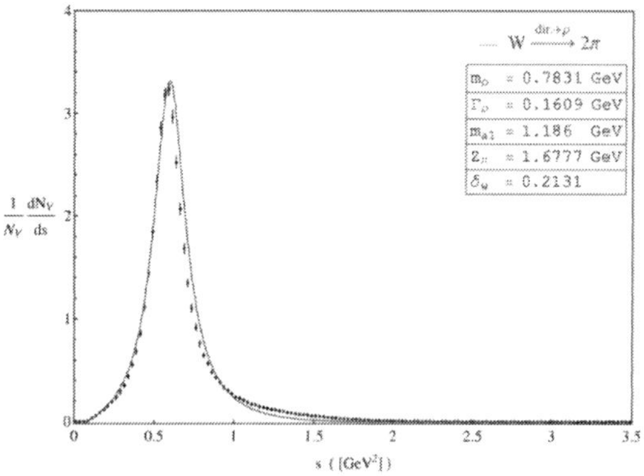

Figure 2: Vector channel spectral function of the coherent sum of the diagrams in Fig. 1a).

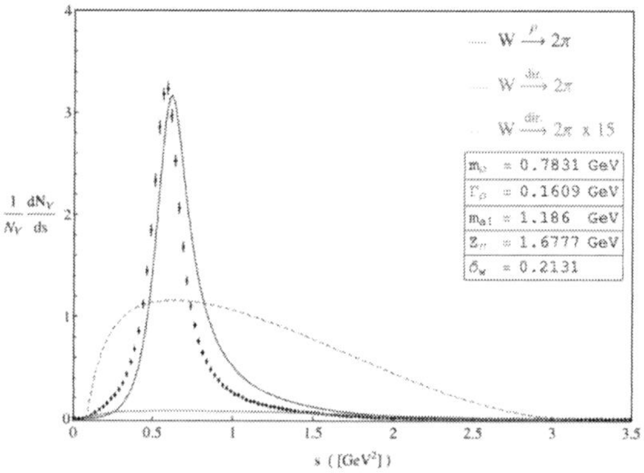

Figure 3: Individual contributions; the blue line represents the W-ρ mixing, the red line $W^- \to \pi^-\pi^0$, and the dashed gray line the direct contribution scaled by 15.

The coupling δ_w that describes the W-ρ mixing is determined by the peak value of the mass distribution $(1/N_V)dN_V/ds$. Apart from a small shift of the peak to higher energies the spectral function nicely describes the experimental distribution. This shift is expected, since if we look at the results of the global fit in [5] we see that also there $\Gamma_{\rho^- \to \pi^-\pi^0}$ deviates slightly from the experimental data. The individual contributions to the coherent sum are shown in Fig. 3. The process $\tau^- \to \pi^-\pi^0$ is clearly dominated by the intermediate ρ meson which constitutes almost the entire strength of the vector spectral function. However, the numerically small contribution of the direct $W^- \to \pi^-\pi^0$ decay has a strong influence on the lineshape, particularly visible e.g. at $s > m_\rho^2$, where due to destructive quantum interference the excess that arises from the ρ resonance contribution disappears. This can be interpreted as a very nice demonstration of VMD. In the low energy region the charged weak interaction fields seem to essentially couple to the pions by forming a vector meson resonance.

The coherent sum in the axial-vector channel shown in Fig. 4 presents a similar picture. While the a_1 meson yields the strongest contribution, the influence of the direct decay $W^- \to \rho\pi$ is clearly visible. Notice that we have not yet included the direct contributions from $W \to 3\pi$ and $a_1 \to 3\pi$. This is probably the reason why the axial-vector spectral function lacks some strength at the peak and above 2 GeV^2. In the non-resonant $a_1 \to 3\pi$ decay there are however unknown parameters that have yet to be determined. Thus, including these contributions would not necessarily render the result more conclusive. We could however use the ALEPH data also to obtain a first estimate on these unknown parameters. It should be stressed once more, that for both spectral-functions we have only one free parameter, the mixing constant δ_w, which is restrained by the peak value of the vector channel spectral function, while all other parameters are taken from the global fit in [5].

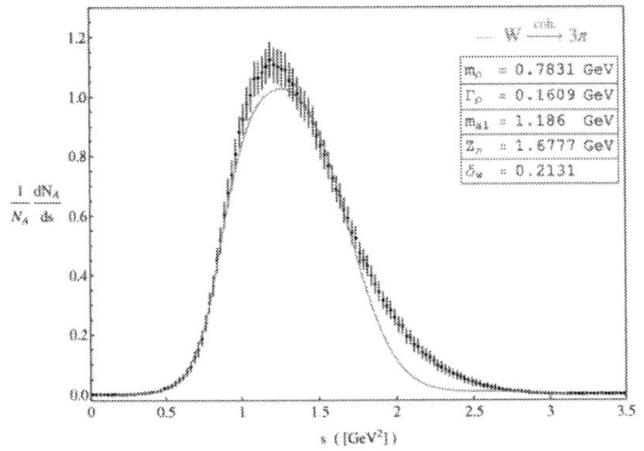

Figure 4: The axial-vector spectral function obtained from the coherent sum of the diagrams in Fig. 1 b).

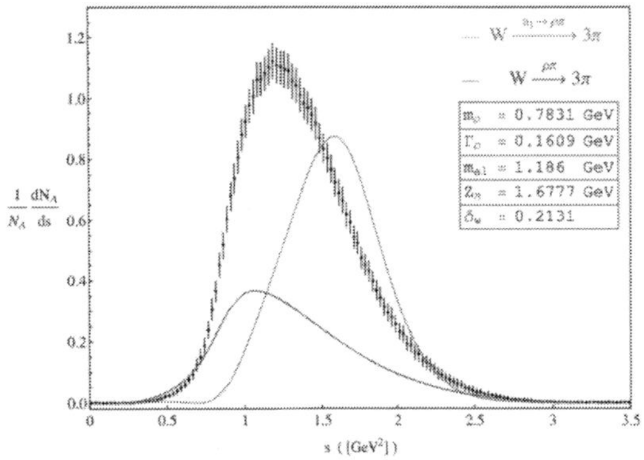

Figure 5: Individual contributions to the axial-vector spectral function. The blue lines represents $W \to a_1 \to \rho\pi$ and the red line $W \to a_1 \to \rho\pi$.

Within this fit all parameters have been determined to ±5% accuracy. We can still adjust the parameters within this range and obtain an even better description.

5. Conclusion

Together with a gauge invariant inclusion of an external electroweak field the eLσM can describe vector and axial-vector channels of the electroweak decay of the τ. The difference in the coupling of the weak bosons to ρ and a_1 arises only from the chiral condensate because of the spontaneously broken chiral symmetry. In our chiral framework the axial-vector resonance can indeed be described as a $\bar{q}q$ meson and as the chiral partner of ρ. In both channels we also obtained a nice demonstration of vector meson dominance. Although, to some extent, the τ decay can be described by ρ and a_1 resonances only, the contributions from W decay are clearly needed. These results can still be improved by adjusting the parameters to this specific process. Moreover, we can still include the direct decays $W \to 3\pi$ and $a_1 \to 3\pi$ into the coherent sum and see how they will influence the spectral function.

Acknowledgments

The authors acknowledge support from the Helmholtz Research School on Quark Matter Studies and thank D. H. Rischke for his collaboration and also M. Buballa, J. Wambach, H. v. Hees, G. Wolf, and Z. Szep for valuable discussions.

References

[1] Schael S et al. 2005 *Phys. Rept.* **421** 191
[2] Wagner M and Leupold S 2008 *Phys. Rev.* D **78** 053001
[3] Roca L, Oset E and Singh J 2006 *AIP Conf. Proc.* **814** 468.
[4] Gasiorowicz S and Geffen D A 1969 *Rev. Mod. Phys.* **41** 531.
[5] Parganlija D, Kovacs P, Wolf G, Giacosa F and Rischke D H 2012 *Phys. Rev.* D **87** 014011
[6] Parganlija D, Giacosa F and Rischke D H 2010 *Phys. Rev.* D **82** 054024
[7] Janowski S, Giacosa F and Rischke D H 2014 arXiv:1408.4921 [hep-ph].
[8] Gallas S, Giacosa F and Rischke D H 2010 *Phys. Rev.* D **82** 014004
[9] O'Connell H B, Pearce C, Thomas A W and Williams A G 1997 *Prog. Part. Nucl. Phys.* **39** 201
[10] Giacosa F and Pagliara G 2007 *Phys. Rev.* C **76** 065204

Multiplicity fluctuations at the quark-hadron phase transition from a fluid dynamical model

Christoph Herold[1], Marlene Nahrgang[2], Yupeng Yan[1], Chinorat Kobdaj[1]

[1] School of Physics, Suranaree University of Technology, 111 University Avenue, Nakhon Ratchasima 30000, Thailand

[2] Department of Physics, Duke University, Durham, NC 27708, USA

E-mail: herold@g.sut.ac.th

Abstract. The region of large net-baryon densities in the QCD phase diagram is expected to exhibit a first-order phase transition. Experimentally, its study will be one of the primary objectives for the upcoming FAIR accelerator. We model the transition between quarks and hadrons in a heavy-ion collision using a fluid which is coupled to the explicit dynamics of the chiral order parameter and a dilaton field. This allows us to investigate signals stemming from the nonequilibrium evolution during the expansion of the hot plasma. Special emphasis is put on an event-by-event analysis of baryon number fluctuations which have long since been claimed to be sensitive to a critical point.

1. Introduction

Although some aspects of the QCD phase structure have been revealed in the past years, still many things are unknown and require more thorough investigation. Regarding the chiral and deconfinement transition, we have learned from lattice QCD data that for small baryochemical potentials, the transition from a hadron gas to a quark-gluon plasma is a smooth crossover rather than a phase transition [1, 2]. A critical endpoint (CEP) and first-order phase transition are expected from effective low-energy models [3, 4, 5, 6] and a Dyson-Schwinger approach [7], though any irrefutable proof is still missing. Experimentally, the dip in the recent measurement of directed flow as a function of beam energy by STAR [8] is sometimes argued to result from the presence of a phase transition. The measurement of proton number fluctuations during the beam energy scan program has revealed some non-monotonic behavior in higher moments such as the kurtosis [9]. The idea hereby is that moments of fluctuations of conserved quantities are sensitive to a CEP as they scale with some powers of the correlation length [10]. However, these predictions rely on the assumption of thermodynamic equilibrium and to understand the observables we need to develop models which are able to describe the full nonequilibrium evolution of the hot and dense matter created in a heavy-ion collision. Effects like critical slowing down near a CEP [11] and spinodal decomposition at a first-order phase transition [12, 13] might influence the obtained signal. Our ansatz for such a model is to couple an ideal fluid of quarks and gluons to the explicit propagation of the relevant order parameters and additionally take into account friction and stochastic fluctuations [14]. We have shown that this model successfully describes supercooling and critical slowing down [15, 16] as well as domain formation due to spinodal dynamics [17]. The goal of this work is to estimate the impact of nonequilibrium effects on an enhancement of net-baryon number fluctuations at a CEP and first-order phase transition [18].

2. Nonequilibrium chiral fluid dynamics

We derive the coupled dynamics of order parameters and quark-gluon fluid from a linear sigma model with dilatons [19]

$$\mathcal{L} = \bar{q}\left(i\gamma^\mu\partial_\mu - g_\mathrm{q}\sigma\right)q + \frac{1}{2}\left(\partial_\mu\sigma\right)^2 + \frac{1}{2}\left(\partial_\mu\chi\right)^2 + \mathcal{L}_A - U_\sigma - U_\chi \,. \tag{1}$$

(2)

It describes the chiral dynamics of the light quarks $q = (u, d)$ coupled to the condensate $\sigma \sim \langle\bar{q}q\rangle$ which melts at high temperatures, thus restoring chiral symmetry. Additionally, gluons are incorporated with a constituent gluon field A whose mass is generated by the glueball condensate $\langle\chi\rangle \sim \langle A_{\mu\nu}A^{\mu\nu}\rangle$, the so-called dilaton field. The quark-meson coupling g_q is fixed to a value of 3.37 to reproduce the vacuum nucleon mass. Fixing the mass of the sigma meson at 900 MeV results in a phase diagram with a CEP at temperature $T_\mathrm{CEP} = 89$ MeV and quark chemical potential $\mu_\mathrm{CEP} = 329$ MeV, with a first-order phase transition for larger chemical potentials.

The mean-field effective thermodynamic potential is given by

$$V_\mathrm{eff} = \Omega_{q\bar{q}} + \Omega_A + U_\sigma + U_\chi + \Omega_0 \,, \tag{3}$$

with the quark and gluon contributions

$$\Omega_{\mathrm{q}\bar{\mathrm{q}}} = -2N_f N_c T \int \frac{\mathrm{d}^3p}{(2\pi)^3}\left\{\ln\left[1+\mathrm{e}^{-\frac{E_\mathrm{q}-\mu}{T}}\right] + \ln\left[1+\mathrm{e}^{-\frac{E_\mathrm{q}+\mu}{T}}\right]\right\} \,, \tag{4}$$

$$\Omega_A = 2(N_c^2-1)T \int \frac{\mathrm{d}^3p}{(2\pi)^3}\left\{\ln\left[1-\mathrm{e}^{-\frac{E_A}{T}}\right]\right\} \,, \tag{5}$$

Here, $E_\mathrm{q} = \sqrt{p^2+m_\mathrm{q}^2}$ and $E_A = \sqrt{p^2+m_A^2}$ denote the quasiparticle energies of constituent quarks and gluons, respectively. Note that we use the quark chemical potential $\mu = \mu_\mathrm{q}$. The constant Ω_0 is added to set the total energy in vacuum to zero. The pressure of the ideal quark-gluon fluid is given by $p = -\Omega_{q\bar{q}} - \Omega_A$.

For the dynamics of the sigma field we adopt the result from [14] for our model, resulting in a Langevin equation containing a dissipative term $\eta_\sigma(T)\partial_t\sigma$ due to the possible decay of a sigma into a quark-antiquark pair and stochastic noise field ξ_σ resembling the back reaction of the fluid

$$\partial_\mu\partial^\mu\sigma + \eta_\sigma(T)\partial_t\sigma + \frac{\delta V_\mathrm{eff}}{\delta\sigma} = \xi_\sigma \,. \tag{6}$$

Due to kinematic reasons, no dissipative processes from interactions between quarks and the dilaton field are possible, therefore we use the classical Euler-Lagrange equation of motion for the dynamics of χ,

$$\partial_\mu\partial^\mu\chi + \frac{\delta V_\mathrm{eff}}{\delta\chi} = 0 \,. \tag{7}$$

Energy-momentum and net-baryon number are conserved through the equations

$$\partial_\mu T^{\mu\nu} = -\partial_\mu\left(T_\sigma^{\mu\nu} + T_\chi^{\mu\nu}\right) \,, \tag{8}$$

$$\partial_\mu N_\mathrm{q}^\mu = 0 \,, \tag{9}$$

which together with the equation of state from the mean-field pressure determine the evolution of the fluid. As can be seen, we allow for a direct transfer of energy-momentum from the order parameters to the quark-gluon fluid through the stochastic source terms T_σ and T_χ.

Journal of Physics: Conference Series **599** (2015) 012012 doi:10.1088/1742-6596/599/1/012012

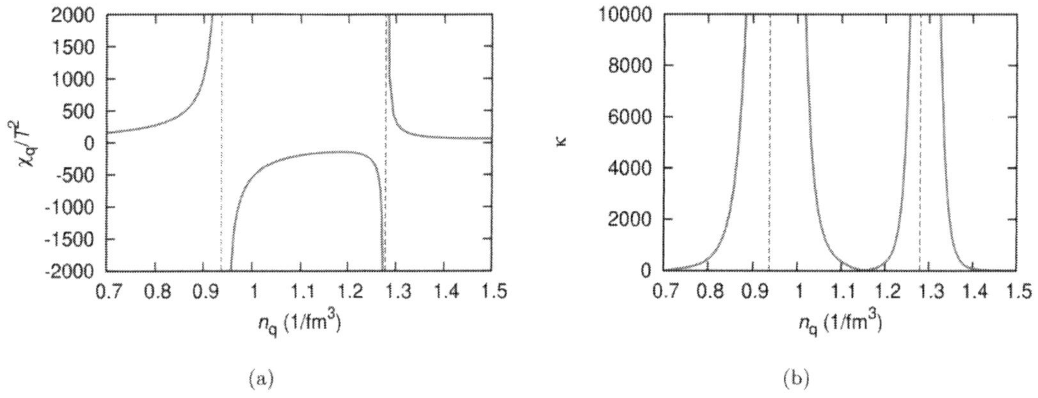

Figure 1. Quark number susceptibility (a) and kurtosis (b) for a nonequilibrium first-order phase transition at $T = 40$ MeV. Figure from [18].

3. Baryon number Fluctuations

Fluctuations of conserved quantities play an important role in identifying the location of a phase transition. They can be studied on several levels. For effective models, we may calculate for instance quark or baryon number susceptibilities according to the formula

$$c_n = \frac{\partial^n(p/T^4)}{\partial(\mu/T)^n} \quad . \tag{10}$$

as some derivative of the pressure with respect to the chemical potential. These coefficients are usually finite and diverge only at a CEP, thus marking a clear signal for a critical structure. However, as shown in [20, 18], divergences occur also at the spinodal lines of a first-order phase transition if instabilities are taken into account. Fig. 1 shows the susceptibility and kurtosis from the linear sigma model with dilatons for a constant temperature of $T = 40$ MeV as a function of the net quark density. We see that both quantities diverge around the spinodal points, with critical indices of $2/3$ and 2, respectively. This result is in agreement with what has been found for the NJL model in [20].

Given this, one may expect a clear enhancement of event-by-event fluctuations in experiment not only for a CEP but also, and probably even stronger, for a first-order phase transition. Note that the important aspect here is that the system is allowed to develop spinodal instabilities after the formation of a supercooled phase. Only in nonequilibrium, enhanced fluctuations can be observed for a first-order transition. We test this assumption by extracting fluctuations of the net-baryon number from dynamical simulations of heavy-ion collisions using the nonequilibrium chiral fluid dynamics model. For this purpose we begin with a spherical droplet of plasma, defined by an initial temperature and chemical potential. We choose these values such that the subsequent evolution proceeds through the desired region of the phase diagram, enabling us to study a crossover, CEP and first-order phase transition. As the total baryon number is conserved throughout the expansion of the hot and dense matter, we need to define some region of acceptance, as we are not able to study fluctuations in a grand canonical ensemble as the susceptibilities would indicate. We rather have to observe some appropriate part of the phase space in the canonical ensemble. As shown in [21], ratios of cumulants significantly depend on the ratio of measured to total baryons N/N_{tot} due to the overall conservation law, making the choice of a suitable range of acceptance a delicate and nontrivial problem. We show results for the variance and kurtosis as a function of time in Fig. 2. The baryon number within a

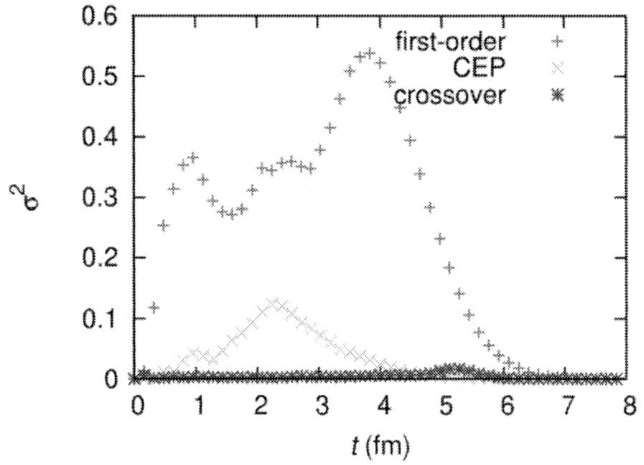

(a)

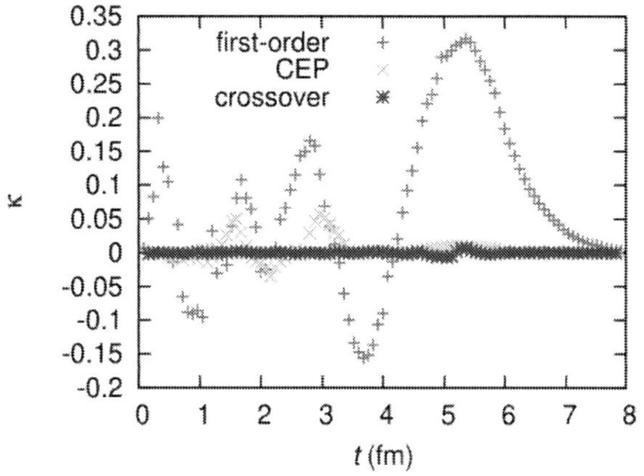

(b)

Figure 2. Variance (a) and kurtosis (b) of the net-baryon number from the fluid dynamical evolution as a function of time. Figure from [18].

single event is calculated over all cells with rapidity $|y| < 0.5$ and transverse momentum density $100\ \mathrm{MeV/fm^3} < p_T < 500\ \mathrm{MeV/fm^3}$. In the upper plot we see the variance clearly enhanced for a CEP and even more for a first-order phase transition in comparison with a crossover scenario. The same holds for the kurtosis as shown in the lower plot. We see that it also becomes negative for both a CEP and a first-order phase transition.

We also see that the enhanced fluctuations vanish for larger times, so the signal gets washed out in the hydrodynamic phase after passing the phase boundary. Therefore it is important to consider a criterion for a freeze-out to determine if the fluctuations in the density get finally imprinted in fluctuations of actual particle numbers as measured by a detector. We expect that due to baryon number conservation the fluctuations remain present even after particlization and final state interactions. However, this assumption has to be verified in the future.

Acknowledgements
This work was funded by Suranaree University of Technology (SUT) and the CHE-NRU (NV.12/2557) project. The authors thank Igor Mishustin and Chihiro Sasaki for fruitful discussions and Dirk Rischke for providing the SHASTA code that was used numerical evolution of the ideal fluid. M. N. acknowledges support from the U.S. Department of Energy under grant DE-FG02-05ER41367 and a fellowship within the Postdoc-Program of the German Academic Exchange Service (DAAD). The computing resources have been provided by the National e-Science Infrastructure Consortium of Thailand, the Center for Computer Services at SUT and the Frankfurt Center for Scientific Computing.

References

[1] Aoki Y, Endrodi G, Fodor Z, Katz S D and Szabo K K 2006 *Nature* **443** 675–678 (*Preprint* hep-lat/0611014)
[2] Borsanyi S *et al.* (Wuppertal-Budapest Collaboration) 2010 *JHEP* **1009** 073 (*Preprint* 1005.3508)
[3] Scavenius O, Mocsy A, Mishustin I N and Rischke D H 2001 *Phys.Rev.* **C64** 045202 (*Preprint* nucl-th/0007030)
[4] Schaefer B J, Pawlowski J M and Wambach J 2007 *Phys.Rev.* **D76** 074023 (*Preprint* 0704.3234)
[5] Fukushima K 2008 *Phys.Rev.* **D77** 114028 (*Preprint* 0803.3318)
[6] Herbst T K, Pawlowski J M and Schaefer B J 2011 *Phys.Lett.* **B696** 58–67 (*Preprint* 1008.0081)
[7] Fischer C S, Luecker J and Welzbacher C A 2014 (*Preprint* 1405.4762)
[8] Adamczyk L *et al.* (STAR Collaboration) 2014 *Phys.Rev.Lett.* **112** 162301 (*Preprint* 1401.3043)
[9] Adamczyk L *et al.* (STAR Collaboration) 2014 *Phys.Rev.Lett.* **112** 032302 (*Preprint* 1309.5681)
[10] Stephanov M A 2009 *Phys.Rev.Lett.* **102** 032301 (*Preprint* 0809.3450)
[11] Berdnikov B and Rajagopal K 2000 *Phys.Rev.* **D61** 105017 (*Preprint* hep-ph/9912274)
[12] Mishustin I N 1999 *Phys.Rev.Lett.* **82** 4779–4782 (*Preprint* hep-ph/9811307)
[13] Randrup J 2010 *Phys.Rev.* **C82** 034902 (*Preprint* 1007.1448)
[14] Nahrgang M, Leupold S, Herold C and Bleicher M 2011 *Phys.Rev.* **C84** 024912 (*Preprint* 1105.0622)
[15] Nahrgang M, Leupold S and Bleicher M 2012 *Phys.Lett.* **B711** 109–116 (*Preprint* 1105.1396)
[16] Herold C, Nahrgang M, Mishustin I N and Bleicher M 2013 *Phys.Rev.* **C87** 014907 (*Preprint* 1301.1214)
[17] Herold C, Nahrgang M, Mishustin I and Bleicher M 2014 *Nuclear Physics A* **925** 14 – 24 ISSN 0375-9474
[18] Herold C, Nahrgang M, Yan Y and Kobdaj C 2014 *J.Phys.* **G41** 115106 (*Preprint* 1407.8277)
[19] Sasaki C and Mishustin I N 2012 *Phys.Rev.* **C85** 025202 (*Preprint* 1110.3498)
[20] Sasaki C, Friman B and Redlich K 2007 *Phys.Rev.Lett.* **99** 232301 (*Preprint* hep-ph/0702254)
[21] Bzdak A, Koch V and Skokov V 2013 *Phys.Rev.* **C87** 014901 (*Preprint* 1203.4529)

Regge behaviour within the Bethe-Salpeter approach

Stanislav Kubrak, Christian S Fischer and Richard Williams
Institut für Theoretische Physik, Justus-Liebig–Universität Giessen, 35392 Giessen, Germany.

Abstract. We present a calculation of the spectrum of light and heavy quark bound states in the rainbow-ladder truncation of Dyson-Schwinger/Bethe-Salpeter equations. By extending the formalism include the case of total angular momentum $J = 3$, we are able to explore Regge trajectories and make prediction of tensor bound states for light and heavy quarkonia.

1. Introduction

The theory of QCD presents many interesting challenges, not least of which that the underlying dynamics described in terms of quarks and gluons is obscurred by the process of hadronization. In this direction, meson spectroscopy remains an interesting topic especially in light of Belle, Babar and BES, where a wealth of new states have been discovered and warrant investigation.

In this work we focus on the functional approach, namely Dyson–Schwinger equations (DSEs) and Bethe–Salpeter equations (BSEs). By extending the formalism to include mesons with higher angular momentum [1], we can make predictions for states with $J = 3$ in the light and heavy quark sector [2]. This enables the investigation of Regge trajectories in this approach for the first time.

2. Framework

The starting point are the Dyson-Schwinger equations (DSEs), the corresponding solutions of which are the one-particle irreducible Green's functions. Throughout this paper we employ the rainbow-ladder approach, a widely used truncation that preserves the chiral nature of the pion. The dressed quark propagator, $S^{-1}(p) = Z_f^{-1}(p^2)(i\not{p} + M(p^2))$, with $Z_f(p^2)$ the quark wave function and $M(p^2)$ its mass function, is obtained by solution of its DSE

$$S^{-1}(p) = Z_2 S_0^{-1}(p) + 4\pi Z_2^2 C_F \int \frac{d^4k}{(2\pi)^4}\gamma^\mu S(k+p)\gamma^\nu(\delta_{\mu\nu} - k_\mu k_\nu/k^2)\,\frac{\alpha_{\text{eff}}(k^2)}{k^2}\,, \tag{1}$$

which we have already specialised to the case of rainbow-ladder by combining the dressed gluon propagator and quark-gluon vertex to form an effective one-gluon exchange model with effective coupling $\alpha_{\text{eff}}(k^2)$ [3, 4, 5]. The bare propagator is $S_0^{-1}(p) = i\not{p} + m_0$, with bare quark mass m_0 related to the renormalized one via $Z_2\, m_0 = Z_2\, Z_m\, m_q$ and the renormalisation factors Z_2, Z_m. The $C_F = 4/3$ is the usual color Casimir.

For our effective interaction we employ the frequently used model of Maris and Tandy [6]. This consists of an ultraviolet part that guarantees agreements with one-loop resummed perturbation theory, and an infrared term whose integrated strength triggers dynamical chiral symmetry breaking. It has the form

$$\alpha_{\text{eff}}(q^2) = \pi\eta^7 x^2 e^{-\eta^2 x} + \frac{2\pi\gamma_m (1 - e^{-y})}{\ln[e^2 - 1 + (1+z)^2]}, \tag{2}$$

with momenta $x = q^2/\Lambda^2$, $y = q^2/\Lambda_t^2$, $z = q^2/\Lambda_{\text{QCD}}^2$. Typically Λ is constrained in the light quark sector to be equal to 0.72 GeV in order to set the scale of an non-pertubative interaction such, that the pion mass and leptonic decay constant are reproduced correctly. In order to incorporate flavor dependence, the η parameter differs in the effective interaction for light and heavy hadron states.

Bound states of a quark and an antiquark are described by the Bethe–Salpeter equation for the corresponding Bethe–Salpeter amplitude $\Gamma(p;P)$

$$[\Gamma(p;P)] = \lambda(P_i^2) 4\pi Z_2^2 C_F \int \frac{d^4k}{(2\pi)^4} \gamma^\mu \left[S_+\Gamma(k;P)S_-\right] \gamma^\nu (\delta_{\mu\nu} - k_\mu k_\nu/k^2) \frac{\alpha_{\text{eff}}(k^2)}{k^2}. \tag{3}$$

This equation can be seen as an eigenvalue problem, whose solutions $\lambda\left(P_i^2\right) = 1$ found at $P^2 = -M_i^2$ form a discrete spectrum. The quantum numbers of the meson follow from the covariant structure of $\Gamma(p;P)$.

3. Results

In Fig. 1 we show the resulting spectrum for $n\bar{n}$ mesons. With the parameters Λ and η to be adjusted such, that the pion mass is fixed to experimental value, the ground state mass in the vector channel ρ agrees with experimental value. However, further on the spectra, for the scalar and axialvector states we see a significant deviation in masses. Nevertheless, the situation greatly improves for the two tensor channels with $J = 2^{++}$ and $J = 3^{--}$.

It is apparent that the sequence of ground states lying on the Regge trajectory with quantum numbers $J^{PC} = 1^{--}, 2^{++}, 3^{--}$ is in good agreement with experiment. The physical interpretation of this is as follows: in the scalar and axial-vector states the tensor force plays a significant role, but for the $J^{PC} = 2^{++}, 3^{--}$ states by contrast it is vanishing or small, whereas the spin-orbit forces are dominant. Translating our observation to a naive quark model

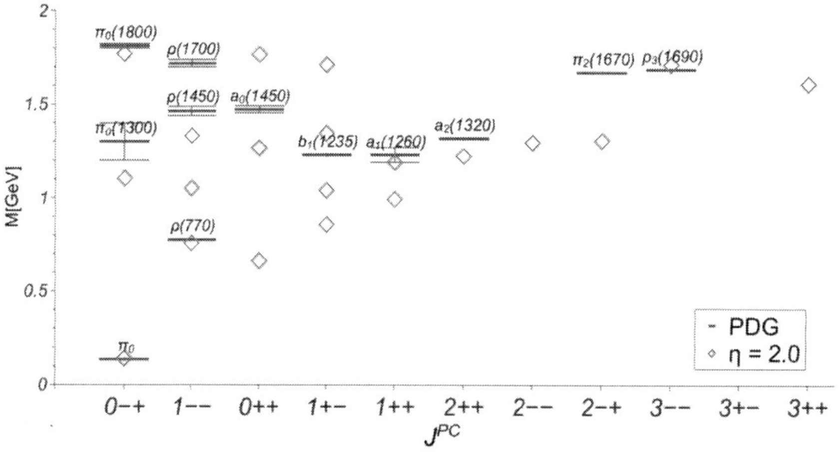

Figure 1. The calculated $n\bar{n}$ spectrum, compared to experiment.

Journal of Physics: Conference Series **599** (2015) 012013 doi:10.1088/1742-6596/599/1/012013

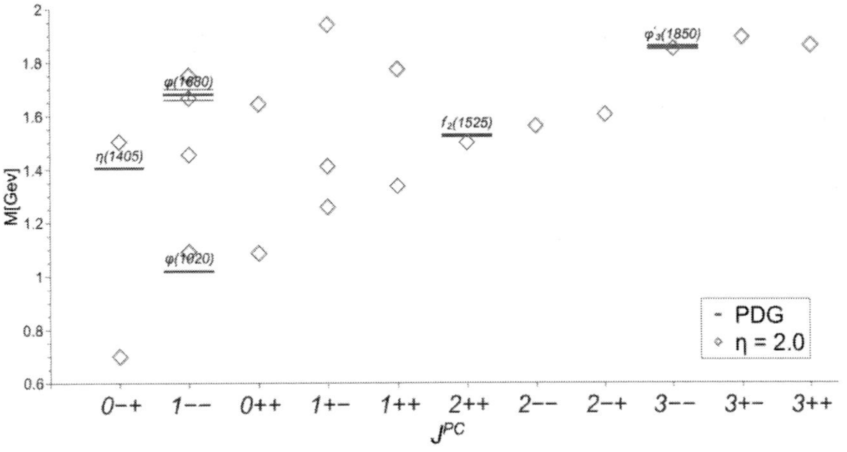

Figure 2. Calculated $s\bar{s}$ spectrum, compared to experiment.

picture, we conclude that in the rainbow-ladder truncation the analogue of the tensor force is not sufficient.

The $s\bar{s}$ spectrum is displayed in Fig. 2. As before one observes a similar picture in that the sequence $J^{PC} = 1^{--}, 2^{++}, 3^{--}$ agrees fairly well with experiment. However the pseudoscalar $s\bar{s}$-state is too light because effects from the $U_A(1)$ anomaly [7] are not taken into account. In the $J^{PC} = 2^{++}$ and $J^{PC} = 3^{--}$ channels, we closely reproduce the experimental values for the $f_2(1525)$ and the $\varphi_3(1850)$.

For the charmonium sector the J/Ψ and X_{c2} states masses were used to fix the two model parameters, η and the quark mass. The remaining spectrum can be viewed as predictions. In addition to the ground states, we observe a good agreement for excited states in the scalar and vector channel. The masses of η_c' and $\Psi(2S)$ states coincide with experiment within a 1% percent accuracy. Even the third state in vector channel, the $\Psi(3S)$, is nicely represented. Unfortunately

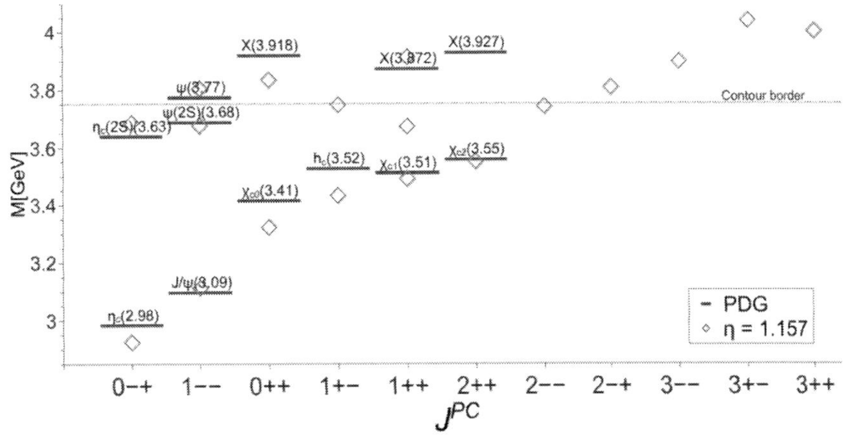

Figure 3. Spectrum of ground and excited charmonium states.

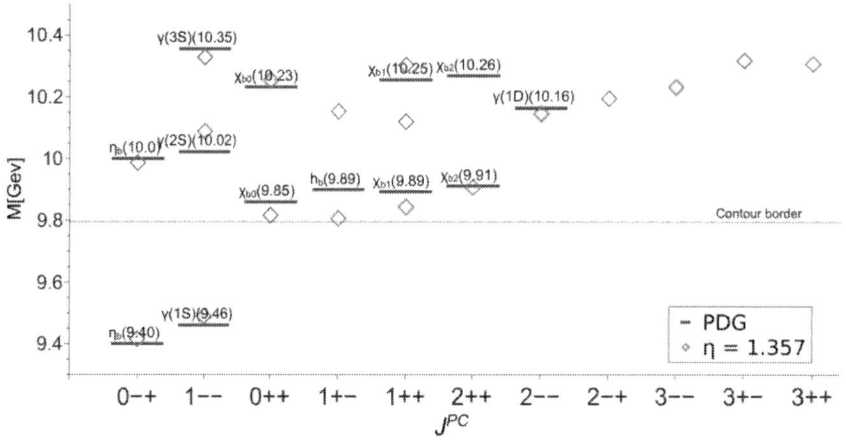

Figure 4. Spectrum of ground and excited bottomonium states.

any further excitations are unreachable due to the non-analytic structure of the quark dressing functions at large M_i^2, as well as the onset of the $D\bar{D}$-threshold.

Similarly, for bottomonium the spectrum, shown in Fig. 4, is also very well represented for the ground-states, agreeing well with experimental data. Note that the η-parameter was changed to $\eta = 1.357$ in comparison to a charmonium calculation, in order to take into account the intrinsic flavour dependence of the quark-gluon interaction [8].

Further, we apply an observation made in light quark sector, that the channels $1^{--}, 2^{++}, 3^{--}$ (the natural Regge trajectory) are well described by the rainbow-ladder truncation, to predict the charmonium and bottomonium 3^{--}-state. We find $m^{3^{--}}_{charm} = 3.896\,\mathrm{GeV}$ and $m^{3^{--}}_{bottom} = 10.169\,\mathrm{GeV}$. The result agrees with the quark model prediction [9] and the lattice QCD results [10, 11].

Finally, we extract the parameters of the Regge-trajectory given by $M^2 = M^2(0) + \beta J$, with the slope parameters $M^2(0)$ and β given in Table 1. That we obtain Regge behaviour is intriguing, because the model that we employ cannot (directly) be related to a linear rising quark potential.

	$n\bar{n}$	$s\bar{s}$	$c\bar{c}$	$b\bar{b}$
$M^2(0)$	−0.42	0.05	2.72	9.12
β	0.99	1.12	0.39	0.371

Table 1. Parameters of Regge trajectories for $n\bar{n}$, $s\bar{s}$, $c\bar{c}$ and $b\bar{b}$ mesons with natural parity.

4. Summary and conclusions

We presented a calculation of the spectra of light and heavy mesons within within the rainbow-ladder truncation using a well-established effective interaction.

In the light quark sector, the results are reliable on the five percent level for channels where tensor force does not play a dominant role, *i.e.* $J^{PC} = 2^{++}, 3^{--}$. As a result, we find that the associated Regge trajectory agrees with experiment. Since our approach does not incorporate a linear rising potential, it is unexpected that we observe Regge trajectories. This fact may cast some doubts on the naive interpretation of Regge-behavior.

In the case of heavy quarkonia, for spectrum of charmonia and bottomonia we obtained good results, as for ground states as well as for first and second radial excitations in the scalar and vector channels. For the tensor states 3^{--} we have enough confidence in our calculation to make a prediction since it lies on the natural Regge trajectory $J^{PC} = 1^{--}, 2^{++}, 3^{--}$ and is therefore well suited to the rainbow-ladder interaction.

However, the comparison with experiment highlights the necessity to extend this framework beyond that of rainbow-ladder . There are a several ways to approaches in this direction: to take into account the effects of the gluon self-interaction [12, 13], pion cloud [14] or through modelling the quark-gluon vertex beyond the leading γ_μ [15, 16, 17, 18].

Acknowledgments

We thank the organizers of FAIRNESS2014 conference for giving an opportunity to present the results. This work was supported by the BMBF under contract No. 06GI7121, the Helmholtz International Center for FAIR within the LOEWE program of the State of Hesse.

[1] C. S. Fischer, S. Kubrak and R. Williams, Eur. Phys. J. A **50** (2014) 126 [arXiv:1406.4370 [hep-ph]].
[2] C. S. Fischer, S. Kubrak and R. Williams, arXiv:1409.5076 [hep-ph].
[3] M. S. Bhagwat and P. Maris, Phys. Rev. C **77** (2008) 025203 [nucl-th/0612069].
[4] M. Blank and A. Krassnigg, Phys. Rev. D **84** (2011) 096014 [arXiv:1109.6509 [hep-ph]].
[5] T. Hilger, C. Popovici, M. Gomez-Rocha and A. Krassnigg, arXiv:1409.3205 [hep-ph].
[6] P. Maris and P. C. Tandy, Phys. Rev. C **60** (1999) 055214 [nucl-th/9905056].
[7] R. Alkofer, C. S. Fischer and R. Williams, Eur. Phys. J. A **38** (2008) 53 [arXiv:0804.3478 [hep-ph]].
[8] R. Williams, arXiv:1404.2545 [hep-ph].
[9] D. Ebert, R. N. Faustov and V. O. Galkin, Eur. Phys. J. C **71** (2011) 1825 [arXiv:1111.0454 [hep-ph]].
[10] G. S. Bali, S. Collins and C. Ehmann, Phys. Rev. D **84** (2011) 094506 [arXiv:1110.2381 [hep-lat]].
[11] L. Liu *et al.* [Hadron Spectrum Collaboration], JHEP **1207** (2012) 126 [arXiv:1204.5425 [hep-ph]].
[12] C. S. Fischer and R. Williams, Phys. Rev. Lett. **103** (2009) 122001 [arXiv:0905.2291 [hep-ph]].
[13] H. Sanchis-Alepuz and R. Williams, *in preparation.*
[14] C. S. Fischer and R. Williams, Phys. Rev. D **78** (2008) 074006 [arXiv:0808.3372 [hep-ph]].
[15] L. Chang and C. D. Roberts, Phys. Rev. Lett. **103** (2009) 081601 [arXiv:0903.5461 [nucl-th]].
[16] L. Chang, Y. X. Liu and C. D. Roberts, Phys. Rev. Lett. **106** (2011) 072001 [arXiv:1009.3458 [nucl-th]].
[17] L. Chang and C. D. Roberts, Phys. Rev. C **85** (2012) 052201 [arXiv:1104.4821 [nucl-th]].
[18] W. Heupel, T. Goecke and C. S. Fischer, Eur. Phys. J. A **50** (2014) 85 [arXiv:1402.5042 [hep-ph]].

Spinodal density enhancements in nuclear collisions at the CBM experiment

J. Steinheimer[1], V. Koch[2], J. Randrup[2] and M. Bleicher[1,3]

[1] Frankfurt Institute for Advanced Studies, Ruth-Moufang-Str. 1, Frankfurt am Main, Germany

[2] Nuclear Science Division, Lawrence Berkeley National Laboratory, Berkeley, California 94720, USA

[3] Institut für Theoretische Physik, Goethe-Universität, Max-von-Laue-Str. 1, Frankfurt am Main, Germany

E-mail: steinheimer@fias.uni-frankfurt.de

Abstract. We discuss a novel approach to describe the evolution of a fireball, created in a high-energy nuclear collision, experiencing spinodal instabilities due to the first-order deconfinement phase transition of quantum chromo dynamics (QCD). We show that initial density fluctuations in these collisions are enhanced in the mechanically unstable region of the QCD phase diagram. In our study we find that the most favorable energy range for observing these density enhancements is at the lower end of the SIS100 accelerator at FAIR, currently under construction. Furthermore we discuss how one can distinguish and constrain different types of QCD phase transitions, one of hadron-quark type and one of liquid-gas type, leading to strong differences in the dynamical evolution of the QCD medium.

1. Introduction

Strongly interacting matter, in particular compressed baryonic matter, is expected to posses a rich phase structure and may exhibit a first-order phase transition that persists up to a certain critical temperature [1]. At vanishing baryon chemical potential, lattice QCD calculations predict a smooth transformation from confined to deconfined matter at a cross-over temperature of $T_\times \approx 150$ - 160 MeV [2, 3]. Unfortunately, the lattice calculations cannot be extended to finite chemical potential due to the so-called sign problem. It is therefore necessary to experimentally determine the phase structure of baryon-rich strongly interacting matter. For such experimental endeavors [4, 5, 6] to be successful, it is important to identify observables that may serve as signals of the expected phase transition. Because the colliding system is relatively small, non-uniform, far from global equilibrium, and rapidly evolving, the connection between experimental observables and the idealized uniform equilibrium matter described by the equation of state are obscured. Consequently it is important to carry out dynamical simulations of the collisions with suitable transport models.

In a novel approach we perform numerical simulations with finite-density fluid dynamics, using a previously developed two-phase equation of state and incorporating a gradient term in the local pressure [7]. This latter refinement emulates the finite-range effects for a proper description of the phase transition physics [8]. Thus we employ a transport model that treats the associated physical instabilities in a numerically reliable manner.

2. A model for the phase transition

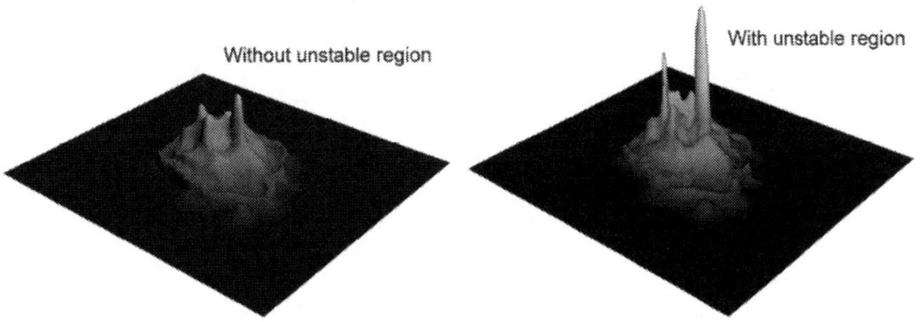

Figure 1. Density profile after 4 fm/c with a single UrQMD initial condition for $E_{lab} = 3A$ GeV. While the initial fluctuations are not amplified in the case without an unstable region (left figure), they are strongly amplified when an unstable region in the EoS is included (right figure).

For our present investigations, we describe the evolution of the colliding system by ideal fluid dynamics. The basic equation of motion in ideal fluid dynamics expresses four-momentum conservation, $\partial_\mu T^\mu = 0$, where the stress tensor is given by

$$T^{\mu\nu}(x) = [p(x) + \varepsilon(x)]u^\mu(x)u^\nu(x) - p(x)g^{\mu\nu} , \tag{1}$$

where $u^\mu(x)$ is the four-velocity of the fluid. When taking account of the baryon current density,$N^\mu(x) = \rho(x)u^\mu(x)$, the basic equation of motion is supplemented by the continuity equation, $\partial_\mu N^\mu = 0$. These equations of motion are solved by means of the code SHASTA [10] in which the propagation in the three spatial dimensions is carried out consecutively.

In order to obtain a suitable equation of state, used as input to the fluid dynamical simulations, we employ the method developed in Ref. [9]. For a given T, we obtain the free energy density $f_T(\rho)$ in the phase coexistence region by performing a suitable spline between two idealized systems (either a gas of pions and interacting nucleons or a bag of gluons and quarks) held at that temperature. For each T the spline points were adjusted so the resulting $f_T(\rho)$ would exhibit a concave anomaly, *i.e.* there would be two densities, $\rho_1(T)$ and $\rho_2(T)$, for which the tangent of $f_T(\rho)$ would be common. This ensures phase coexistence because then the chemical potentials $\mu_T = \partial_\rho f_T(\rho)$ match, $\mu_T(\rho_1) = \mu_T(\rho_2)$, as do the pressures, $p_T(\rho_1) = p_T(\rho_2)$. The resulting two-phase EoS (HQ-EoS) therefore exhibits a transition from a hadronic to a deconfined quark phase.

As mentioned above, a proper description of spinodal decomposition requires that finite-range effects be incorporated [8]. Therefore we write the local pressure as

$$p(\boldsymbol{r}) = p_0(\varepsilon(\boldsymbol{r}), \rho(\boldsymbol{r})) - a^2 \varepsilon_s \frac{\rho(\boldsymbol{r})}{\rho_s} \nabla^2 \frac{\rho(\boldsymbol{r})}{\rho_s} , \tag{2}$$

where we recall that $p_0(\varepsilon, \rho)$ is the equation of state, the pressure in uniform matter characterized by ε and ρ. With $\rho_s = 0.153/\mathrm{fm}^3$ being the nuclear saturation density and $\varepsilon_s \approx m_N \rho_s$ the associated energy density, the gradient term is normalized such that its strength is conveniently governed by the length parameter a which was chosen to take the value $a = 0.033$ fm [11].

We have shown that our model produces both meaningful interface properties, including the temperature-dependent strength of the associated tension, and reasonable spinodal growth rates, including the emergence of an optimal phase separation scale [7]. The model is therefore suitable for addressing dynamical scenarios involving phase-transition instabilities.

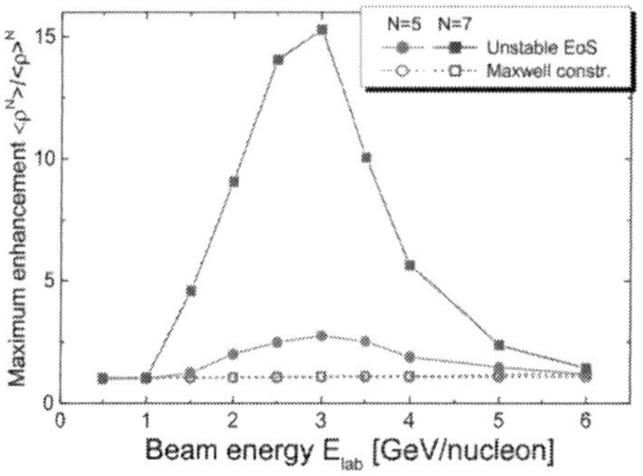

Figure 2. Maximal enhancement of the fifth and seventh moments of the net baryon density distributions as function of the beam energy $E_{\rm lab}$, for central collisions of Pb+Pb. We compare a two phase equation of state including the effects of the spinodal instabilities (solid lines) with a single phase equation of state (dashed lines) obtained via a Maxwell construction.

For a first study, we have considered central collisions of lead nuclei bombarded onto a stationary lead target at various kinetic energies, $E_{\rm lab}$. For each energy, an ensemble of several hundred separate evolutions are generated, each one starting from a different initial configuration generated by (the cascade mode of) the UrQMD model [12, 13, 14] which treats the non-equilibrium dynamics during the very early stage of the collision.

Figure 1 shows the effect of the spinodal instabilities on a single initial density profile, obtained with UrQMD. Here the initial fluctuations are not amplified in the case without an unstable region, they are strongly amplified when an unstable region in the EoS is included. When considering an ensemble of many events, a convenient quantitative measure of the resulting degree of "clumping" in the system is provided by the moments of the baryon density density $\rho(\boldsymbol{r})$,

$$\langle\rho^N\rangle \equiv \frac{1}{A^N}\int \rho(\boldsymbol{r})^N \rho(\boldsymbol{r})\, d^3\boldsymbol{r} \ , \tag{3}$$

where $A = \int\rho(\boldsymbol{r})d^3\boldsymbol{r} = \langle\rho^0\rangle$ is the total (net) baryon number. The corresponding normalized moments, $\langle\rho^N\rangle/\langle\rho\rangle^N$, are dimensionless and increase with the order N, for a given density distribution $\rho(\boldsymbol{r})$; the normalized moment for $N = 1$ is unity.

Figure 2 shows the (ensemble average) maximum enhancement (for $N = 5$ and 7) achieved, as a function of the beam energy for the two equations of state. The existence of an optimal collision energy is clearly brought out. While the presently employed equation of state suggests that this optimal range is $E_{\rm lab} \approx 2 - 4\,A\,\text{GeV}$, it should be recognized that others may lead to different results.

3. The effect of the equation of state

As we pointed out, the effective EoS used in the dynamical simulations for the QCD phase transition may have a strong impact on the resulting observable signals, as well as the optimal collisional beam energy needed to see signals of the phase transition. In this section we will discuss possible ways to qualitatively distinguish different types of phase transitions in QCD and how they influence the dynamical evolution of nuclear collisions in the region of large baryon densities, where the corresponding baryon-number chemical potential μ is of the order of the temperature or higher.

Among the many effective models for QCD thermodynamics, those that couple constituent quarks to mesonic fields and the Polyakov loop have become popular because they include chiral-

Journal of Physics: Conference Series **599** (2015) 012014 doi:10.1088/1742-6596/599/1/012014

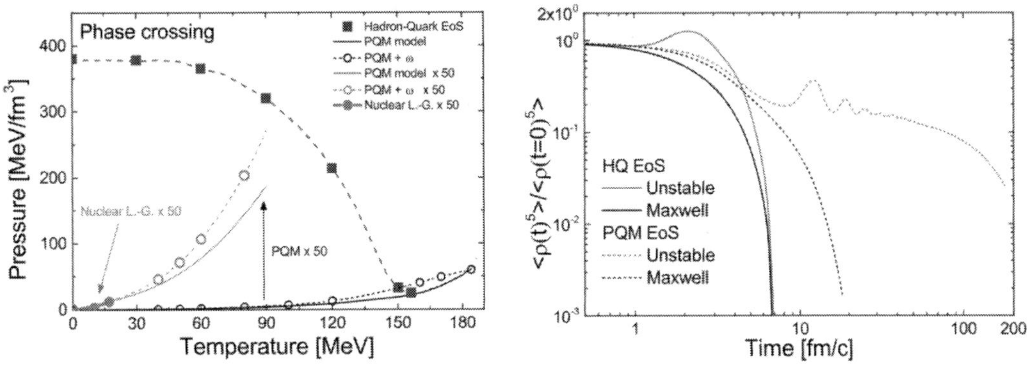

Figure 3. Left: Pressure along the coexistence line for the two different equations of state discussed in the text. Right: Time evolution of the fifth moment of the net baryon density for the two different equations of state discussed in the text, and an initial state that locates the bulk of the system within the unstable region of the phase diagram.

symmetry breaking and a "thermal" confinement of quarks. Such approaches are the Polyakov Nambu–Jona-Lasino (PNJL) [15, 16] and the Polyakov Quark-Meson (PQM) [17] models. These models can be brought into agreement with lattice results at vanishing net baryon density but their results for the finite-density domain depend significantly on the parameterizations used. A particular weakness in most of these models is the absence of explicit hadronic degrees of freedom in the confined phase. There exist only few examples where the equations of state (EoS) includes an approximately realistic description of the hadronic phase. These equations of states are constructed either by matching a purely hadronic model to an MIT-bag EoS or by combining a PNJL/PQM-type model with a chiral hadronic model [18, 19]. Obviously, these models need to be constrained both by lattice QCD results at zero and small chemical potential and by known properties of nuclear matter at small temperatures and densities up to about twice nuclear matter density [20], as well as by astrophysical observations on neutron star properties [21].

It is instructive to consider the temperature dependence of the *pseudo-critical* pressure, $p_{\rm pc}$, the pressure at which the phase transformation occurs for the different types of EoS. Figure 3 (left) shows the temperature dependent pseudo-critical pressure for the two models considered here, the PQM model and the constructed Hadron-Quark EoS, as discussed in section 2. For a specified temperature T, $p_{\rm pc}^{\rm HQ}$ is obtained by increasing the chemical potential μ until the ideal hadron-gas and the quark-gluon bag have the same pressure, while $p_{\rm pc}^{\rm PQM}$ is taken as the pressure along the inflection point of the chiral condensate [22]. While the resulting pseudo-critical pressures are comparable at the highest temperatures (corresponding to $\mu \approx 0$), and consistent with lattice results [2], they deviate strongly at large chemical potentials: As the temperature is reduced, $p_{\rm pc}^{\rm HQ}$ increases steadily, as was already noted in previous studies employing models that describe the hadron-quark transition [23, 24]. By contrast, $p_{\rm pc}^{\rm PQM}$ decreases steadily and vanishes at $T = 0$, a behavior that is a robust feature of the PQM models and persists also in the presence of a repulsive quark interaction [25, 26]. This generic liquid-gas behavior differs qualitatively from the HQ result, for which the pressure *decreases* steadily with temperature.

We find that the PQM model shows a transition that is very similar to that of the liquid-gas transition in nuclear matter and thus this model differs qualitatively from the HQ model with regard to the thermodynamic properties near the phase coexistence line. The qualitative differences between the two equations of state examined in this work lead to significant quantitative differences in the time evolution of fireballs that expand through the respective unstable region of the phase diagram. In dynamical simulations with the PQM model the lifetime of quark clusters is orders of magnitude longer than what is usually expected for the timescales of heavy-ion collisions and it predicts stable dense quark matter droplets at zero temperature, i.e. coexisting with the vacuum [22]. In figure 3 (right) the time evolution of the density moments for the two equations of state discussed above are shown. One can clearly observe, that the moments drop rapidly in the case of the HQ EoS which has a large coexistence pressure, while in the case of the PQM model the density clusters survive for a very long time as they can only slowly decay in the vacuum.

4. Conclusion

In conclusion, we have obtained a transport model that is suitable for simulating nuclear collisions in the presence of a first-order phase transition: it describes both the tension between coexisting phases and the dynamics of the unstable spinodal modes. Applying this model to lead-lead collisions, we have found that the associated instabilities may cause significant amplification of initial density irregularities. Furthermore we have shown that qualitative differences between the PQM and a HQ equations of state lead to considerable differences in the dynamical evolution of the system. We believe that these qualitative differences persist independent of the specific EoS model adopted but are rather characteristic for either a liquid-gas type or a more plausible hadron-quark type transition.

5. Acknowledgments

This work was supported by GSI and the Hessian initiative for excellence (LOEWE) through the Helmholtz International Center for FAIR (HIC for FAIR) and by the Office of Nuclear Physics in the U.S. Department of Energy's Office of Science under Contract No. DE-AC02-05CH11231.

6. References

[1] M. A. Stephanov, K. Rajagopal and E. V. Shuryak, Phys. Rev. Lett. **81**, 4816 (1998).
[2] S. Borsanyi *et al.*, JHEP **1011**, 077 (2010).
[3] A. Bazavov and P. Petreczky [HotQCD collaboration], J. Phys. Conf. Ser. **230**, 012014 (2010).
[4] B.I. Abelev *et al.*, Internal STAR Note SN0493, 2009: drupal.star.bnl.gov/STAR/starnotes/public/sn0493.
[5] B. Friman, C. Höhne, J. Knoll, S. Leupold, J. Randrup, R. Rapp, and P. Senger, *The CBM Physics Book*, (Springer, Berlin, 2010).
[6] A.N. Sissakian and A.S. Sorin, J. Phys. G **36**, 064069 (2009).
[7] J. Steinheimer and J. Randrup, Phys. Rev. Lett. **109**, 212301 (2012).
[8] Ph. Chomaz, M. Colonna, and J. Randrup, Phys. Rep. **389**, 263 (2004).
[9] J. Randrup, Phys. Rev. C **79**, 054911 (2009).
[10] D. H. Rischke, S. Bernard and J. A. Maruhn, Nucl. Phys. A **595**, 346 (1995).
[11] J. Steinheimer and J. Randrup, Phys. Rev. C **87**, no. 5, 054903 (2013).
[12] S. A. Bass *et al.*, Prog. Part. Nucl. Phys. **41**, 255 (1998) [Prog. Part. Nucl. Phys. **41**, 225 (1998)].
[13] M. Bleicher *et al.*, J. Phys. **G25**, 1859 (1999).
[14] H. Petersen, J. Steinheimer, G. Burau, M. Bleicher and H. Stocker, Phys. Rev. C **78**, 044901 (2008).
[15] K. Fukushima, Phys. Lett. B **591**, 277 (2004).
[16] C. Ratti, M. A. Thaler and W. Weise, Phys. Rev. D **73**, 014019 (2006).
[17] B. J. Schaefer, J. M. Pawlowski and J. Wambach, Phys. Rev. D **76**, 074023 (2007).
[18] J. Steinheimer, S. Schramm and H. Stöcker, J. Phys. **G38**, 035001 (2011).
[19] L. Turko, D. Blaschke, D. Prorok and J. Berdermann, J. Phys. Conf. Ser. **455**, 012056 (2013).
[20] J. Steinheimer, S. Schramm and H. Stöcker, Phys. Rev. C **84**, 045208 (2011).
[21] V. Dexheimer, J. Steinheimer, R. Negreiros and S. Schramm, Phys. Rev. C **87**, 015804 (2013).
[22] J. Steinheimer, J. Randrup and V. Koch, Phys. Rev. C **89**, no. 3, 034901 (2014).
[23] J. Randrup, Phys. Rev. C **82**, 034902 (2010)
[24] M. Hempel, V. Dexheimer, S. Schramm and I. Iosilevskiy, Phys. Rev. C **88**, no. 1, 014906 (2013).
[25] O. Lourenco, M. Dutra, A. Delfino and M. Malheiro, Phys. Rev. D **84**, 125034 (2011).
[26] M. B. Pinto, V. Koch and J. Randrup, Phys. Rev. C **86**, 025203 (2012).

Approaching the QCD phase diagram for $N_f = 2+1$ and $N_f = 2+1+1$ quark flavors

Christian A Welzbacher[1], **Christian S Fischer**[1] **and Jan Luecker**[2,3]

[1]Institut fuer Theoretische Physik, Justus-Liebig-Universitaet Giessen, Heinrich-Buff-Ring 16, D-35392 Giessen, Germany.

[2]Institut fuer Theoretische Physik, Universitaet Heidelberg, Philosophenweg 16, D-69120 Heidelberg, Germany.

[3]Institut fuer Theoretische Physik, Goethe-Universitaet Frankfurt, Max-von-Laue-Strasse 1, D-60438 Frankfurt/Main, Germany

E-mail: `Christian.A.Welzbacher@theo.physik.uni-giessen.de`

Abstract. Recent results for the QCD phase structure at finite temperature and light-quark chemical potential are summarized, where the cases of $N_f = 2+1$ and $N_f = 2+1+1$ dynamical quark flavors are considered. Order parameters for the chiral and deconfinement transitions are obtained from solutions of a coupled set of truncated Dyson-Schwinger equations for the quark and gluon propagators of Landau gauge QCD. Based on very good agreement with lattice QCD for zero chemical potential the phase diagram in the whole T-μ plane and the appearance of a putative critical end-point at large chemical potential are studied.

Keywords: Critical end point, QCD phase diagram, Dyson-Schwinger equations

1. Introduction

In recent years, experimental and theoretical interest in heavy ion collisions and the phase diagram of QCD grew rapidly. On the experimental side facilities such as RHIC at BNL, ALICE at LHC and CBM at the future FAIR facility (will) search for possible signals of phase transitions. On the theoretical side lattice QCD has established a crossover transition at small chemical potential, see e.g. [1, 2] for $N_f = 2+1$ flavors. In our work we approach the QCD phase diagram with the functional method of Dyson-Schwinger equations and put particular emphasis on the possible influence of the charm quark [3]. The contribution of the charm quark to the equation of state starts to become significant at top LHC energies. However, possibly even at smaller temperatures close or above the light-quark crossover region, effects of charm quarks, although predicted to be small by perturbation theory [4], may not be entirely negligible. Within lattice QCD preliminary results on calculations with $N_f = 2+1+1$ flavors for transition temperatures and the equation of state using staggered [5, 6] and Wilson type quarks [7] are available. Here we summarize recent results for the phase diagram at zero and finite chemical potential within the framework of Dyson-Schwinger equations. At zero chemical potential we compare with lattice results for the quark condensate and the unquenched gluon propagator and discuss our results for the critical end point for $N_f = 2+1$ and $N_f = 2+1+1$ quark flavors.

Journal of Physics: Conference Series **599** (2015) 012015 doi:10.1088/1742-6596/599/1/012015

2. Framework

Dyson-Schwinger equations (*DSEs*) are the equations of motion for the n-point functions of QCD. In Fig. 1 we show the DSE for the quark propagator, which is the foundation of an infinite tower of coupled integral equations. These are in principle ab initio, if solved completely and self-consistently. In practice, however, in most cases one needs to impose a truncation using input from other sources such as symmetries, constraints or other methods like lattice QCD. The truncation used in this work has been developed over the past few years (see [3, 8–11]) and consists of two key points. The first part of the truncation is depicted in Fig. 2, showing the approximation in the DSE for the Landau gauge gluon propagator. In order to calculate this quantity we use input from lattice QCD for the quenched gluon propagator (denoted by the yellow dot), replacing diagrams with internal gluons and Faddeev-Popov ghosts. To this we add a quark loop for each flavor.

Figure 1. The quark Dyson-Schwinger equation.

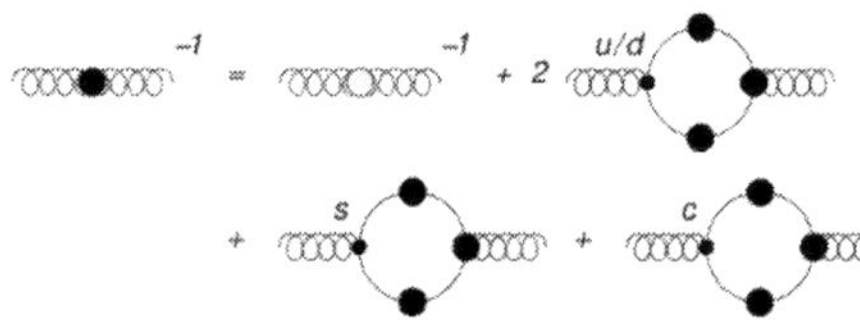

Figure 2. The truncated gluon Dyson-Schwinger equation.

This approach neglects all the second order unquenching effects in the Yang-Mills diagrams but ensures that we take the important temperature effects of the unquenched gluon into account. Additionally the unquenched gluon propagator inherits a dependence on the temperature and the chemical potential via the quark loops, giving a contribution to the thermal mass. The gluon propagators at finite temperature T and quark-chemical potential μ is given by

$$D_{\alpha\beta}(p) = P^L_{\alpha\beta}(p)\frac{Z^L(p)}{p^2} + P^T_{\alpha\beta}(p)\frac{Z^T(p)}{p^2}\,, \tag{1}$$

with momentum $p = (\omega_n, \vec{p})$, $\omega_n = 2n\pi T$ for bosons. The projectors $P^{L,T}_{\alpha\beta}$ are longitudinal (L) and transverse (T) with respect to the heat bath and given by

$$P^T_{\alpha\beta} = (1-\delta_{\alpha 4})(1-\delta_{\beta 4})\left(\delta_{\alpha\beta} - \frac{p_\alpha p_\beta}{\vec{p}^{\,2}}\right), \tag{2}$$

$$P^L_{\alpha\beta} = P_{\alpha\beta} - P^T_{\alpha\beta}\,, \tag{3}$$

where $P_{\alpha\beta}$ is the transverse projector with respect to the four momentum.
The second element of our truncation is an approximation for the full quark-gluon vertex. Information on this vertex can be gathered from lattice QCD or from solutions to its DSE. Corresponding studies at zero temperature and chemical potential are under way; for recent works see e.g. [14, 15] in the DSE framework and [16] in the functional renormalization group approach. At finite temperature, however, information from these sources is not yet available. We therefore use a functional form for this vertex which combines information from the well-known perturbative behavior at large momenta and an approximate form of the Slavnov-Taylor identity at small momenta studied long ago by Ball and Chiu [12].

The explicit form of this approximate expression for the vertex is discussed in Refs. [3, 9, 10] and shall not be repeated here for brevity. It contains one open parameter, called d_1 in the following, which controls the infrared strength of the vertex and sets the temperature scale.

Furthermore, via the Ball-Chiu construction, it contains dressing functions of the two attached quarks. Therefore the vertex is flavor, as well as temperature and chemical potential dependent. To investigate the influence of the charm quark, we developed two approaches to set the IR-strength of the quark-gluon vertex as well as the physical masses of the quarks. The first one (called Sets A_{Nf}) connects our calculations to lattice QCD, since the IR-strength and light-quark mass is fixed to reproduce the (pseudo-) critical temperature $T_C \approx 155$ MeV from [1] for N_f=2+1 flavors at zero chemical potential. Since in a crossover region there are different definitions of a critical temperature, we point out that the chosen one is defined via the inflection point of the chiral condensate. The mass of the strange quark is chosen to be 27 times the light-quark mass. To this setup we merely added a charm quark with a mass of 300 MeV at a renormalization point of 80 GeV, without adjusting the IR-strength. In our second approach (called Sets B_{Nf}) we set the IR-strength in the quark-gluon vertex self-consistently by fixing the pion decay constant, as well as the pion, kaon and η_C mass in the very same truncation via the Bethe-Salpether equation, separately for N_f=2+1 and N_f=2+1+1. This was possible due to recent progress in the Bethe-Salpether framework [13].

To describe the (pseudo-) critical temperature of our system, we calculate the chiral condensate

$$\langle\bar{\psi}\psi\rangle_f = Z_2 Z_m N_c T \sum_n \int \frac{d^3p}{(2\pi)^3} \mathrm{Tr}_D \left[S^f(p)\right], \tag{4}$$

where $S^f(p)$ is the calculated quark propagator for flavor f, Z_2 and Z_m are the corresponding wave-function and quark mass renormalization constants, respectively, and $N_c = 3$ is the number of colors. Any quark flavor with non-zero bare quark mass leads to a condensate which is quadratically divergent and needs to be regularized. To circumvent this divergence one considers the regularized chiral condensate

$$\Delta_{l,s} = \langle\bar{\psi}\psi\rangle_l - \frac{m_l}{m_s}\langle\bar{\psi}\psi\rangle_s . \tag{5}$$

In our work we extract T_C either by considering the maximum of the derivative of this quantity with respect to temperature (*inflection point*) or with respect to the mass of the light quarks (*chiral susceptibility*).

3. Results

In this section we show the results calculated within the framework discussed in the previous section. In Fig. 3 we compare the electric part of the gluon propagator for N_f=2 and a pion mass of 316 MeV, fixed in the manner of Set B, with lattice data from [17]. Note that our result, determined first in [10], has been a prediction, verified later by the lattice results. Therefore the quite good agreement for the unquenched gluon, featuring an inversion of the temperature ordering when going from the quenched to the unquenched case serves as a further justification for our truncation scheme. In Fig. 4 the same quantity is shown, in this case for Set A_{2+1}. We observe the same inversion of the temperature ordering as well as the suppression of the maximum after the unquenching.

In Fig. 5 we show the regularized chiral condensate at μ=0 MeV for all sets of interest, compared to lattice data taken from [1]. We see a nice matching of Set A_{2+1} (dashed red line) with the lattice data, not only in the scale (T_C; controlled by the choice of d_1 as shown in Fig. 6) but also in the steepness of the regularized condensate, which is a non-trivial result. Additionally we observe that by merely adding the charm quark in Set A (dash-dotted blue line), we find a shift of the scale to lower temperatures, which is an effect caused by adding the charm quark to the system without adjusting the physical scale. To investigate this, by looking at Sets B (solid black line) we find that by self-consistently readjusting the scale, the charm quark has no effect.

Journal of Physics: Conference Series **599** (2015) 012015 doi:10.1088/1742-6596/599/1/012015

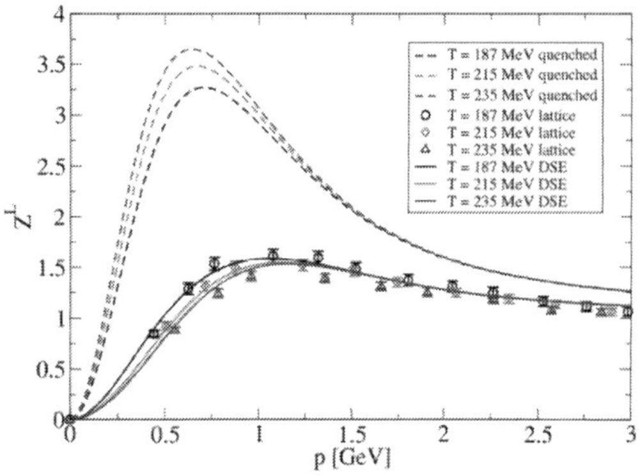

Figure 3. Electric part of the gluon dressing functions for N_f=2, m_π=316 MeV.

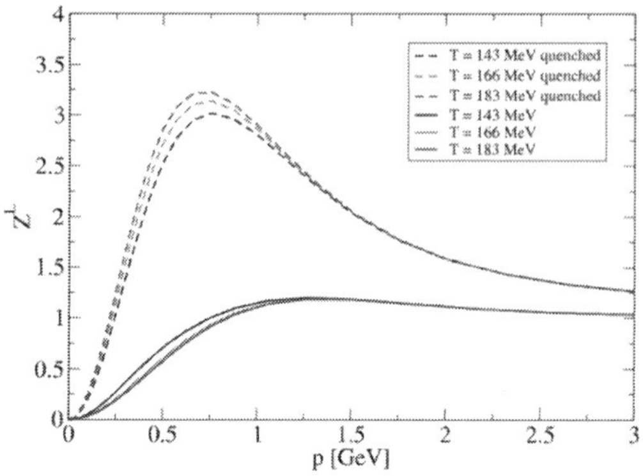

Figure 4. Electric part of the gluon dressing functions for Set A_{2+1}.

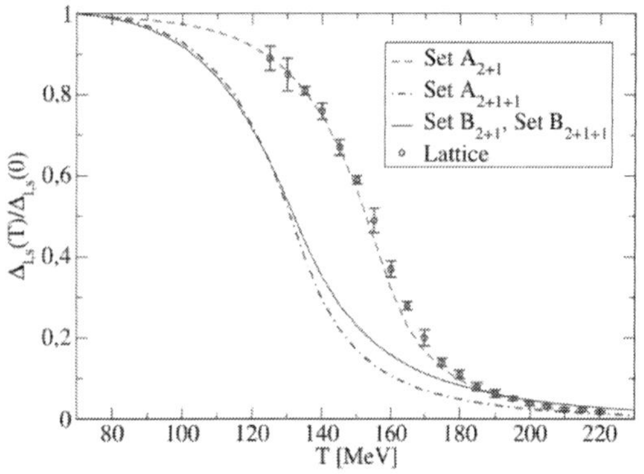

Figure 5. $\Delta_{l,s}$ at μ=0 MeV [3].

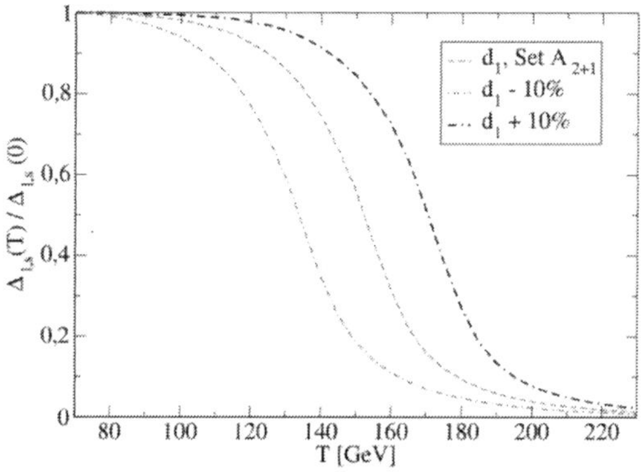

Figure 6. Influence of d_1 on $\Delta_{l,s}$.

Fig. 7 shows the phase diagrams for Sets B. It immediately becomes apparent, that the negligible contribution of the charm quark which we observed for μ=0 MeV in Fig. 5, continues to be small for the whole T-μ-plane for both, chiral and deconfinement transition. The deconfinement transition was fixed via the Polyakov loop of the minimum of the background field potential (see [11] for more details), where the inflection point is used. On the other hand, the critical temperature in the chiral crossover region was defined by the maximum of the chiral susceptibility. This partly explains the gap for small chemical potential between the crossover lines.

Our results for Fig. 7 suggest that there is almost no influence of the charm quark on the transition temperatures, provided the internal scales are controlled by external input. For the QCD phase diagram shown in Fig. 8 we therefore return to lattice input for this task (Set A_{2+1}). In this plot, the critical temperature for the crossover region is in both cases, chiral and deconfinement transition, defined by the inflection point method. One can see, that the chiral crossover (dashed black line) becomes ever steeper with increasing chemical potential and turns into a critical end point (CEP) at

$$(T^c, \mu_q^c) = (115, 168)\,\text{MeV}. \tag{6}$$

To guide the eye we also show the lines for μ_B/T=2 and μ_B/T=3 and added predictions for the curvature of the chiral transition line from lattice extrapolations for $N_f = 2+1$ of different

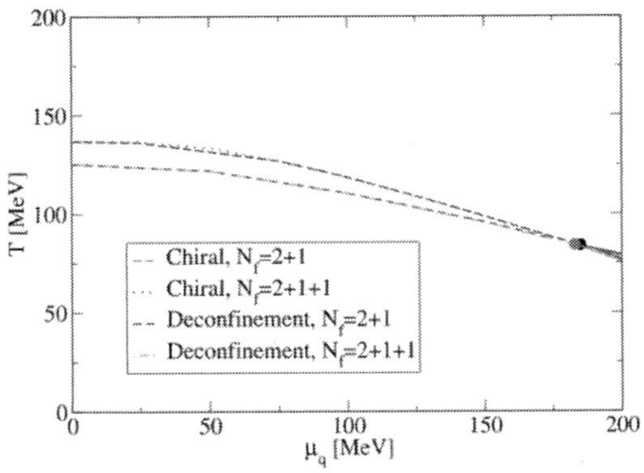

Figure 7. Phase diagram Sets B [3].

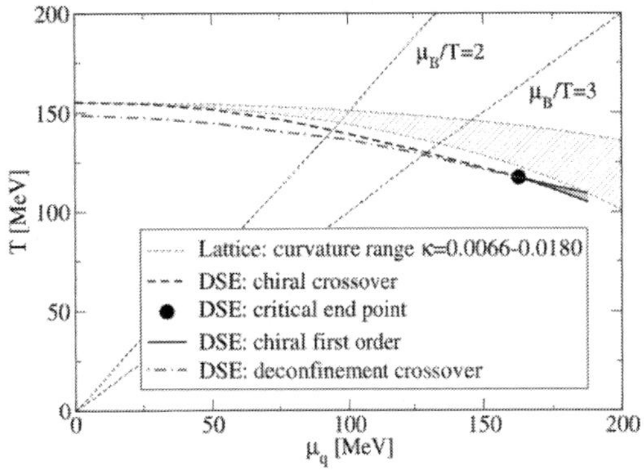

Figure 8. Phase diagram Set A_{2+1} [3].

groups at imaginary and zero chemical potential into the real chemical potential region [18–20]. The agreement between the lattice extrapolation and our DSE results is quite satisfactory. We close with the remark that potential effects of baryons onto the location of the critical endpoint have not yet been taken into account, neither in the lattice extrapolations of the curvature, nor in our DSE approach. In order to turn this qualitative study into a quantitative one, these need to be addressed. Thus the close agreement of both approaches, although interesting, may very well not be the final word.

Acknowledgements
This work has been supported by the Helmholtz Alliance HA216/EMMI and by ERC-AdG-290623 as well as the Helmholtz International Center for FAIR within the LOEWE program of the State of Hesse.

References
[1] Borsanyi S et al 2010 *JHEP* **1009** 073, *arXiv:1005.3508.*
[2] Bazavov A et al (2012) *Phys. Rev.* D **85** 054503 *arXiv:1111.1710.*
[3] Fischer C S, Luecker J and Welzbacher C A 2014 *Phys. Rev.* D **90** 034022, *arXiv:1405.4762.*
[4] Laine M and Schroder Y (2006) *Phys. Rev.* D **73** 085009 *arXiv:0603.048.*
[5] Bazavov A et al (2013) *PoS* **LATTICE2013** 154 *arXiv:1312.5011.*
[6] Borsanyi S et al (2010) *JHEP* **1011** 077, *arXiv:11007.2580.*
[7] Burger F, Hotzel G, Muller-Preussker M, Ilgenfritz E-M and Lombardo M P (2013) *arXiv:1311.1631.*
[8] Fischer C S 2009 *Phys.Rev.Lett.* **103** 052003. *arXiv:0904.2700.*
[9] Fischer C S, Maas A and Mueller J A 2010 *Eur.Phys.J.* C **68** 165-81. *arXiv:1003.1960.*
[10] Fischer C S and Luecker J 2013 *Phys.Lett.* B **718** 1036-43. *arXiv:1206.5191.*
[11] Fischer C S, Fister L, Luecker J and Pawlowski J M 2014 *Phys.Lett.* B **732** 273-77. *arXiv:1306.6022.*
[12] Ball J S and Chiu T-W 1980 *Phys. Rev.* D **22** 2542.
[13] Heupel W, Goecke T and Fischer C S 2014 *Eur. Phys. J.* A **50** 85, *arXiv:1402.5042.*
[14] Williams R 2014, *submitted to Eur. Phys. J. A, arXiv:1404.2545.*
[15] Aguilar A C, Binosi D, Ibanez D and Papavassiliou J 2014 *Phys. Rev.* D **90** 065027 *arXiv:1405.3506.*
[16] Mitter M, Pawlowski J M and Strodthoff N, *in preparation.*
[17] Aouane R, Burger F, Ilgenfritz E M, Muller-Preussker M and Sternbeck A 2013 *Phys. Rev.* D **87** 114502 *arXiv:1212.1102.*
[18] Endrodi G, Fodor Z, Katz S and Szabo K (2011) *JHEP* **1104** 001, *arXiv:1102.1356.*
[19] Kaczmarek O, Karsch F, Laermann E, Miao C, Mukherjee S, et al.(2011) *Phys. Rev.* D **83** 014504 *arXiv:1011.3130.*
[20] Cea P, Cosmai L and Papa A (2014) *Phys. Rev.* D **89** 074512 *arXiv:1403.0821.*

FAIRNESS 2014 IOP Publishing
Journal of Physics: Conference Series **599** (2015) 012016 doi:10.1088/1742-6596/599/1/012016

Prospects for dilepton rates from lattice QCD

Anthony Francis
Helmholtz Institut Mainz, Johannes Gutenberg-Universität Mainz, D-55099 Mainz, Germany
E-mail: `afrancis@uni-mainz.de`

Abstract. We discuss the prospects of computing thermal dilepton rates from first principles lattice QCD. The focus lies in the determination of the meson vector-vector current spectral function to estimate the electrical conductivity, heavy quark diffusion and quarkonium dissociation. We review and compare recent results from continuum-extrapolated, quenched calculations, as well as dynamical two-flavor setups.

1. Electromagnetic spectral function in heavy-ion collisions

Heavy-ion collision experiments such as those carried out at RHIC, LHC and the future facilities at FAIR probe nuclear matter under extreme conditions, i.e. at high temperatures and/or baryon densities. Given the possible applications there has been a large effort from the phenomenological and theoretical communities to work out the properties of relevant observables in QCD, for recent reviews see [1, 2]. One of these is the spectral function of the electromagnetic current $J^{\rm em}_\mu = \frac{2}{3}\bar{u}\gamma_\mu u - \frac{1}{3}\bar{d}\gamma_\mu d - \frac{1}{3}\bar{s}\gamma_\mu s + \ldots$:

$$\rho_{\mu\nu}(\omega,\vec{p}) = \int d^4x\, e^{i\omega\cdot x_0 - i\vec{p}\cdot\vec{x}} \left\langle 0\left|\left[J^{\rm em}_\mu(x), J^{\rm em}_\nu(0)\right]\right|0\right\rangle \quad , \tag{1}$$

in vacuum it is accessible to experiments from the R-ratio via the optical theorem

$$\frac{\sigma(e^+e^- \to \text{hadrons})}{48\pi^3\alpha(s)^2/(3s)} = \frac{R(s)}{12\pi^2} = \frac{1}{6\pi}\frac{\rho_{\mu\mu}(\omega^2)}{\omega^2} \quad . \tag{2}$$

Conversely the spectral function determines the production rate of lepton-antilepton pairs:

$$\frac{dN_{l^+l^-}}{d\omega d^3p} = \sum_{f=u,\ldots} Q_f^2 \, \frac{\alpha^2_{\rm em}}{6\pi^3} \, \frac{\rho_{\mu\mu}(\omega,\vec{p},T)}{(\omega^2-\vec{p}^2)(e^{\omega/T}-1)} \quad . \tag{3}$$

The low frequency part of the spectral function is related to the transport properties of the medium. For example, in the high temperature phase and for light vector mesons, the spatial components give the response of the quark-gluon plasma (QGP) to electromagnetic fields and the soft emission rate of photons

$$\sigma_{\rm em}(T) = \sum_{f=u,d} Q_f^2 \, \frac{1}{6}\lim_{\omega\to 0}\frac{\rho_{ii}(\omega,\vec{0},T)}{\omega} \quad \text{and} \quad \lim_{\omega\to 0}\omega\frac{dR_\gamma}{d^3p} = \frac{3}{2\pi^2}\sigma_{\rm em}(T)T\alpha_{\rm em} \quad . \tag{4}$$

For heavy quark vector mesons the same limit gives the heavy quark diffusion coefficient D:

$$\frac{2T^2}{\kappa} = D = \frac{1}{6\chi_q} \lim_{\omega \to 0} \frac{\rho_{ii}(\omega, \vec{0}, T)}{\omega}, \tag{5}$$

where χ_q is the quark number susceptibility. For heavy quarkonia the spectral function around threshold of not only the vector but also the other possible channels is especially interesting, since the dissociation pattern can be used to define a QGP-thermometer [3].

The spectral function is related to Euclidean current-current correlation functions by the integral transform:

$$G_{\mu\nu}(x_0, \vec{p}) = \int_0^\infty \frac{d\omega}{2\pi} \, \rho_{\mu\nu}(\omega, \vec{p}, T) \, \frac{\cosh[\omega(\beta/2 - x_0)]}{\sinh(\omega\beta/2)}, \tag{6}$$

where $\beta = 1/T$ and we define the kernel $K(\omega, x_0, T) = \cosh[\omega(\beta/2 - x_0)]/\sinh(\omega\beta/2)$. This connection opens the possibility to study the spectral function indirectly via lattice QCD. The central question is, how well the numerically ill-posed inverse transform to obtain the spectral function can be controlled. In this talk, I review a selection of recent results from different lattice approaches to tackle this issue and the insights on the underlying physics phenomena gained in this way.

2. Meson correlation functions from lattice QCD

In lattice QCD, observables are computed on a set of gauge field configurations generated via Monte-Carlo methods and the discretized QCD action. The correlation functions of two mesonic (vector) currents at vanishing momentum is

$$G_{\mu\nu}(x_0, T) = \int d^3x \langle J_\mu(x_0, \vec{x}) J_\nu(0, \vec{0})^\dagger \rangle \quad , \tag{7}$$

whereby here $J_\mu(x) = \frac{1}{\sqrt{2}}\left(\bar{u}(x)\gamma_\mu u(x) - \bar{d}(x)\gamma_\mu d(x)\right)$ is the isospin current with only quark-line connected diagrams contributing[1]. In practice, computing this quantity on the lattice involves setting up an appropriate interpolating creation operator at the source $(0, \vec{0})$ and correlating it with a corresponding annihilation operator at the sink $(x_0, \vec{x})$ located within the four-volume of the background gauge-field configurations. To obtain a physically meaningful result the so-called "gauge average" has to be taken:

$$\langle J_\mu(x_0, \vec{x}) J_\nu(0, \vec{0})^\dagger \rangle = \frac{\int dU \; J_\mu(x_0, \vec{x}) J_\nu(0, \vec{0})^\dagger \; e^{S_{QCD}^{lat}[U]}}{\int dU \; e^{S_{QCD}^{lat}[U]}} \quad , \tag{8}$$

in the limit of infinitely many configurations this corresponds to computing the path integral over all possible gauge field backgrounds.

3. Spectral function reconstruction

The process of reconstructing the spectral function from lattice correlator data is made more difficult by the kernel $K(\omega, x_0, T)$. As such it highly suppresses the features of the spectral function itself and renders the correlator largely insensitive to them. For example, in [5, 6] mock data studies could show the difference in the correlator at the midpoint of the lattice between a

[1] Note, a recent estimate of the disconnected contribution [4] could show the isospin and electromagnetic correlators to be the same for distances up to $a \lesssim 1.5$ fm

Journal of Physics: Conference Series **599** (2015) 012016 doi:10.1088/1742-6596/599/1/012016

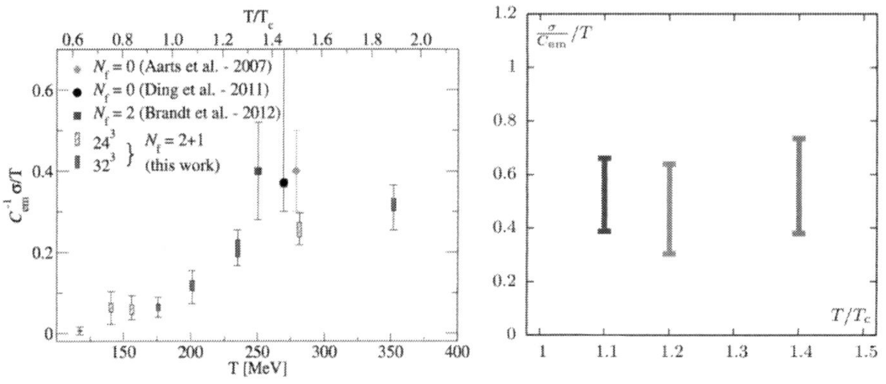

Figure 1. Left: Collected results on the electrical conductivity of the QGP across the deconfinement phase transition, provided in [8]. Right: A recent update on the electrical conductivity in the continuum limit of quenched QCD [20].

strongly peaked, i.e. bound state, and a flat, i.e. un-bound, spectral function is at the 5%-level. Taking this as an optimistic lower bound, it follows that, the lattice data has to be more precise than this value.

With precision data available, two approaches have been widely adapted to reconstruct the spectral function. The first is the "Maximum Entropy Method" (MEM) [7]. Here Bayes' theorem is invoked to determine the most probable spectral function given the data and a so-called default model. Commonplace algorithms maximize an entropy term to determine the coefficients $c(k)$ of a set of transformed basis functions $\mathcal{L}(B(k))$ to obtain an approximation of the spectral function based on the correlator data,

$$G(x_0) = \sum_k c(k)\mathcal{L}\Big(B(k,w), x_0\Big) \quad \Rightarrow \quad \rho(\omega) = \sum_k c(k)B(k,\omega) \ , \tag{9}$$

since the inverse transformation of the basis functions is known, one thereby obtains also an estimate of the spectral function. The greatest caveat of this method is the dependence on the default model. Alternatively, instead of fixing the basis functions and determining their weights, one can define the basis functions by a set of Ansätze $F(a_k)$ for the spectral function, i.e.

$$G(x_0) = \sum_k \mathcal{L}\Big(F(a_k,\omega), x_0\Big) \quad \Rightarrow \quad \rho(\omega) = \sum_k F(a_k,\omega) \ . \tag{10}$$

Naturally, in this case the Ansätze introduce a model-dependence and have to be justified. Generally phenomenological considerations enter the analysis at this point. Since both methods introduce a difficult to estimate degree of systematic uncertainty, due to the ill-posed nature of the inverse transformation, cross-checks using multiple approaches are mandatory.

3.1. Electrical conductivity across deconfinement and in the quenched continuum

In the case of the light vector meson spectral function and especially the electrical conductivity of the QGP a number of calculations is available, using both MEM and Ansätze for the reconstruction. In Fig. 1(left) we show a collection of results on $\sigma_{\rm em}(T)/T$ recently published[2] in [8]. Here, the first determination using staggered fermions in quenched QCD and using

[2] Note, the charge factor $\sum_f Q_f^2$ is neglected in the figure.

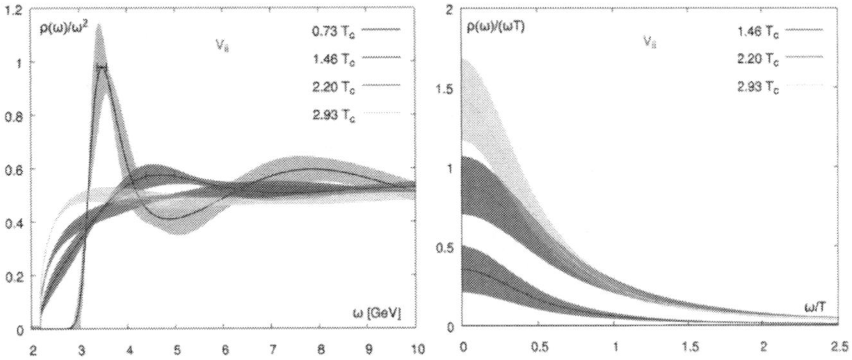

Figure 2. The vector charmonium spectral function over ω^2 (left) and ω (right) computed on large, quenched ensembles using MEM [14]. Left: The existence/non-existence of peaks is indicative of dissociation. Right: The linear intercept gives the heavy quark diffusion coefficient.

MEM for reconstruction [9] is shown in brown. Also in the quenched approximation, but using Wilson-Clover fermions, results using the Ansatz-method extrapolated to the continuum limit are given in black [10]. In the dynamical sector results obtained using ensembles with two light quark flavors of Wilson-Clover fermions and a pion mass $m_\pi \simeq 270\,\text{MeV}$ are given in blue [11]. This calculation made use of the Ansatz-method and additionally incorporated QCD sum rules for further constraints on the spectral function. Finally, in yellow and red, results on anisotropic ensembles with two light and one strange quark flavor of Wilson-Clover fermions at $m_\pi \simeq 400\,\text{MeV}$ are shown [8]. This study also uses MEM to reconstruct the spectral function. Recently, also the analysis using the Ansatz-method in the continuum limit of quenched QCD was updated [13, 20] to cover a temperature range of $T/T_c \in [1.1 : 1.5]$, see Fig. 1(right). Throughout, a consistent picture is emerging towards the value of σ_{em}, especially in the $T/T_c \simeq 1.5$ region. In the quenched continuum limit no clear temperature dependence can be observed, while there is a consistent drop in the dynamical case. This might be due to to different natures of the deconfinement phase transitions in these two theories and is under current investigation.

3.2. Heavy quark diffusion and quarkonium dissociation

In the case of heavy mesons, [14] used MEM to study the dissociation Fig. 2(left) and the diffusion coefficient 2(right) of charmonia on large, quenched lattices. Here, the J/Ψ was found to dissociate below $T/T_c \simeq 1.5$, while the heavy quark diffusion coefficient was found to be $\kappa/T^3 \in [4 : 7]$. A determination based on effective field theory methods [15] in the continuum limit [16] reconstructing using an Ansatz, shown in Fig. 3(left), found $\kappa/T^3 = 2.5(4)$, in line also with [17, 18, 19, 20]. This discrepancy is currently under investigation. Using NRQCD to determine the correlation functions for bottomonia [21, 22], the Υ-peak was observed for temperatures up to $T/T_c \simeq 1.9$, whereby using MEM for reconstruction, see Fig. 3(right). The emerging picture from these studies, and others not shown here e.g. [23, 24], indicates a faster dissociation of charmonia above T_c than previously thought, while bottomonia persist to larger temperatures, as expected.

4. Conclusions

In this talk I outlined the basic steps required to reconstruct spectral functions from Euclidean, lattice QCD, current-current correlation functions. The two most widely used methods for reconstruction used were also commented on. Focusing on vector mesons, the applications in heavy-ion collision physics were highlighted and a selection of studies was discussed aiming at

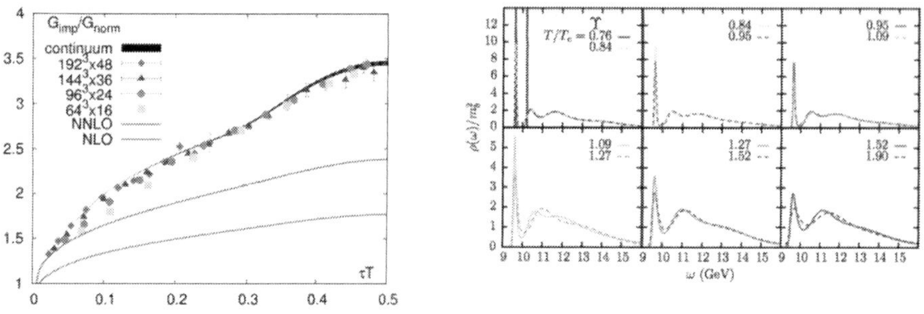

Figure 3. Left: The electric force correlator can be used to estimate heavy quark diffusion [15]. The black curve shows the result extrapolated to the continuum [16]. Right: Bottomonium spectral functions determined using NRQCD across the deconfinement phase transition [21].

providing an overview to the current status from lattice QCD. We observe an emerging agreement on the value of the electrical conductivity at large temperatures from different approaches and a fast dissociation of charmonia above the deconfinement temperature. At the same time bottomonia are seen to persist to larger temperatures. In the case of heavy quark diffusion and the electrical conductivity around the deconfinement transition work is still needed to pin down any additional systematic uncertainties. These studies are currently underway.

Acknowledgments
I wish to thank the organizers of "FAIRNESS 2014" for giving me the opportunity to present this review and for hosting such a broad and stimulating conference. Additionally I want to extend my gratitude to Gert Aarts of Swansea University for kindly sharing his results. Finally, I want to thank Harvey B. Meyer of Mainz University for looking over this draft.

References
[1] R. Rapp, *et al.*, Lect. Notes Phys. **814**, 335 (2011).
[2] R. Rapp, D. Blaschke and P. Crochet, Prog. Part. Nucl. Phys. **65**, 209 (2010) [arXiv:0807.2470 [hep-ph]].
[3] T. Matsui and H. Satz, Phys. Lett. B **178**, 416 (1986).
[4] A. Francis, V. Gülpers, B. Jäger, H. B. Meyer, G. von Hippel, H. Wittig; PoS LATTICE **2014** (2014) 128.
[5] F. Karsch, *et al.*, Nucl. Phys. A **715**, 701 (2003) [hep-ph/0209028].
[6] S. Wissel, phD-Thesis, Bielefeld University 2006.
[7] M. Asakawa, T. Hatsuda and Y. Nakahara, Prog. Part. Nucl. Phys. **46**, 459 (2001) [hep-lat/0011040].
[8] A. Amato, *et al.*, Phys. Rev. Lett. **111**, 172001 (2013) [arXiv:1307.6763 [hep-lat]].
[9] G. Aarts, *et al.*, Phys. Rev. Lett. **99**, 022002 (2007) [hep-lat/0703008 [HEP-LAT]].
[10] H.-T. Ding, *et al.*, Phys. Rev. D **83**, 034504 (2011) [arXiv:1012.4963 [hep-lat]].
[11] B. B. Brandt, A. Francis, H. B. Meyer and H. Wittig, JHEP **1303**, 100 (2013) [arXiv:1212.4200 [hep-lat]].
[12] A. Francis, *et al.*, PoS LATTICE **2013**, 453 (2013) [arXiv:1311.3759 [hep-lat]].
[13] O. Kaczmarek, *et al.*, PoS ConfinementX , 185 (2012) [arXiv:1301.7436 [hep-lat]].
[14] H. T. Ding, *et al.*, Phys. Rev. D **86**, 014509 (2012) [arXiv:1204.4945 [hep-lat]].
[15] S. Caron-Huot, M. Laine and G. D. Moore, JHEP **0904**, 053 (2009) [arXiv:0901.1195 [hep-lat]].
[16] O. Kaczmarek, arXiv:1409.3724 [hep-lat].
[17] H. B. Meyer, New J. Phys. **13**, 035008 (2011) [arXiv:1012.0234 [hep-lat]].
[18] D. Banerjee, *et al.*, Phys. Rev. D **85**, 014510 (2012) [arXiv:1109.5738 [hep-lat]].
[19] A. Francis, *et al.*, PoS LATTICE **2011**, 202 (2011) [arXiv:1109.3941 [hep-lat]].
[20] A. Francis, *et al.*, PoS LATTICE **2013**, 453 (2013) [arXiv:1311.3759 [hep-lat]].
[21] G. Aarts, *et al.*, JHEP **1111**, 103 (2011) [arXiv:1109.4496 [hep-lat]].
[22] G. Aarts, *et al.*, JHEP **1407**, 097 (2014) [arXiv:1402.6210 [hep-lat]].
[23] C. Nonaka, M. Asakawa, M. Kitazawa and Y. Kohno, J. Phys. G **38**, 124109 (2011).
[24] S. Borsanyi, *et al.*, JHEP **1404**, 132 (2014) [arXiv:1401.5940 [hep-lat]].

Journal of Physics: Conference Series **599** (2015) 012017 doi:10.1088/1742-6596/599/1/012017

Cooper-Frye Negative Contributions in a Coarse-Grained Transport Approach

D. Oliinychenko[1,3], **P. Huovinen**[1,2], **H. Petersen**[1,2]

[1]Frankfurt Institute for Advanced Studies, D-60438 Frankfurt am Main, Germany

[2]Institut für Theoretische Physik, Goethe-Universität, D-60438 Frankfurt am Main, Germany

[3]Bogolyubov Institute for Theoretical Physics, Kiev 03680, Ukraine

E-mail: `oliiny@fias.uni-frankfurt.de`

Abstract. Many models of heavy ion collisions employ relativistic hydrodynamics to describe the system evolution at high densities. The Cooper-Frye formula is applied in most of these models to turn the hydrodynamical fields into particles. However, the number of particles obtained from the Cooper-Frye formula is not always positive-definite. Physically negative contributions of the Cooper-Frye formula are particles that stream backwards into the hydrodynamical region.

We quantify the Cooper-Frye negative contributions in a coarse-grained transport approach, which allows to compare them to the actual number of underlying particles crossing the transition hypersurface. It is found that the number of underlying inward crossings is much smaller than the one the Cooper-Frye formula gives under the assumption of equilibrium distribution functions. The magnitude of Cooper-Frye negative contributions is also investigated as a function of hadron mass, collision energy in the range $E_{\rm lab} = 5 - 160A$ GeV, and collision centrality. The largest negative contributions we find are around 13% for the pion yield at midrapidity at $E_{\rm lab} = 20A$ GeV collisions.

1. Introduction

In the hot and dense system of strongly interacting matter created in heavy ion collisions the mean free path of the particles is much smaller than the size of the fireball. This fact together with the assumption of fast thermal equilibration allows to apply relativistic hydrodynamics for the dynamical description of heavy ion collisions. At later times of the evolution and at its edges the system is dilute, the mean free path is larger than system size, and hydrodynamics is not applicable. An adequate description of the system in those circumstances is provided by kinetic (transport) equations, such as the Boltzmann equation or its modifications for the quantum case. State of the art simulations of heavy ion collisions couple hydrodynamics for the early stage of the evolution to hadron transport for the late stage. Such approaches are called hybrid approaches [1].

The transition from hydrodynamics to transport (so-called particlization) is non-trivial, because hydrodynamics contains no microscopic information that is needed for transport. Currently, most of the models implement particlization in the following way. Hydrodynamical equations are solved in the whole forward light cone, including those regions, where hydrodynamics is not applicable. Then the particlization hypersurface is found from the hydrodynamical evolution. Usually a hypersurface of constant time, temperature or energy density is taken. Particle distributions to be fed into the kinetic model are generated on this hypersurface according to the Cooper-Frye formula [2]:

$$p^0 \frac{dN}{d^3p} = p^\mu d\sigma_\mu f(p) \qquad (1)$$

Here $\frac{dN}{d^3p}$ is a spectrum of particles emerging from an element of the hypersurface, $d\sigma_\mu$ is the normal four-vector to the hypersurface, its length being equal to the area of hypersurface element. In ideal fluid calculations the distribution function $f(p)$ is taken as an equilibrium Fermi or Bose distribution: $f(p) = \left[exp\left(\frac{p^\mu u_\mu - \mu}{T}\right) \pm 1\right]^{-1}$. An advantage of the Cooper-Frye formula is that it respects conservation laws. The disadvantage is that for space-like elements of the hypersurface (*i.e.* where $d\sigma^\mu d\sigma_\mu < 0$) there exist particle momenta p^μ such that $p^\mu d\sigma_\mu < 0$ and thus $\frac{dN}{d^3p} < 0$ in the Cooper-Frye formula. This is clearly unphysical, because the number of particles must be positive in every element of phase space. However, in all practical cases after integration over the hypersurface one gets positive definite spectra. One can split the integrated Cooper-Frye formula into positive and negative parts:

$$\frac{dN}{d^3p} = \int_\sigma \frac{p^\mu}{p^0} d\sigma_\mu f(p)\Theta(p^\mu d\sigma_\mu) + \int_\sigma \frac{p^\mu}{p^0} d\sigma_\mu f(p)\Theta(-p^\mu d\sigma_\mu)\,. \qquad (2)$$

The second term is called Cooper-Frye negative contributions and the first one - positive. For the applicability of the Cooper-Frye formula the negative contributions must be much smaller than the positive ones. This is usually true for simulations of high energy collisions. Negative contributions of around 9% relative to the positive ones were reported for RHIC top energy [4]. At the same time negative contributions at $E_{\rm lab} = 160A$GeV were found to be around 13%. This suggests that at lower collision energies negative contributions might become large and Cooper-Frye formula will be inapplicable. Does it really happen? To answer this question we systematically investigate the negative Cooper-Frye contributions differentially in rapidity and transverse momentum against collision energy, centrality and particle species.

Several ways to circumvent the problem of negative contributions were suggested (see [5] for a short summary). One practical way to do it in hybrid models would be to simultaneously neglect negative contributions and remove the particles from the transport calculation, if they fly to the hydrodynamic region. If the distributions in hydrodynamics and cascade are the same, then conservation laws will be fulfilled. Such an approach has never been implemented. To check if it is feasible, we compare negative contributions from the Cooper-Frye formula to distributions of backscattered particles from the cascade. We take advantage of the coarse-grained approach, which allows to calculate both within one framework.

2. Methodology

We aim at two goals: systematic estimation of Cooper-Frye negative contributions and comparing them to distributions of transport particles flying inwards to the hydrodynamics region. As a transport model we take Ultra-Relativistic Quantum Molecular Dynamics (UrQMD 3.3p2) [3]. UrQMD allows to simulate heavy ion collisions as a sequence of elementary particle collisions. Included processes are 2 to 2 scattering, resonance formation and decays, string excitation and fragmentation.

We generate an ensemble of UrQMD Au+Au collision events and average them on a rectangular grid to obtain the energy momentum tensor $T^{\mu\nu}$ and the baryon current j^μ in each cell. In each cell we find the Landau rest frame (LRF, a frame, where energy flow is zero: $T^{0i}_{LRF} = 0$). We obtain the energy density $\epsilon_{LRF} = T^{00}_{LRF}$, flow velocity u^μ and baryon density in the LRF $n_{LRF} = j^\mu_B u_\mu$. Knowing ϵ_{LRF} for each grid cell we construct the hypersurface

Σ of constant ϵ_{LRF} and find the normal vectors $d\sigma_\mu$ for each piece of Σ. The latter is done using the Cornelius subroutine [4], that provides a continuous surface without holes and avoids double counting of hypersurface pieces. The hypersurface Σ mimics the transition hypersurface in hybrid models. When Σ is obtained we perform a Cooper-Frye calculation on it and compare to distributions of underlying UrQMD particles that cross Σ.

We perform our calculations with time step $\Delta t = 0.1$ fm/c, grid spacing in the beam direction $\Delta z = 0.3$ fm, and grid spacings in transverse direction $\Delta x = \Delta y = 1$ fm. For collision energy $E_{\rm lab} = 160A$ GeV we take $\Delta z = 0.1$ fm, and $\Delta x = \Delta y = 0.3$ fm. We have checked that for such a choice of grid spacing conservation laws on the surface are fulfilled with an accuracy better than 1%. In other words $\int_\Sigma T^{\mu 0} d\sigma_\mu$ and total energy flowing out of the hypersurface calculated by particles differ by no more than 1%. To create a smooth hypersurface and obtain reproducible results we employ a Gaussian smearing procedure. For the construction of the hypersurface every UrQMD particle with coordinates (t_p, x_p, y_p, z_p) and 4-momentum p^μ is substituted by 300 marker particles with coordinates distributed with the probability density $f(x,y,z) \sim \exp\left(-\frac{(x-x_p)^2}{2\sigma^2} - \frac{(y-y_p)^2}{2\sigma^2} - \gamma_z^2\frac{(z-z_p)^2}{2\sigma^2}\right)$, where $\gamma_z = (1 - p^z/p^0)^{-1/2}$. In this way every particle contributes to $T^{\mu\nu}$ and j^μ not only of the cell, where it is located, but also to the adjacent cells. We take as number of events $N = 1500$ and the Gaussian width $\sigma = 1$ fm. Choice of N, σ, grid spacing and sensitivity of results to these choices are discussed in [5].

3. Results

If the distribution of UrQMD particles is thermal on some closed hypersurface and the system is in chemical equilibrium, then the Cooper-Frye formula should give results identical to explicit particle counting. This would allow to compensate negative contributions by removing particles from UrQMD, if they cross the hypersurface inwards. In such a treatment conservation laws on the surface would be respected. We check, if this can be done on the hypersurface Σ of constant energy density in Landau frame, $\epsilon = 0.3$ GeV/fm^3.

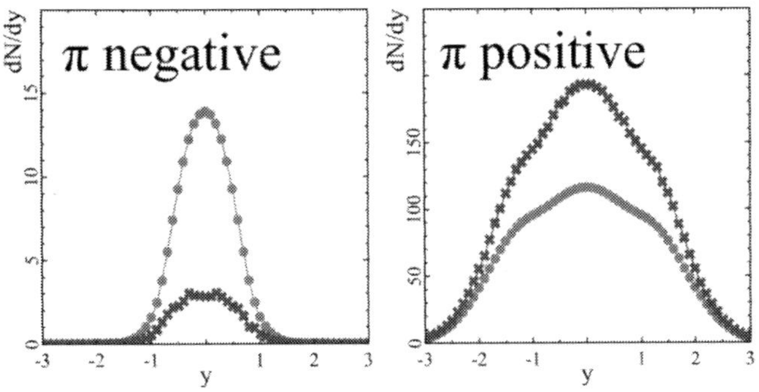

Figure 1. Cooper-Frye rapidity spectra for pions on Σ (red circles) are compared to distribution of UrQMD pions crossing Σ (blue crosses). Left panel: negative contributions and inward crossings. Right panel: positive contributions and outward crossings. Collision energy $E_{\rm lab} = 40$ AGeV, central collisions. Note the very different scale of negative and positive contributions.

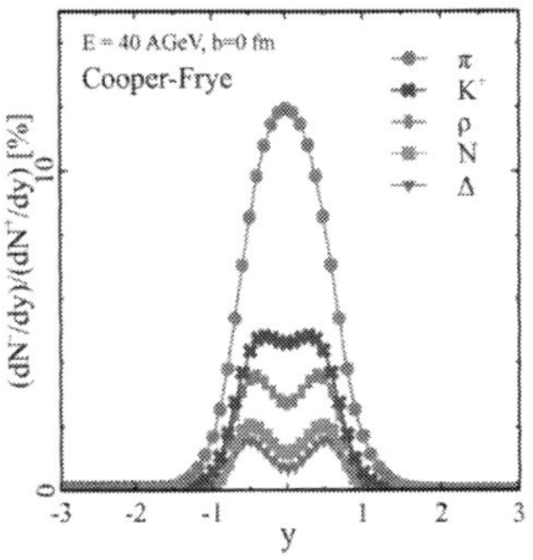

Figure 2. Rapidity distribution of the ratio of negative to positive contributions for different hadrons: pions (red circles), K^+ (blue crosses), ρ (grey bars), nucleons (magenta rectangles) and Δ (green triangles). Cooper-Frye calculation in central Au+Au collisions at $E_{\rm lab} = 40$ AGeV.

When calculating the net number of pions passing through the surface, one finds that the number of pions in UrQMD is larger than in the equilibrated Cooper-Frye calculation [5]. It

Journal of Physics: Conference Series **599** (2015) 012017 doi:10.1088/1742-6596/599/1/012017

might be possible to explain this as a sign of chemical non-equilibrium in UrQMD, but when one looks at the positive and negative contributions to the pion distributions shown in Fig. 1, one sees that a difference in the pion density only is not sufficient to explain the differences in the contributions. The positive contribution is much larger in UrQMD than in the Cooper-Frye calculation, whereas the negative contribution depicts the opposite behaviour: in UrQMD it is much smaller than in the Cooper-Frye scenario. This kind of distributions may indicate that the collective flow velocity of pions is much larger than the collective velocity of other particles [6, 7], or that the dissipative corrections are very large.

Since neither negative contributions coincide with inward crossings, nor positive contributions coincide with outward crossings, the above mentioned idea of compensating negative contributions will not work in our case. Instead we concentrate on finding when negative contributions play the most prominent role, and evaluate the ratio of negative contributions to positive ones to estimate the error they bring into hybrid calculations. For that we vary hadron sort, collision energy and centrality. As a relevant variable we consider the ratio of negative to positive contributions integrated over the hypersurface, $(dN^-/dy)/(dN^+/dy)$.

From Fig. 2 one can see that negative contributions become smaller for larger particle mass. It is simple to understand this result in the rest frame of the fluid element where the surface is moving. If the hypersurface Σ moves inwards, as is usually the case in fluid-dynamical calculations, a particle must be faster than the surface to cross it inwards. For larger particle mass the average velocities of the thermal motion are smaller and the probability to catch up the hypersurface is also smaller. Therefore, to find maximal negative contributions it is sufficient to consider pions only. For them we find out that if negative contributions are binned according to rapidity, then the largest contribution occurs at midrapidity.

We show the negative to positive contribution ratio for the pion yield at midrapidity for collision energies $E_{\rm lab} = 5$–$160A$ GeV in Fig. 3. The ratio in UrQMD calculation is much smaller than in the equilibrated Cooper-Frye calculation at all collision energies, but the maximum lies in the region of $20A$ GeV in both approaches. The value of the maximum is about 13% in Cooper-Frye calculation, and about 4% in UrQMD calculation.

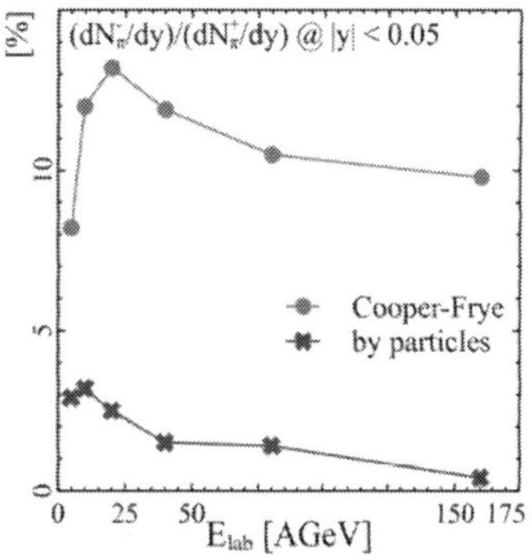

Figure 3. The ratio of negative to positive contributions on the $\epsilon(t, x, y, z) = \epsilon_c = 0.3$ GeV/fm^3 surface for pions at midrapidity. Red circles depict the ratio in a Cooper-Frye calculation assuming thermal equilibrium, and blue crosses in explicit calculation of UrQMD particles.

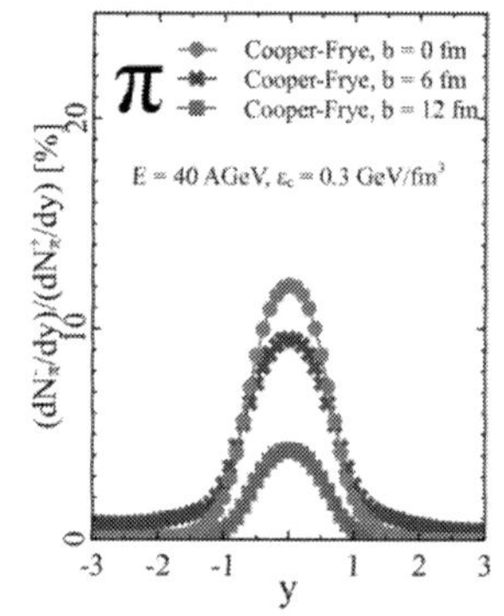

Figure 4. Rapidity distribution of the ratio of negative to positive contributions for pions in Au+Au collisions at $E_{\rm lab} = 40$ AGeV at various centralities: $b = 0$ (red circles), $b = 6$ fm (blue crosses) and $b = 12$ fm (green rectangles).

The dependency of negative to positive contribution ratio on collision energy is non-monotonous. This is not obvious, because several factors influence the ratio: the temperature on the hypersurface, the relative velocities between the flow and the surface, and the relative amounts of volume and surface emission, *i.e.* emission from the time- and space-like parts of the surface. Larger temperature results in larger negative contributions because the thermal velocities increase. Larger relative velocity leads to smaller negative contributions. The larger the relative amount of volume-emission, the smaller the negative contributions. With increasing collision energies the temperature saturates and thus the changes in the last two factors make negative contributions fall with increasing collision energy. The decrease of negative contributions at higher energy is predictable, because the relative amount of volume emission and relative velocity between flow and surface increase with collision energy. However, the behaviour of negative contributions at lower energy is caused by the interplay of all three factors.

We plot negative contributions for pions versus collision centrality in Fig. 4, and find that for peripheral collisions negative contributions are smaller than for central collisions. The relative amount of surface and volume emission plays the most prominent role here. For peripheral collisions volume emission dominates due to the short lifetime of the system and negative contributions are small.

After varying the collision energy and the centrality, we can conclude that in the worst scenario negative contributions can hardly exceed 15%. Changing the criterion of hypersurface to $\epsilon_c = 0.6$ GeV/fm^3 does not change this conclusion (see Ref. [5]).

4. Conclusions

We have investigated negative Cooper-Frye contributions and backscattering using a coarse-grained molecular dynamics approach. Au+Au collisions at $E_{\text{lab}} = 5$–160 A GeV energies have been simulated using UrQMD, and a hypersurface Σ of constant Landau rest frame energy density has been constructed. On this surface we have calculated two quantities: The ratio of Cooper-Frye negative to positive contributions (r_{eq}), which assumes local thermal equilibrium, and the ratio of UrQMD particles crossing Σ inward to crossing Σ outward (r_{neq}), which does not assume equilibrium.

We found that at all collision energies $r_{eq} \gg r_{neq}$. We explain this by a deviation of pions in UrQMD simulation from equilibrium. A non-monotonous dependency of r_{eq} and r_{neq} on collision energy was found with a maximum at 10-20 A GeV, maximal r_{eq} being around 13%. The size of the negative contributions is a result of an interplay of several factors: the temperature on the hypersurface, the relative velocities between flow and surface, and the relative amounts of volume and surface emission.

Acknowledgments

This work was supported by the Helmholtz International Center for the Facility for Antiproton and Ion Research (HIC for FAIR) within the framework of the Landes-Offensive zur Entwicklung Wissenschaftlich-Oekonomischer Exzellenz (LOEWE) program launched by the State of Hesse. DO and HP acknowledge funding of a Helmholtz Young Investigator Group VH-NG-822 from the Helmholtz Association and GSI, and PH by BMBF under contract no. 06FY9092. DO acknowledges support of HGS-HIRe. Computational resources have been provided by the Center for Scientific Computing (CSC) at the Goethe University of Frankfurt.

References

[1] T. Hirano, P. Huovinen, K. Murase and Y. Nara, Prog. Part. Nucl. Phys. **70**, 108 (2013).
[2] F. Cooper and G. Frye, Phys. Rev. D **10**, 186 (1974). nd W. Greiner, Phys. Rev. C 59 (1999) 1
[3] S. A. Bass, M. Belkacem et al., Prog. Part. Nucl. Phys. **41**, 225 (1998);
M. Bleicher, E. Zabrodin, et al., J. Phys. G: Nucl. Part. Phys. **25**, 1859 (1999).
[4] P. Huovinen, H. Petersen, Eur. Phys. J. A **48**, 171 (2012).
[5] D. Oliinychenko, P. Huovinen and H. Petersen, arXiv:1411.3912 [nucl-th].
[6] H. Sorge, Phys. Lett. B **373**, 16 (1996).
[7] S. Pratt and J. Murray, Phys. Rev. C **57**, 1907 (1998).

Charmonium suppression in a baryon rich quark-gluon plasma

Partha Pratim Bhaduri
Variable Energy Cyclotron Centre,
1/AF, Bidhan Nagar, Kolkata 700 064, India
E-mail: partha.bhaduri@vecc.gov.in

Abstract. We have investigated the survival probability of different charmonium states, in a high baryon density parton plasma, expected to be produced in nuclear collisions at FAIR. Charmonia are assumed undergo complete dissociation by color screening, if the in-medium Debye radius becomes comparable to the spatial size of the corresponding bound state. Results indicate a non-trivial dependence of the suppression pattern on the plasma evolution dynamics.

1. Introduction

Bound states of charm (c) and anti-charm ($\bar{c}$) quark, which are stable under strong decay are collectively known as charmonia. Among different charmonium resonances, the most extensively studied state is J/ψ meson. In relativistic nuclear collisions, J/ψ suppression has long been predicted as an unambiguous and experimentally viable signature to indicate the possible occurrence of phase transition to quark-gluon plasma (QGP) [1]. Till date no measurement exists on charmonium production in heavy-ion collisions below the top SPS energy ($E_b = 158$ A GeV), primarily due to the extremely low production cross sections. This in turn demands accelerators delivering extremely high intensitiy heavy-ion beams and detectors with high rate handling capability. The upcoming Compressed Baryonic Matter (CBM) experiment at FAIR [2], in GSI, Germany, for the first time, is aiming at the measurement of charmonium production in low energy nuclear collisions, over a range E_b=10-40 A GeV. In this energy domain, highest possible baryon densities are expected to be produced at the center of the collision zone [3]. This might lead to a density driven QCD phase transition of the nuclear matter to a baryon rich QGP. In a baryonic plasma, as expected to be produced in the FAIR energy collisions, different charmonium states might get suppressed due to Debye screening. In this present paper, we aim to calculate the survival probability of different charmonium states, suffering dissociation due to plasma screening, at FAIR energy domain. For this purpose we have developed a variant of the threshold model, where Debye screening mass as a function of temperature and baryon chemical potential $m_D(T, \mu_B)$, in a dynamically evolving plasma is used to decide the fate of different charmonium states implanted in the medium.

2. Theoretical formulation

In literature, quarkonium suppression in nuclear collisions, due to color screening inside the plasma, are generally treated within threshold model scenario [4, 5, 6, 7]. The basic assumption is the existence of a characteristic threshold/critical dissociation temperature (T_d) or equivalent

Journal of Physics: Conference Series **599** (2015) 012018 doi:10.1088/1742-6596/599/1/012018

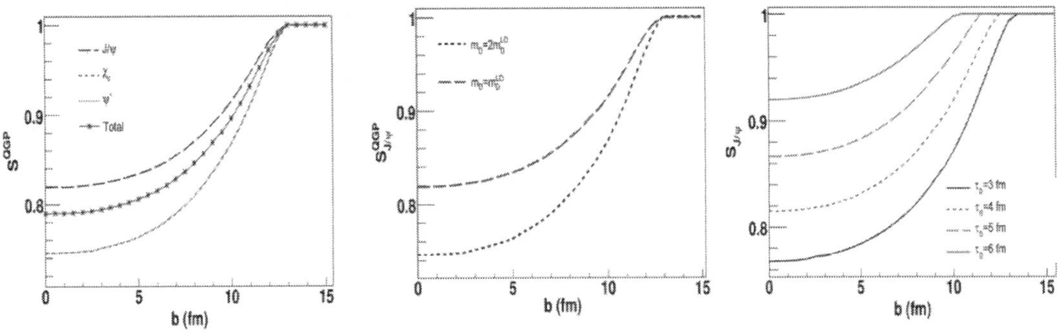

Figure 1. (Left) Centrality dependence of the inclusive survival probability of different charmonium states, in 30 A GeV Au+Au collisions. Suppression is induced by Debye screening mechanism only. (Middle) Sensitivity of the inclusive suppression pattern on non-perturbative effcets in screening mass. (Right) Sensitivity of the inclusive suppression pattern on the plasma formation time.

threshold/critical dissociation energy density ($\epsilon_d \simeq T_d^4$),that encloses the plasma volume where the screening length is shorter than the characteristic Bohr radius of a particular resonance state. This prevents the formation of the bound state for all $c\bar{c}$ pairs inside the region at the corresponding resonance formation time (t_F) in the plasma frame. At mid-rapidity, competition between t_F, the finite volume and life time of the plasma, leads to the characteristic p_T dependent survival probability. Finite lifetime of the plasma sets an upper limit on the p_T at which charmonia are suppressed. The deconfined medium produced in the collisions, is believed to reach thermalization within a time comparable to the formation time of the primordial $c\bar{c}$ pairs in the collision frame. However the situation might be different at FAIR owing to different kinematic conditions. Due to much smaller collision energies, much longer time will be required for the medium to attain thermalization compared to that at higher energies. It might thus be reasonable to consider that at FAIR, different charmonium states will be formed in the pre-equilibrium stage and the plasma would encounter the already formed physical bound states rather than their precursors (see [8] for details). If the local temperature $T(x)$ and chemical potential $\mu(x)$ inside the plasma are such that the screening radius $r_D(T(x), \mu(x)) \leq r_i$, the ith state will melt, where r_i denotes the separation radius of the ith charmonium state. At an impact parameter $\mathbf{b}$, the instantaneous survival probability for the ith state, at a transverse position $\mathbf{s}$, can be expressed as

$$S^i_{QGP}(\mathbf{b}, \mathbf{s}, \tau) = \Theta(r_i - r_D(\mathbf{b}, \mathbf{s}, \tau)) \tag{1}$$

Determination of survival probability thus boils down to the estimation of Debye screening mass m_D (note that m_D is inverse of r_D) at finite T and μ_B. In leading order perturbation theory, the μ_B dependence of the screening mass reads as, $m_D(T, \mu_q) = gT\sqrt{\frac{N_c}{3} + \frac{N_f}{6} + \frac{N_f}{2\pi^2}(\frac{\mu_q}{T})^2}$, where $\mu_q = \mu_B/3$ is the quark chemical potential and the other symbols have their usual meaning [9]. The NLO level estimates of Debye mass suffer from large uncertainties as well as the presently available lattice estimates are still at their infancy [10]. In our present phenomenological study, we thus refrain from using lattice estimates and make use of the LO pQCD estimations only. The model now demands the determination of local T and μ_B of the fluid, as a function of collision energy and collision centrality. Central collisions of gold nuclei have been simulated using various dynamical models, based on transport or hydrodynamical equations, in the FAIR energy regime [3]. For each case, central values of net baryon density (ρ_B) and the total energy density (ϵ) are extracted as a function of time, which exhibit a large degree of mutual agreement.

For our present calculations we make use of the results from UrQMD model to get $\rho_B(t)$ and $\epsilon(t)$, extracted for a central cell of unit thickness ($\Delta z = 1$). To account for the spatial inhomogenity of the medium we set the initial profiles of ρ_B and ϵ in the transverse plane of the collision, in proportion to the participant density ($n_{part}(\mathbf{b}, \mathbf{s})$), a quantity calculable using Glauber model. Space time dependent density values (ρ_B, ϵ) are then simultaneously plugged in phenomenological QGP equations of state (EOS) proposed by Kapusta [11] to solve for the corresponding values of local T and μ_B, as a function of collision energy and centrality. One input parameter of the EOS, T_0, can be identified with T_c the pseudo-critical temperature at $\mu_B \approx 0$. The critical energy density (ϵ_c), required for deconfinement can then be estimated at $\mu_B \approx 0$. Since T_c as a function of μ_B is believed to follow a nearly constant density curve, we assume ϵ_c, to remain constant with increase in μ_B. Incorporation of ϵ_c endows the plasma with finite space-time extent. Debye screening remains operational on a bound state, over the time it spends inside the plasma. The lower limit of the duration of screening can be taken as the same as the thermalization time (t_{th}) of the medium, which we assume to coincide with the passing time of the two colliding nuclei. For beam kinetic energy $E_b = 30$ A GeV, and for a central cell of longitudinal width 1 fm, as used in our present analysis, $t_{th} \simeq 3.7$ fm/c. Again a particle of mass m, inserted at a point $\mathbf{r}$, in the transverse plane ($z = 0$) of the fireball, with velocity $\mathbf{v}$ will travel a distance d ($d = -rcos\phi + \sqrt{R^2 - r^2(1 - cos^2\phi)}$), in the time interval $t_b = M_T d/p_T$, before it escapes from the system of transverse extension R, ϕ being the angle between the vectors $\mathbf{r}$ and $\mathbf{v}$ and $M_T = \sqrt{m^2 + p_T^2}$ denotes the transverse mass. Thus minimum between the two time values t_I and t_b should fix the upper limit of duration, a bound state suffers dissociation due to screening, where t_I denotes the plasma extinction time. Since the chramonia produced in the central rapidity region would have very small p_T, they remain inside the system till the plasma is alive.

3. Numerical Results

We now have all the ingredients to calculate the anomalous J/ψ suppression pattern due to plasma screening in a deconfined medium. Our aim is the estimation of the inclusive J/ψ survival probability as a function of collision centrality. Since J/ψ mesons are produced in the initial hard collisions, we distribute the J/ψs in the transverse plane, following the transverse density of binary collisions $n_{coll}(\mathbf{b}, \mathbf{s})$, as obtained in Glauber model. At a given impact parameter, the total inclusive survival probability suffering dissociation inside a QGP is then obtained by integrating Eq. 1. Results are shown in the left panel of Fig. 1, for different resonance states. As inputs to the EOS, we have chosen $N_f = 2$ and $T_0 = 154$ MeV guided by the current state-of-the-art lattice simulations of T_c [12]. Excited states suffer larger suppression. Magnitude of suppression for ψ' and χ_c are visibly indistinguishable. In practice, it has been found that only about 60% of the observed J/ψ originate directly in hard collisions while 30% of them come from the decay of χ_c and 10% from the ψ'. Hence, the total survival probability of J/ψ becomes, S_{tot}=$0.6S_{J/\psi} + 0.3S_{\chi_c} + 0.1S_{\psi'}$. As evident from the figure, at FAIR energies, color screening effect alone leads to a maximum $15 - 20\%$ suppression in most central collisions. It might now be interesting to test the sensitivity of suppression pattern on various model inputs. Let us first check how S_{tot} quantitatively depends on the enumeration of screening mass. In finite temperature lattice guage theory, at $\mu_B = 0$, the screening behavior is found to be in good agreement with leading order perturbation theory for $T \geq 1.5T_c$. However for lower temperature values, large non-perturbative corrections are found to arise in the estimation of Debye mass. They are quantified through a multiplicative constant A ($A = 1$, for LO pQCD), with $A = 1.39$. Such non-perturbative effects might also be prevalent at finite baryon density regime accessible at FAIR. To find out its effect on the inclusive survival probability, we have calculated $S_{J/\psi}$ with $m_D(T, \mu_B) = 2m_D^0(T, \mu_B)$, where m_D^0 denotes the LO pQCD estimate. Results are depicted in

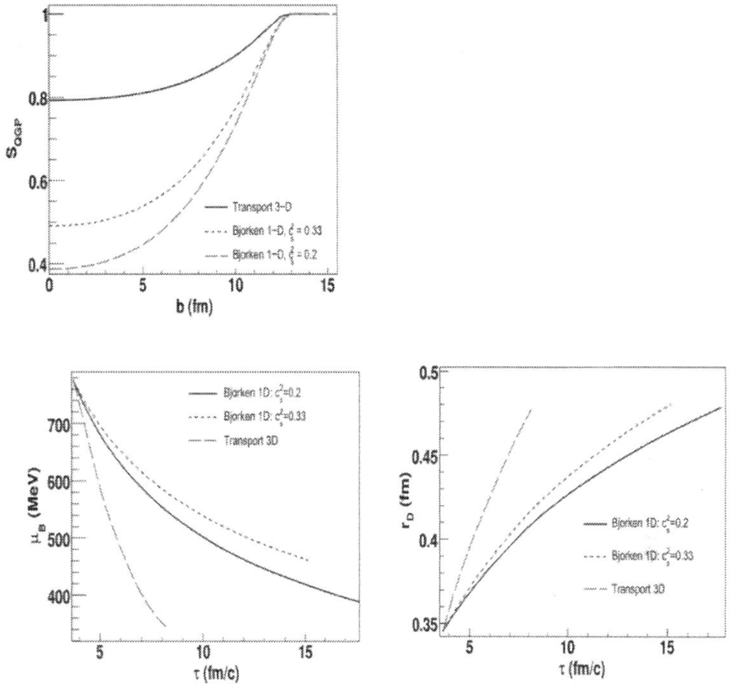

Figure 2. (Top) Sensitivity of the inclusive survival probability on the medium expansion dynamics. (Bottom) Time evolution of local temperature (T), baryon chemical potential (μ_B) and screening radius (r_D) under different expansion scenarios.

the middle panel of Fig. 1. Even for a non-perturbative correction factor as large as $K = 2$, difference in $S_{J/\psi}$ appears to be less than 10%. Perturbation theory might thus do a reasonable job so far as the estimation of the inclusive survival probability is concerned. It might also be useful to test the sensitivity of the suppression pattern on the choice of medium thermalization time. As discussed earlier, we assume thermalization time to be equal to the passing time of the two colliding nuclei, which is commonly believed to set the lower bound of t_{th}. In right panel of Fig. 1, we have calculated the survival probability of the directly produced J/ψs for different values of t_{th} between 3 to 6 fm/c. In case of late thermalization of the deconfined matter, all the J/ψ would possibly escape the suppression due to plasma screening. Our calculations indicate that for $t_{th} \geq 7$ fm, there is no suppression at all.

Finally, let us explore the sensitivity of the suppression pattern on the medium expansion dynamics. Most of the calculations, performed so far, generally assume the plasma to expand following Bjorken 1-D boost invariant scaling solutions. It is basically driven by the expectation that the suppression occurs before the transverse expansion could set in. For a quantitative comparison it might be useful to find the effect of the transverse expansion on the observed suppression pattern. In our present calculations, we have followed the time evolution of the densities from UrQMD which includes the full 3-D expansion of the fireball. Thus in order to find the effect of transverse dynamics, we have also calculated the suppression for a case where the fluid dynamics is governed by Bjorken 1-D hydrodynamical expansion. For such cases the net baryon and the energy densities would evolve as $\rho_b \tau = constant$ and $\epsilon \tau^{1+c_s^2} = constant$, where c_s denotes the speed of sound in the plasma. For ideal massless relativistic gas $c_s^2 = 0.33$. For a massive interacting system c_s^2 would be less than that. We have chosen here two illustrative values of $c_s^2 = 0.33, 0.2$, in accord with [5]. In Figure 2, we have shown comparative suppressions of directly produced J/ψ, between 3-D and 1-D expansion of the partonic fluid. Suppression strongly depends on the medium expansion dynamics. In absence of transverse expansion, plasma would expand slowly and hence screening would operate for a longer time. The rate of expansion, in case of longitudinal dynamics, would be governed by the value of c_s^2. Among the three cases considered, slowest possible expansion occurs for $c_s^2 = 0.2$, giving largest suppression effects. On the other hand presence of the transverse expansion accelerates the rate of expansion thereby shortening the duration of the screening which would increase the chances of J/ψ survival. It can be understood on a more quantitative basis if we look the the evolution of the medium. Hence in Fig. 2, we have also plotted the time evolution of local T, μ_B and r_D, for the central fluid cell in central Au+Au collisions. Fluid is allowed to expand till the energy density drops down to ϵ_c, the critical energy density for deconfinement to occur. Clearly the fluid gets diluted and cold much faster if it follows UrQMD 3-D expansion dynamics. For 1-D expansion, rate of dilution and cooling is controlled by the value of c_s, leading minimum survival for slowest expansion. A static medium would thus generate maximum amount of suppression.

4. Summary

In summary, we have demonstrated the possible charmonium suppression due to color screening, inside a hot baryonic plasma, anticipated to be produced at FAIR energy domain. Screening effects alone are found to generate a maximum of about 20% suppression in most central Au+Au collisions. Centrality dependence of the inclusive survival probability is found to be sensitive on the different parameters of the plasma EOS. Debye mass as a function of T and μ_B, is used to fix the deconfining zone. In the deconfined QGP, quark contribution to the screening mass is dominant and the gluonic contribution is small at low temperature. At high temperature, both quarks and gluon contribute to the screening. In our present work, we have assumed that the deconfinement sets in the nuclear collisions which eventually leads to the formation of a partonic medium in thermal and chemical equilibrium. Chemical equilibration is assumed to occur simultaneously with thermal equilibrium which allows us to estimate the screening mass in terms of T and μ_B. Debye screening could also set in inside a thermally equilibrated but chemically equilibrating plasma. Moreover color screening seems not to be the sole source of charmonium suppression in a deconfined medium. Suppression can also result from collisional dissociation of the charmonia in the pre-resonance or resonance stage with the quarks and gluons co-moving with the evolving charmonium states. Such microscopic dissociation mechanisms do not require equilibration to set in and can be operational in the pre-equilibrium stage as well, as along as the medium remains in deconfined state. Work is under progress for a quantitative estimation of these effects.

[1] T. Matsui and H. Satz, Phys. Lett. B **178**, 416 (1986).
[2] P. Senger, Nucl. Phys. A **862-863**, 139 (2011).
[3] I. C. Arsene *et al.*, Phys. Rev. C **75**, 034902 (2007).
[4] J. P. Blaizot and J. Y. Ollitrault, Phys. Lett. B **199** 499 (1987); M. C. Chu and T. Matsui, Phys. Rev. D. **37**, 1851 (1988); F. Karsch and R. Petronzio, Phys. Lett. B **193** 105 (1987); R. Vogt, Phys. Lett. B **430**,15 (1998).
[5] D. Pal, B. K. Patra and D. K. Srivastava, Eur. Phys. J. C **17**,179 (2000); B. K. Patra and D. K. Srivastava, Phys. Lett. **B 505**, 113 (2001).
[6] O. Linnyk, E. L. Bratkovskaya, W. Cassing and H. Stocker, Nucl. Phys. A **786**, 183 (2007).
[7] M. Mishra, C. P. Singh, V. J. Menon and Ritesh Kumar Dubey, Phys. Lett. **B 656**, 45 (2007); P.K.Srivastava, M. Mishra and C. P. Singh, Phys. Rev. C. **87**, 034903 (2013); T. Gunji, H. Hamagaki, T. Hatsuda and T. Hirano, Phys. Rev. **C 76**, 051901(R) (2007); A. K. Chaudhuri, Phys. Rev. **C 80**, 047901 (2009);
[8] P. P. Bhaduri, A.K. Chaudhuri and S. Chattopadhyay, Phys. Rev. **C 88**, 061902 (2013) .
[9] T. Toimela, Phys. Lett. **B 124**, 407 (1983).
[10] M. Doring, S. Ejiri,O. Kaczmarek,F. Karsch and E. Laermann, Eur. Phys. J. C **46** (2006) 179.
[11] J. I. Kapusta, Phys. Rev. C **81**, 055201 (2010).
[12] A. Bazavov *et al.*, HotQCD Collaboration, Phys. Rev. D **85**,(2012) 054503.

Photon emission within the linear sigma model

F Wunderlich and B Kämpfer
Helmholtz-Zentrum Dresden-Rossendorf 01328 Dresden, Germany and
Institut für Theoretische Physik, TU Dresden 01062 Dresden, Germany
E-mail: f.wunderlich@hzdr.de, b.kaempfer@hzdr.de

Abstract. Soft-photon emission rates are calculated within the linear sigma model. The investigation is aimed at answering the question to which extent the emissivities map out the phase structure of this particular effective model of strongly interacting matter.

1. Introduction

Despite of the increasing success of QCD in describing a large variety of phenomena, both in the perturbative as well as in the non-perturbative regimes, some fundamental questions remain unsolved. Prominent examples are the very nature and detailed properties of the strongly coupled quark gluon plasma which is the conjectured state of QCD matter at temperatures comparable and larger than the QCD energy scale $\Lambda_{\rm QCD}$. Furthermore, the nature and properties of the chiral and deconfinement phase transition as well as the position of a conjectured critical point (CP) in the QCD phase diagram are among the still challenging issues [1]. To answer these questions experimentally, a number of large scale experiments is currently running (ALICE, ATLAS and CMS at the LHC, and STAR and PHENIX at RHIC), planned (MPD at NICA) or under construction (CBM and HADES at FAIR).

On the theory side, lattice QCD yields a smooth crossover from the hadronic phase to the quark-gluon phase for small chemical potential at temperatures of about 150 MeV. At sufficiently large net baryon density the crossover may turn into a first order phase transition at a CP. At non-zero net densities (*i.e.* non-zero quark chemical potential) there is no first-principle approach to the phase diagram. Therefore, one has to rely on effective models or on truncation schemes. Nevertheless, many of these approaches seem to point to a first order phase transition connected to the spontaneous breaking of chiral symmetry at densities a few times the nuclear density.

The end point of this transition line has interesting properties on its own. From macroscopic examples, the phenomenon of critical opalescence, *i.e.* the diverging scattering strength of transparent media in the vicinity of a critical point, has been known for a long time [2]. Quite common also is the phenomenon of critical slowing down, *i.e.* the diverging relaxation time at criticality [3]. These two examples as well as most of the special properties of critical points have their reason in the diverging correlation length making the system scale free.

To understand the mass generation connected to chiral symmetry breaking, several effective models have been constructed with the Nambu-Jona-Lasinio (NJL) model [4, 5] and the linear sigma model (LσM) [6, 7] being the most prominent ones.

Our present investigation is motivated by the question whether penetrating probes reflect directly the phase structure of strongly interacting matter. We focus here on real photons and

Journal of Physics: Conference Series **599** (2015) 012019 doi:10.1088/1742-6596/599/1/012019

select the LσM to mimic the above anticipated phase structure. The LσM contains quark and meson (pion and sigma) fields as basic degrees of freedom, where the fluctuations of the latter ones are accounted for in linear approximation, as in [8–10] and the photon field is minimally coupled to the strongly interacting components of the LσM.
There is a large difference in the time scales concerning the strong and the electromagnetic interactions, respectively. This makes possible separating the two interactions involved. The strong interaction is responsible for the relaxation towards a local thermal equilibrium as well as to the mass generation via the spontaneous breaking of chiral symmetry. The electromagnetic interaction with a perturbative radiation field contributes little to this, because its effects are $\mathcal{O}(\alpha_{\rm em}/\alpha_s)$ suppressed. Therefore we might calculate the thermodynamics without regarding electromagnetism and use the thermodynamic properties as well as the effective masses of the dressed quarks and mesons later on in the photon emission calculations.

2. Thermodynamics of the LσM with linearized meson fluctuations

The LσM is a widely used effective model of QCD and has been applied often for studying various aspects of thermodynamics of strongly ineracting matter. It was suggested by Gell-Mann and Levy in 1960 [11] for studying chiral symmetry breaking. In absence of an explicit symmetry breaking term, the model has a $SU(2) \times SU(2) \simeq O(4)$ symmetry and therefore belonging to the same universality class as $N_f = 2$ QCD in the chiral limit [12]. This symmetry present at high temperatures is spontaneously broken to a residual $SU(2)$ symmetry with the three pseudoscalar π mesons being the Goldstone modes. Breaking chiral symmetry explicitly, the pions acquire non-zero masses. Besides these satisfying properties there is a close connection to the non-linear σ model, which in turn is equivalent to leading order chiral effective field theory of QCD. Compared to the NJL model the LσM has the advantage of including the mesons directly as dynamic field quanta, making it easier to address their properties.
The LσM Lagrangian reads

$$\mathcal{L}_{\rm L\sigma M} = \bar{\psi}(i\gamma^\nu\partial_\nu - g(\sigma + i\gamma^5\vec{\tau}\vec{\pi}))\psi + \frac{1}{2}\partial_\rho\sigma\partial^\rho\sigma + \frac{1}{2}\partial_\kappa\vec{\pi}\partial^\kappa\vec{\pi} + \frac{\lambda}{4}(\sigma^2 + \vec{\pi}^2 - v^2)^2 - H\sigma, \quad (1)$$

where the Dirac field ψ describes a doublet of quarks, σ corresponds to an iso-scalar and Lorentz-scalar field, and $\vec{\pi}$ describes an iso-vector and Lorentz-pseudoscalar field, the latter ones conveniently interpreted as the σ and π mesons. From the Lagrangian the thermodynamic potential Ω is constructed via the path integral of the exponential of the Euclidean action and evaluated following the procedure described in [8–10] for including linearized fluctuations. First, one integrates over the fermionic fields ψ and $\bar{\psi}$. The remaining path integral corresponds to a purely mesonic theory with a complicated interaction potential, which is approximated by a quadratic one to account for small fluctuations. The parameters of this quadratic potential are identified with the masses and thermodynamic averages of the meson fields. This leads to self consistency relations for the masses.
The parameters are fixed by the following requirements: The mass of the pions is set to 138 MeV in vacuum ($T = \mu = 0$) and the sigma meson mass to 700 MeV. The effective quark mass in the vacuum is fixed to one third of the nucleon mass $m^0_{\rm eff} = gv = 312$ MeV, and the parameter v is identified with the pion decay constant in vacuum, $v = 92.4$ MeV.
With these parameters one obtains the results depicted in Fig. 1. Figs. 1(a)-1(c) show contour plots of the masses over the phase diagram. One notes that the pion mass (Fig. 1(a)) increases with temperature and chemical potential with the strongest change at the phase boundary. The sigma meson mass (Fig. 1(b)) on the other hand exhibits a valley of low mass values around the phase boundary and with a global minimum at the critical point. The quark mass plotted in Fig. 1(c) drops from its vacuum value to about 30 MeV. The most drastic change, again, is at the phase boundary, signaling that the mechanism for mass generation is indeed the spontaneous

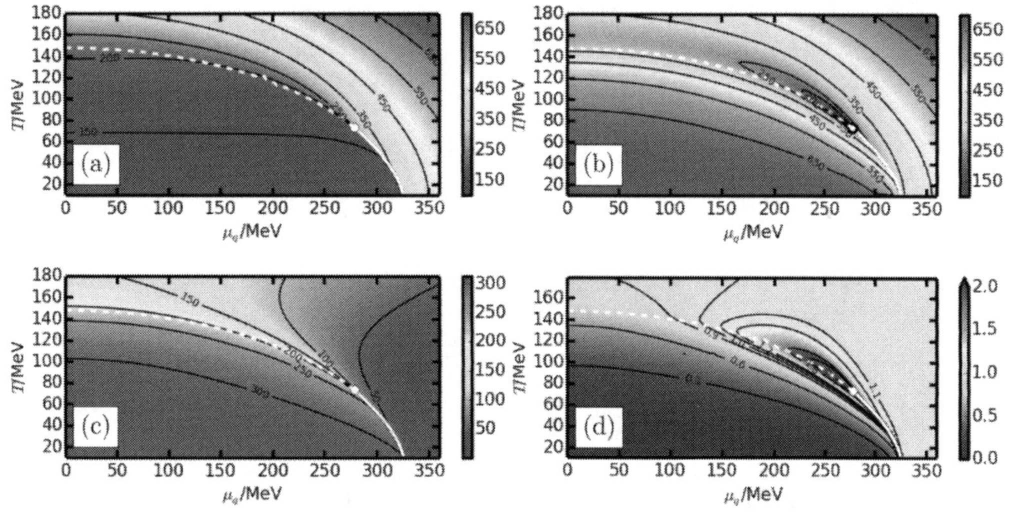

Figure 1. Contour plots of dynamically generated masses for the pion (a) and sigma (b) mesons as well as the quarks (c) (in MeV). The quark number susceptibility (d), normalized to the susceptibility of a ideal massless Fermi-gas, is increased around the CP (white circle). The solid white curve denotes the coexistence curve for the 1st order phase transition and the dashed line estimates the pseudocritical temperature. The later one is defined as the the position of the extremal normalized heat capacity $\bar{c} = -c_0^{-1}\partial_T^2\Omega(\mu, T)$ with $-c_0$ being the second derivative w.r.t. temperature of the grand canonical potential Ω_0 of an ultrarelativistic free Fermi gas.

breaking of chiral symmetry within the LσM. Because the chiral symmetry is also explicitly broken by a nonzero H in the Lagrangian, the quark mass does not drop to zero, but stays finite in the high temperature phase. Comparing the meson masses (*cf.* Figs. 1(a) and 1(b)), one realizes that they are degenerate above the 1st order phase transition curve and the crossover region, respectively, but very different below. This behavior of the mass difference of these chiral partners is another sign of the chiral symmetry breaking and restoration.
For quantifying the size of the critical region the quark number susceptibility $\chi := -\partial^2\Omega/\partial\mu^2$ is chosen, since the susceptibility scales with the correlation length whose divergence causes many of the special features of a CP. In Fig. 1(d), χ is normalized to the susceptibility χ_0 of a massless ideal fermion gas to scale out trivial contributions.

3. Photon emission rates within the LσM

For calculating photon emission rates, the LσM Lagrangian is extended by an electromagnetic sector coupled minimally (*cf.* [13]) to the strongly interacting part.

$$\mathcal{L}_{\gamma\mathrm{L}\sigma\mathrm{M}} = \mathcal{L}_{\mathrm{L}\sigma\mathrm{M}} + \mathcal{L}_\gamma + \mathcal{L}_{\mathrm{int}}, \tag{2}$$

$$\mathcal{L}_{\mathrm{int}} = -eQ_f\bar{\psi}A\!\!\!/\psi + \frac{1}{2}e^2\pi^+\pi^-A^\nu A_\nu + \frac{1}{2}eA_\nu(\pi^-\partial^\nu\pi^+ + \pi^+\partial^\nu\pi^-), \tag{3}$$

where $\mathcal{L}_\gamma = -\frac{1}{4}F^{\mu\nu}F_{\mu\nu}$ is the free photon Lagrangian and A^μ denotes the photon field.
Photon emission rates are, in a kinetic theory approach, convolutions of squared matrix elements $|M|^2$ and phase space distribution functions $f_\pm$, the latter ones explicitly depending on T and μ. Superimposed are implicit T and μ dependencies from the effective masses of the involved fields, as displayed in Figs. 1(a)-1(c). Given the marked variations of these masses one can expect an

pronounced impact on the emission rates

$$\omega\frac{d^7N}{dx^4dk^3}=\frac{\mathcal{N}}{(2\pi)^5}\int\frac{dp^3}{2p^0}\int\frac{dq^3}{2q^0}\int\frac{dz^3}{2z^0}|\mathcal{M}|^2 f_\pm(p^0)f_\pm(q^0)(1\mp f_\pm(z^0))\delta^{(4)}(p+q-z-k). \quad (4)$$

Owing to the weakness of the electromagnetic interaction we restrict the calculations to first order in the electromagnetic coupling. Since we expect to have captured the dominant part of the strong interaction in the calculation of the thermodynamic potential and the effective masses, the residual interaction is expected to be relatively weak. Therefore we restrict our calculation to 1st order processes in the quark-meson coupling. Within this approximation the contributing processes are the tree-level processes in the s, t and u channels.
In (4), four of the nine integrations can be carried out exactly applying the delta distribution. Another (angular) integration drops out by symmetry reasons, so one is left with four integrals, which have to be executed numerically resulting in the rates depicted in Fig. 2. When photon

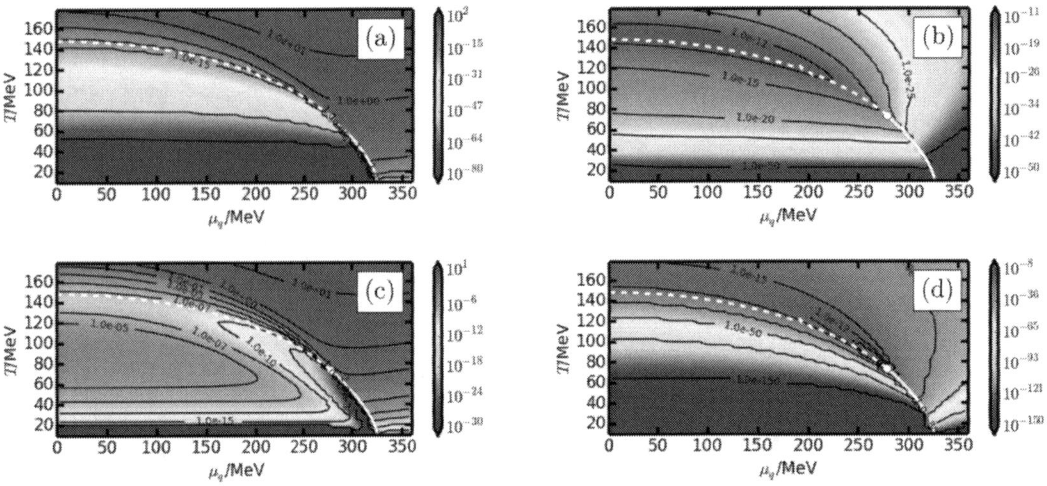

Figure 2. Contour plots of the photon emissivity $\omega\frac{d^7N}{d^4xd^3k}$ in units of MeV^2 as functions of T and μ at photon energies $\omega = 10\,\mathrm{MeV}$ for the processes $\overline{\psi}\psi \to \pi\gamma$ (a), $\psi\pi \to \psi\gamma$ (b), $\overline{\psi}\psi \to \sigma\gamma$ (c) and $\psi\sigma \to \psi\gamma$ (d). Phase contour definitions as in Fig. 1

energies ω are much larger than the respective masses, it is not expected to see much of the details of the phase structure. Contrary, at lower energies there are huge differences in available phase space and matrix elements squared leading to pronounced patterns which reflect phase diagram features, in particular the effective masses. For this reason $\omega = 10\,\mathrm{MeV}$ is chosen.
Figure 2 shows contour plots of the photon rates for the different contributing processes over the phase diagram. In Figs. 2(b) and 2(d) we see an enhancement in the crossover region and in Fig. 2(d) a global maximum in the critical region. In Figs. 2(a) and 2(c) one notices large rates in the chirally restored phase and much less photon emission in the chirally broken phase, which in case of 2(c) is superimposed by an island of enhanced rates for $T \sim 100\,\mathrm{MeV}$ and $\mu \lesssim 200\,\mathrm{MeV}$. Figures 2(a) and 2(b) show photon rates from processes involving pions. Pions exhibit a large mass difference between the two phases, but contrary to the sigma meson whose mass has a global minimum at the CP the pion mass does not show special features at this point. This leads to a large difference in the emissivity between the phases but no features characteristic for the CP itself. For the pion-involving Compton process (Fig. 2(b)) there is

an enhancement in the crossover region. This is probably due to a combination of phase space effects and the (comparatively) large probability for the internally propagating pion to get on-shell. A better channel for obtaining signatures of a CP are sigma involving processes. This is expected, since the sigma meson is precisely the mode getting massless at the CP making long range interactions possible and thus driving the critical processes. Unfortunately, the inclusion of linearized fluctuations increases the sigma mass, so it is not clear whether the endpoint of the 1st order phase transition shows correctly the critical behavior. But linearizing fluctuations anyhow restricts to small fluctuations making it not adequate very near the CP. Nevertheless the sigma mass drops to small values in the critical region, which has a notable effect on the corresponding processes, *e.g.* the excess of the photon rate in the critical region in Fig. 2(d).
There is a large difference in the rates for the processes under consideration, even between the corresponding Compton (Figs. 2(b) and 2(d)) and annihilation (Figs. 2(a) and 2(c)) processes. These can be understood in terms of available phase space in combination with thermal suppression. Within a Boltzmann approximation two of the remaining integrals in (4) can be solved to obtain

$$\omega \frac{d^7N}{dx^4dk^3} \overset{\omega \ll m_i}{\sim} \int_{s_0} \frac{ds}{s-z^2} \int dt |M(s,t)|^2 \exp\{-(s-z^2)/(4\omega T)\}. \tag{5}$$

The difference between the minimal kinematically allowed value of the center of mass energy $\sqrt{s_0} = \max\{m_1 + m_2, m_3\}$ for the different processes, together with a small value of ω leads to the huge thermal suppression at small T seen in Figs. 2 (a) and (d).

4. Summary

Focusing on soft-photon emission rates we demonstrate that some features of the phase diagram provided by the linear sigma model are nicely mapped out. Being aware of some limitations, such as the restriction to linearized fluctuations (*cf.* [14] for a proper account of fluctuations) and the need to implement more complete rates in a model of space-time evolution of the matter, we hope that improved calculations can provide useful complementary information on strongly interacting matter produced in the course of relativistic heavy-ion collisions at various energies, system sizes and centralities.

References

[1] Friman B et al (eds.) *The CBM Physics Book: Compressed Baryonic Matter in Laboratory Experiments*, Lect. Notes Phys. **814** (Springer, Berlin, 2011)

[2] von Smoluchowski M 1908, *Ann. Phys.* **330**, 205

[3] Hohenberg P C and Halperin B I 1977, *Rev. Mod. Phys.* **49**, 435

[4] Klevansky S P 1992, *Rev. Mod. Phys.* **64**, 649

[5] Asakawa M and Yazaki K 1989, *Nucl. Phys.* **A504**, 668

[6] Bochkarev A and Kapusta J I 1996, *Phys. Rev.* **D54**, 4066

[7] Jungnickel D U and Wetterich C 1996, *Phys. Rev.* **D53**, 5142

[8] Mocsy A, Mishustin I N, and Ellis P J 2004, *Phys. Rev.* **C70**, 015204

[9] Bowman E S and Kapusta J I 2009, *Phys. Rev.* **C79**, 015202

[10] Ferroni L, Koch V, and Pinto M B 2010, *Phys. Rev.* **C82**, 055205

[11] Gell-Mann M and Levy M 1960, *Nuovo Cim.* **16**, 705

[12] Pisarski R D and Wilczek F 1984, *Phys. Rev.* **D29**, 338

[13] Mizher A J, Chernodub M N, and Fraga E S 2010, *Phys. Rev.* **D82**, 105016

[14] Tripolt R-A, Strodthoff N, von Smekal L, and Wambach J 2014, *Phys. Rev.* **D89**, 034010

In-medium Spectral Functions in a Coarse-Graining Approach

Stephan Endres[1,2], Hendrik van Hees[1,2], Janus Weil[1,2] and Marcus Bleicher[1,2]

[1]Institut für Theoretische Physik, Universität Frankfurt, Max-von-Laue-Straße 1, 60438 Frankfurt, Germany
[2]Frankfurt Institute for Advanced Studies, Ruth-Moufang-Straße 1, 60438 Frankfurt, Germany

E-mail: endres@th.physik.uni-frankfurt.de

Abstract. We use a coarse-graining approach to extract local thermodynamic properties from simulations with a microscopic transport model by averaging over a large ensemble of events. Setting up a grid of small space-time cells and going into each cell's rest frame allows to determine baryon and energy density. With help of an equation of state we get the corresponding temperature T and baryon-chemical potential μ_B. These results are used for the calculation of the thermal dilepton yield. We apply and compare two different spectral functions for the ρ meson, firstly a calculation from hadronic many-body theory and secondly a calculation from experimental scattering amplitudes. The results obtained with our approach are compared to measurements of the NA60 Collaboration. A relatively good description of the data is achieved with both spectral functions. However, the hadronic many-body calculation is found to be closer to the experimental data with regard to the in-medium broadening of the spectral shape.

1. Introduction

Although it is widely accepted nowadays that the dynamics of strong interactions are governed by quantum chromodynamics (QCD), we still lack a full understanding of the QCD phase structure. At low energies and densities, the relevant degrees of freedom are hadrons, i.e., composite objects of quarks and gluons. However, for sufficiently large energies and/or densities the creation of a deconfined phase with free quarks and gluons is expected. Also the symmetry pattern of QCD is assumed to change - e.g., chiral symmetry which is broken in the vacuum but expected to be restored at some finite temperature. In consequence, the study of the in-medium properties of hadrons plays an important role especially for a better understanding of the non-perturbative part of QCD [1, 2]. Experimentally, heavy-ion collisions at ultra-relativistic energies are a good possibility to create hot and dense matter in the laboratory. However, the measurement of in-medium modifications is difficult, as the fireball has a lifetime of only a few fm/c, and all the hadronic observables only give information on the final freeze-out. Here dileptons offer the big advantage that they do not interact strongly and can leave the hot and dense fireball unscathed after their production. But this has a drawback, too. As lepton pairs are produced via many different processes at all stages of the reaction, their spectra reflect the whole evolution of the collision. What one can measure are only time integrated spectra. This is a challenge for theory, as it requires a reliable description of the entire reaction dynamics.

In these proceedings a coarse-graining approach [3] is used (following a similar ansatz as proposed by Huovinen et al. [4]), which allows to combine in-medium spectral functions with an underlying microscopic description of the reaction dynamics. We argue that this is a good and realistic compromise between transport approaches, where an inclusion of medium modifications of spectral functions is difficult, and fireball models that use a simplified description of the reaction evolution but where the application of spectral functions is straightforward. In contrast to common hydrodynamic models, its advantage is the unified description of the whole reaction dynamics, from the first hadron-hadron collisions to the final freeze-out (except for a different treatment of the very low temperature cells, see section 2).

2. The Model

The coarse-graining is based on the time-step-wise output from simulations with the Ultra-relativistic Quantum Molecular Dynamics model (UrQMD). It is a non-equilibrium transport approach and describes a nuclear reaction in terms of classically propagated hadrons combined with subsequent elastic and inelastic binary scatterings [5–7]. To extract thermodynamic properties, we take an ensemble of UrQMD events (≈ 1000) and average over them, such that we get a relatively smooth phase-space distribution function. Hereby the UrQMD output is set on a space-time grid with $\Delta t = 0.2$ fm/c and $\Delta x = 0.8$ fm. The choice of the cell size is a compromise between the highest possible resolution and the memory and computational requirements. However, we checked that a smaller size does not improve or change the results.

In each cell the baryon four-flow is then calculated and Eckart's definition [8] is used to perform a Lorentz boost into the local rest frame of the cell. This enables to calculate local baryon and energy densities. To extract the temperature T and baryon-chemical potential μ_{B} we use the equation of state of a hadron resonance gas [9] that includes the same hadronic degrees of freedom as the UrQMD model. To account also for dilepton emission from the quark-gluon plasma phase, we use a second equation of state for temperatures above 170 MeV that has been fitted to Lattice calculations [10]. As the values of T of the two EoS agree almost perfectly in the temperature range from 150 to 170 MeV, a smooth transition is guaranteed. Note, however, that the chemical potential is set to zero for $T > 170$ MeV, as the Lattice fit is only done for vanishing quark chemical potential. In consequence we can not have a smooth transition for μ_B. (Besides T and μ_{B}, we also calculate the effective baryon density, ρ_{eff}, in each cell, as it is the input for one of the spectral functions implemented here, see section 3 below). When we know the thermodynamic properties of the cell, the dilepton emission per four-volume and four-momentum is calculated according to [11]

$$\frac{\mathrm{d}N_{ll}}{\mathrm{d}^4x\mathrm{d}^4p} = -\frac{\alpha_{\mathrm{em}}^2}{\pi^3 M^2} L(M) f^{\mathrm{B}}(E;T)\,\mathrm{Im}\,\Pi_{\mathrm{em}}^{(\mathrm{ret})}(E,p;\mu_{\mathrm{B}},T) \tag{1}$$

with the dilepton phase-space function L and the Bose-distribution function f^{B}. In case of the thermal ρ emission the retarded electromagnetic current-current correlator is given by the in-medium propagator of the meson, $\mathrm{Im}\,\Pi_{\mathrm{em}} = (m_\rho^2/g_\rho)^2\,\mathrm{Im}\,D_\rho$ (where m_ρ respectively g_ρ denote the bare mass and the coupling strength of the ρ).

Note that the original dilepton emission rate, as derived by McLerran and Toimela [12], assumes chemical equilibrium. However, here we use a generalization for an off-equilibrium situation with finite pion chemical potentials. In this case an additional squared fugacity factor $z_\pi^2 = \exp(2\mu_\pi/T)$ enters in equation (1) for the thermal ρ emission. Within our approach the pion chemical potential is extracted in each cell in the relativistic Boltzmann approximation. A positive value for μ_π means that the corresponding number of produced pions is larger than one would obtain in complete chemical equilibrium [13]. This is what we also find when using the input from the UrQMD model. As even small values of μ_π give a strong increase on the dilepton yield [14], it is important to include this effect in the calculations.

Although the focus is on the in-medium effects of the ρ in the present study, we need to consider also the other sources that contribute to the dilepton excess yield. For a full description one has to calculate the emission from the QGP, i.e., in our case for temperatures above 170 MeV. For this we use the rates from lattice calculations, which are extrapolated for finite momenta [15]. As we deal with extremely high densities of hadronic matter in the heavy-ion collisions under consideration, multi-pion interactions will also give a significant contribution for the dilepton yield. The rates used in our approach take vector-axialvector mixing into account [16, 17].

Finally, we have to consider the contribution from those cells, where the density (respectively temperature) is very low and therefore the assumption of a thermal emission is not reasonable. For these cells we simply take the ρ mesons from the UrQMD calculation directly and apply a shining procedure [18, 19]. As this is only done for the cells in which no thermal emission is considered, we avoid double counting.

For more details on the coarse-graining approach, the reader is referred to reference [3].

3. Spectral Functions

Only few approaches exist which model the influence of both finite temperature and baryon density on the ρ spectral function. We here apply and compare two widely used spectral functions: On the one hand, the hadronic many-body calculation by Rapp and Wambach (Rapp SF) [20] which has previously proven to be a good description of experimental results [11, 16, 17]. And secondly the more model-independent approach by Eletsky and others (Eletsky SF) [21] for which the self-energies of the ρ are calculated using empirical scattering amplitudes from resonance dominance. In the Rapp SF the propagator of the ρ takes the form

$$D_\rho(E,p) = \frac{1}{M^2 - M_\rho^2 - \Sigma_{\rho\pi\pi}(M) - \Sigma_{\rho M}(E,p) - \Sigma_{\rho B}(E,p)}, \tag{2}$$

which includes interactions of the pion cloud with the ρ ($\Sigma_{\rho\pi\pi}$) and direct scatterings off mesons and baryons ($\Sigma_{\rho M}$, $\Sigma_{\rho B}$). The meson-gas effects include interactions with the most abundant π, K and ρ mesons and are saturated with resonances up to 1.3 GeV [20, 22]. Similarly, the effects of the baryonic matter enter via direct $\rho N \to B$ scattering and resonance hole excitations [23, 24]. For the Eletsky SF the ρ-meson self-energy is calculated within a heat bath of nucleons and pions. The in-medium modifications are then consequently determined by ρN and $\rho\pi$ scatterings. In the low-energy regime the calculation of the scattering amplitudes $f_{\rho a}(s)$ includes the resonances from the partial wave analysis by Manley [25], while at higher energies a Regge parametrization is applied. The contribution to the self-energy from the scattering of a ρ meson with a hadron is calculated in low-density expansion as

$$\Sigma_{\rho a}(E,p) = -4\pi \int \frac{\mathrm{d}^3k}{(2\pi)^3} n_a(\omega) \frac{\sqrt{s}}{\omega} f_{\rho a}(s) \tag{3}$$

with $\omega^2 = m_a^2 + k^2$ and the occupation number n_a. Note that in this model the scattering amplitudes are evaluated on the mass shell and therefore only depend on p here.

When comparing the two approaches for the spectral function, there are some caveats one has to consider. Especially it is important to know that the implementation of baryonic effects is not the same. In the parametrized version of the Rapp SF [26], which we use for our calculations, the baryonic effects enter via an effective baryon density $\rho_{\text{eff}} = \rho_{\text{N}} + \rho_{\bar{\text{N}}} + 0.5\,(\rho_{\text{B}^*} + \rho_{\bar{\text{B}}^*})$. Firstly, this takes into account that the interaction of the ρ meson is the same with baryons and anti-baryons. And secondly it reflects the fact, that the scattering with excited baryons has less effect on the spectral shape [17]. As the Eletsky SF is determined for a gas of π and N only, not considering the effects of anti-baryons and excited resonance states, the baryon effects enter via a nucleon chemical potential μ_N. In the non-interacting hadron gas EoS, which we apply when determining

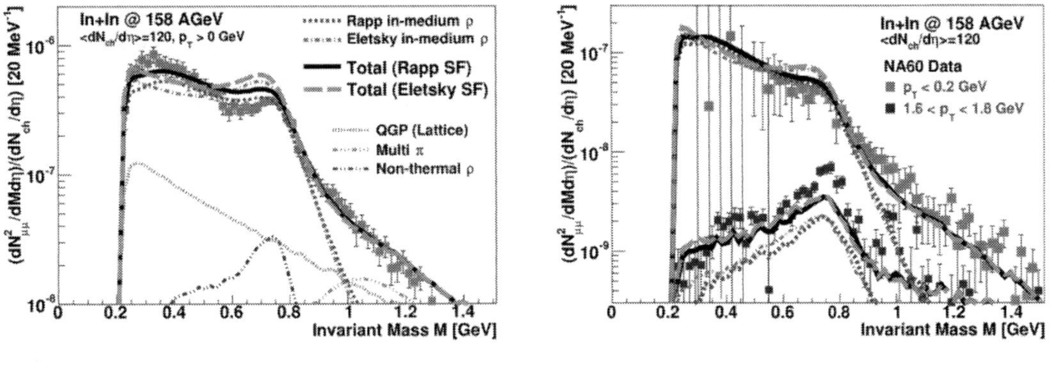

Figure 1. (Color online) Invariant mass spectra of the dilepton excess for In+In collisions at $E_{\rm lab} = 158\,A$GeV for the low-mass region for all p_t (left plot) and two distinct transverse momentum bins (right plot). The results are compared to the experimental data from the NA60 Collaboration [27–29].

T and $\mu_{\rm B}$, this μ_N is identical with the baryon-chemical potential $\mu_{\rm B}$. However, it is important to bear in mind that the degrees of freedom are different for the underlying UrQMD model and the hadron gas EoS on the one side and the πN gas, for which the scattering amplitudes of the Eletsky SF were calculated, on the other side. At SPS energies, a significant fraction of the baryons is found in an excited state in the UrQMD calculations, in conflict with the assumption of a gas of pions and nucleons only. Here a significantly higher degree of consistency is given with the hadron-resonance gas used for the hadronic many-body calculations by Rapp.

4. Results

The resulting invariant mass spectra from our coarse-graining calculation are presented in Figure 1. The left plot shows the total p_t-integrated yield. In general we achieve a rather good description of the measured spectra [27–29] with both spectral functions. A somewhat stronger melting of the peak at the ρ pole mass is however visible in case of the Rapp SF. It only slightly overshoots the NA60 data by about 20% in the mass range from 0.6 to 0.8 GeV, while the Eletsky SF gives a yield that is almost a factor of two above the data. This finding is in line with the fact that the Eletsky approach resembles a virial expansion, that is valid only for low densities. The broadening is mainly due to the presence of baryonic matter which in such heavy-ion collisions can easily reach several times normal matter density. Consequently it is clear that the low-density approximation will underestimate the effect. On the contrary, we get more dilepton yield in the mass range from 0.3 to 0.5 GeV with the Rapp SF, in better agreement with the data than the Eletsky SF. The latter additional enhancement was partly attributed to the inclusion of off-shell ρ-N resonances [2], whereas in the Eletsky SF the scattering amplitudes are calculated on-shell only. The overall shape of the invariant-mass spectrum is clearly better described when applying the Rapp SF; we find a stronger broadening here, in good agreement with the data. The other thermal dilepton sources, i.e., the quark-gluon plasma and the multi-pion interactions, only give relatively small contributions at low masses. However, they become significant in the mass range above 1 GeV, which is not considered in detail here. A relatively small contribution stems also from the non-thermal ρ.

In the right plot of Figure 1 we see the results for two transverse-momentum bins. The upper curves (with red data points) denote the results for $p_t < 0.2$ GeV, while the lower results (blue data points) are for momenta between 1.6 and 1.8 GeV. One can see from the comparison, that the most dominant modification of the spectral functions is observed for low momenta. Here - as in the p_t-integrated spectrum - we find a stronger broadening for the Rapp SF, better describing the data around the pole mass. For the higher momentum window, the spectral shape of the ρ resembles more its vacuum shape. Consequently, the differences between the spectral functions become less dominant. However, for high p_t there remains some excess of the data above the model results around the ρ pole mass. Obviously we miss some of the "freeze-out" ρ contribution here. Note also that the thermal ρ contribution clearly dominates for invariant masses up to 1 GeV at low transverse momenta, while for the higher p_t bin it makes up only roughly 50% in this mass region, i.e., the QGP contribution and the non-thermal ρ become more relevant here (not shown in the plot for reasons of lucidity).

5. Summary & Conclusions

In general, the coarse-graining approach has proven its applicability for the description of experimental dilepton yields. Although the two spectral functions used in the present work stem from two rather different calculations, both give a reasonable description of the experimentally measured dilepton yields at SPS energy within the coarse-graining approach. In comparison, however, the description of the broadening (which is mainly due to scattering of the ρ with baryons) is more realistic in the hadronic many-body calculation by Rapp and Wambach.

Acknowledgments

We thank Ralf Rapp for providing the parametrization of the spectral function and many fruitful discussions. This work was supported by BMBF, HIC for FAIR and H-QM.

References

[1] Leupold S, Metag V and Mosel U 2010 *Int. J. Mod. Phys. E* **19** 147–224
[2] Rapp R and Wambach J 2000 *Adv. Nucl. Phys.* **25** 1
[3] Endres S, van Hees H, Weil J and Bleicher M 2014 (*Preprint* `arXiv:1412.1965 [nucl-th]`)
[4] Huovinen P, Belkacem M, Ellis P J and Kapusta J I 2002 *Phys. Rev. C* **66** 014903
[5] Bass S A, Belkacem M, Bleicher M, Brandstetter M, Bravina L *et al.* 1998 *Prog. Part. Nucl. Phys.* **41** 255–369
[6] Bleicher M, Zabrodin E, Spieles C, Bass S A, Ernst C *et al.* 1999 *J. Phys. G* **25** 1859–1896
[7] Petersen H, Bleicher M, Bass S A and Stöcker H 2008 (*Preprint* `arXiv:0805.0567 [hep-ph]`)
[8] Eckart C 1940 *Phys. Rev.* **58** 919–924
[9] Zschiesche D, Schramm S, Schaffner-Bielich J, Stöcker H and Greiner W 2002 *Phys. Lett. B* **547** 7–14
[10] He M, Fries R J and Rapp R 2012 *Phys. Rev. C* **85** 044911
[11] Rapp R 2013 *Adv. High Energy Phys.* **2013** 148253
[12] McLerran L D and Toimela T 1985 *Phys. Rev. D* **31** 545
[13] Bandyopadhyay D, Gorenstein M, Stoecker H, Greiner W and Sorge H 1993 *Z. Phys. C* **58** 461–464
[14] Koch P 1992 *Phys. Lett. B* **288** 187–194
[15] Ding H T, Francis A, Kaczmarek O, Karsch F, Laermann E *et al.* 2011 *Phys. Rev. D* **83** 034504
[16] van Hees H and Rapp R 2006 *Phys. Rev. Lett.* **97** 102301
[17] van Hees H and Rapp R 2008 *Nucl. Phys. A* **806** 339–387
[18] Li G Q, Ko C M, Brown G E and Sorge H 1996 *Nucl. Phys. A* **611** 539–567
[19] Schmidt K, Santini E, Vogel S, Sturm C, Bleicher M *et al.* 2009 *Phys. Rev. C* **79** 064908
[20] Rapp R and Wambach J 1999 *Eur. Phys. J. A* **6** 415–420
[21] Eletsky V L, Belkacem M, Ellis P and Kapusta J I 2001 *Phys. Rev. C* **64** 035202
[22] Rapp R and Gale C 1999 *Phys. Rev. C* **60** 024903
[23] Rapp R, Urban M, Buballa M and Wambach J 1998 *Phys. Lett. B* **417** 1–6
[24] Urban M, Buballa M, Rapp R and Wambach J 1998 *Nucl. Phys. A* **641** 433–460
[25] Manley D and Saleski E 1992 *Phys. Rev. D* **45** 4002–4033
[26] Rapp R 2014 private communication
[27] Arnaldi R *et al.* (NA60 Collaboration) 2009 *Eur. Phys. J. C* **59** 607–623
[28] Specht H J (NA60 Collaboration) 2010 *AIP Conf. Proc.* **1322** 1–10
[29] Arnaldi R *et al.* (NA60 Collaboration) 2009 *Eur. Phys. J. C* **61** 711–720

Journal of Physics: Conference Series **599** (2015) 012021 doi:10.1088/1742-6596/599/1/012021

Determination of freeze-out conditions from fluctuations in the Hadron Resonance Gas model

P Alba[1], W Alberico[1], R Bellwied[2], M Bluhm[3], V Mantovani Sarti[1], M Nahrgang[4] and C Ratti[1,2]

[1] Department of Physics, Torino University and INFN, Sezione di Torino, via P. Giuria 1, 10125 Torino, Italy
[2] Department of Physics, University of Houston, Houston, TX 77204, USA
[3] Department of Physics, North Carolina State University, Raleigh, NC 27695, USA
[4] Department of Physics, Duke University, Durham, NC 27708-0305, USA

E-mail: `pgalba@to.infn.it`

Abstract. Fluctuations of conserved charges measured in Heavy-Ion Collisions (HICs) received increasing attention in recent years, because they are good candidates to explore the phase diagram of QCD matter. During the last year, net-electric charge and net-proton moments of multiplicities measured at RHIC have been published by the STAR collaboration, for a range of collision energies which spans a region of the phase diagram at finite chemical potential. Here we present a new freeze-out curve obtained using the Hadron Resonance Gas (HRG) model approach to fit these experimental data. The HRG model is modified in order to have a realistic description of the HICs: kinematic cuts, resonance feed-down and resonance regeneration are taken into account. Our result is in agreement with preliminary studies by the ALICE collaboration, and is supported by a recent lattice analysis of the same quantities.

1. Introduction

Various studies on Quantum ChromoDynamics (QCD) converge to the idea that for arbitrarily high temperature and/or densities the strongly interacting confined matter undergoes a transition to a phase in which the fundamental degrees of freedom, namely quarks and gluons, are deconfined [1]. With joined efforts, physicists all around the world were able to obtain first insight on the formation of such a state by means of Heavy-Ion Collisions (HICs). The quest for the right set of observables in order to get a clean signal of the quark liberation has reached the first step, mainly thanks to the possibility of performing simulations on a lattice (lQCD); on the lattice they are able to exactly solve QCD, and consequently to find probes of the free-quark phase from first principles. From the lattice we know that in the pure gauge sector the Polyakov loop is the order parameter for the confinement first order transition, and that in a system with dynamical quarks the transition is a smooth crossover [2]. Unfortunately, lattice simulations are available only for small chemical potentials, due to the so called sign problem which does not allow to use Monte Carlo techniques. In order to complement lattice simulations many options are available from effective theories, like the quasi-particle approach for the deconfined phase [3, 4, 5, 6, 7]. To study the confined thermal-equilibrated hadron phase, the most used tool is the Hadron-Resonance Gas (HRG) model, which is based on the idea that a system of interacting hadrons in the ground state can be described in terms of a non-interacting gas of hadrons and resonances. This model is very successful in describing the hadron production in HICs for a wide range of collision energies, ranging from AGS to the LHC [8, 9]. In the following we will show how to describe the moments of particle multiplicity distributions measured in experiments, by means of the fluctuations of conserved charges calculated in the HRG model. A first matching with lattice calculations shows a good agreement with the HRG predictions [10] , which potentially can provide the link between experiment and first principle calculations. One has to keep in mind that several sources of non statistical fluctuations might potentially affect the experimental measurements: for a discussion on these issues see e.g. [11, 12]

The paper is organized as follows: in Section 2 we will introduce the main concepts about the HRG model, in Section 3 we will introduce the concept of fluctuations and relate them to the moments of multiplicity distributions, in Section 4 we will describe the main corrections applied to the HRG, in Section 5 we will show the main results, and in the end the conclusions.

2. The Hadron Resonance Gas model

The idea at the basis of the HRG model is that a system of interacting hadrons can be described by a non interacting gas of hadrons and resonances. The resonance formation accounts for the strong interaction among particles.
The thermally equilibrated system created after the hadronization process can be described by the sum of independent particle contributions, e.g. for the pressure:

$$p(T,\{\mu_R\}) = \sum_R (-1)^{B_R+1} \frac{d_R T}{(2\pi)^3} \int d^3p \ln\left[1+(-1)^{B_R+1} e^{-(\sqrt{\vec{p}^2+m_R^2}-\mu_R)/T}\right], \tag{1}$$

where B_R, d_R, m_R are respectively baryon number, degeneracy factor and mass of the particle R. The chemical potentials μ_R are given by a linear combination of all the conserved charges times the correspondent chemical potential, for example in full chemical equilibrium where we consider baryon number, electric charge and strangeness:

$$\mu_R = B_R \mu_B + Q_R \mu_Q + S_R \mu_S \,. \tag{2}$$

The other quantities follow from the usual thermodynamic identities, e.g. the particle number densities:

$$n_R(T,\mu_R) = \left(\frac{\partial p}{\partial \mu_R}\right)_T = \frac{d_R}{(2\pi)^3} \int d^3p \frac{1}{1+(-1)^{B_R+1} e^{-(\sqrt{\vec{p}^2+m_R^2}-\mu_R)/T}}. \tag{3}$$

We use a PDG list [13] with particles with masses up to 2 GeV. Due to its simple ideal-gas nature it is easy in the HRG to modify a selected set of particles in order to fulfill the experimental constraints, as showed below.

3. Fluctuations and higher order moments

From Eq. 1 it is possible to extract fluctuations of conserved charges as follows:

$$\chi_{lmn}^{BSQ} = \frac{\partial^{l+m+n}\left(p/T^4\right)}{\partial\left(\mu_B/T\right)^l \partial\left(\mu_S/T\right)^m \partial\left(\mu_Q/T\right)^n}. \tag{4}$$

The mean M, the variance σ, the skewness S and the kurtosis κ of the multiplicity distribution of a net-quantity (e.g. electric charge Q) are related to the above fluctuations through the following equations:

$$M = \langle N \rangle = VT^3\chi_1 \quad , \quad \sigma^2 = \langle (\delta N)^2 \rangle = VT^3\chi_2,$$
$$S = \frac{\langle (\delta N)^3 \rangle}{\sigma^3} = \frac{VT^3\chi_3}{(VT^3\chi_2)^{3/2}} \quad , \quad \kappa = \frac{\langle (\delta N)^4 \rangle}{\sigma^4} - 3 = \frac{VT^3\chi_4}{(VT^3\chi_2)^2}. \tag{5}$$

Ratios of them are volume independent (to leading order), and so are in principle easily comparable to lattice simulations and HRG model results:

$$\sigma^2/M = \chi_2/\chi_1\,, \quad S\sigma = \chi_3/\chi_2\,, \quad S\sigma^2/M = \chi_3/\chi_1\,, \quad k\sigma^2 = \chi_4/\chi_2\,. \tag{6}$$

First experimental data on net-charge and net-proton number have been recently published [15] and compared to lattice QCD results [16].

4. Corrections to match the experimental situation

Many corrections can be applied to the HRG model in order to get closer to the experimental situation. For example, in order to reproduce the initial conditions one usually imposes $N_Q = 0.4\,N_B$ and $N_S = 0$; this connects the temperature T and the three chemical potentials to each other, namely μ_Q and μ_S are functions of T and μ_B.

In the experiment it is not possible to cover the entire phase space, but just a portion of it. This leads to the fluctuations of conserved charges: in the entire phase space, such charges would be strictly conserved and there would be no fluctuations at all.

In accordance with the experimental kinematic setup the formulae modify as follow:

$$n_R(T,\mu_R) = \frac{d_R}{4\pi^2}\int_{-y_{MAX}}^{y_{MAX}} \mathrm{d}y \int_{p_T^{MIN}}^{p_T^{MAX}} \mathrm{d}p_T \frac{p_T\sqrt{p_T^2+m_R^2}\cosh(y)}{(-1)^{B_R+1}+\exp((\cosh(y)\sqrt{p_T^2+m_R^2}-\mu_R)/T)}, \tag{7}$$

$$n_R(T,\mu_R) = \frac{d_R}{4\pi^2}\int_{-\eta_{MAX}}^{\eta_{MAX}} \mathrm{d}\eta \int_{p_T^{MIN}}^{p_T^{MAX}} \mathrm{d}p_T \frac{p_T^2\cosh(\eta)}{(-1)^{B_R+1}+\exp((\sqrt{p_T^2(\cosh(\eta))^2+m_R^2}-\mu_R)/T)}. \tag{8}$$

The higher order cumulants are obtained accordingly. The effects of kinematic cuts on χ_2^Q/χ_1^Q are shown in Fig. 1 (left). For a systematic study see [17].

At a certain point in the evolution of the system it is reasonable to expect that inelastic scatterings between hadrons cease to happen, and hadrons will only interact elastically or quasi-elastically, through the formation and subsequent strong decay of resonances. This is called chemical freeze-out. Only the particles with a lifetime longer than the hadron phase one will be detected; the effective number $\bar{N}_i$ of these *stable* particles, given by their primordial thermal number plus the contribution from the resonance decays (Eq. 9), is thus conserved.

$$\bar{N}_i = N_i + \sum_r d_{r\to i}N_r. \tag{9}$$

This situation is different from the full chemical equilibrium described by the lattice simulations, and so we will call it Partial Chemical Equilibrium (PCE).

As a consequence, after chemical freeze-out, the resonance chemical potentials can be expressed in terms of the stable particle ones, using the branching ratios of the decays.

The contribution to the higher order cumulants for every stable particle in PCE is easily obtained using the partial derivative with respect to its effective chemical potential. See

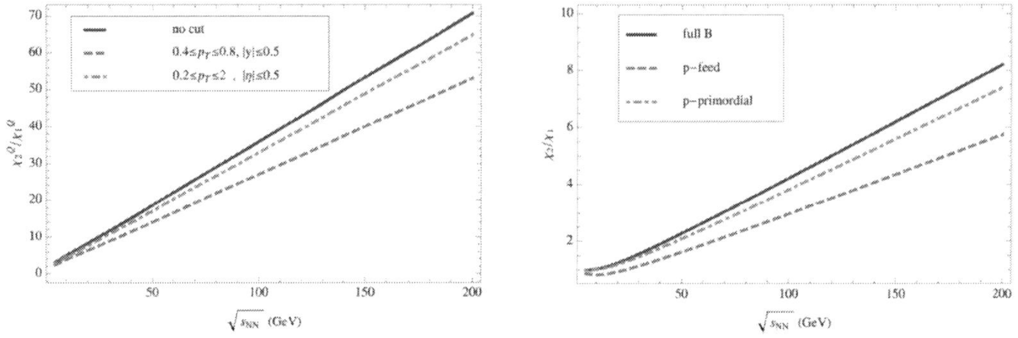

Figure 1. (Color online) Left panel: effects of the kinematic cuts on χ_2/χ_1 for the net-electric charge compared to the situation of no cuts (blue full line); the different cuts read: $0.4 \leq p_T \leq 0.8$ (GeV) and $|y| \leq 0.5$ (red dashed line), $0.2 \leq p_T \leq 2$ (GeV) and $|\eta| \leq 0.5$ (green dot-dashed line). Right panel: feed down correction to χ_2/χ_1 for the net-proton number (red dashed line), compared to the result for the net-baryon (blue full line) and primordial net-proton (green dot-dashed line). The curves are obtained using the freeze-out parametrization from [21].

Fig. 1 (right) to see the effect of the feed down correction. Experimentally, pions, kaons and (anti)protons are the most abundant particles in the energy range of interest, but we extended the set of stable particles in order to exclude the contribution from weak decays [14].

The multiplicity distribution of neutrons has not been measured. Nevertheless, the contribution of neutrons on the distribution of their isospin partner can still be sizable, due to some particular processes. Consider for example the formation cycle of a Delta resonance:

$$p(n) + \pi^0(\pi^+) \rightarrow \Delta^+ \rightarrow n(p) + \pi^+(\pi^0)\,; \tag{10}$$

this regeneration process during the hadron stage realizes an isospin randomization, namely at the detection it is impossible to distinguish between a primordial proton, or a proton coming from the interaction of a neutron with a pion. The lifetime of a Delta resonance, for the temperature range of interest, is big enough to perform a complete isospin randomization for collision energies $\sqrt{s_{NN}} \geq 10$ GeV. The prescription to include such a process has been defined in [18], and effectively this correction has a well established effect on the proton distribution. See [19] for more details.

5. Results and conclusions

Using the experimental data from [15] on σ^2/M for net-charge and net-protons, we are able to calculate for each collision energy (11.5, 19.6, 27, 39, 62.4 and 200 GeV) the temperature and the μ_B at the chemical freeze-out. For further details on the calculation see [20]. In Fig. 2 (right) we show a comparison to the previous results by [21].

In our approach, we were able to find a common freeze-out surface which describes simultaneously the net-charge and net-proton low order fluctuations. The agreement with previous results is satisfactory for the μ_B, while our temperature is lower by about 20 MeV. This difference may follow from the fact that the observables we are analyzing are dominated by light quarks, namely pions, kaons and protons, while the former result is obtained from a simultaneous fit of particle multiplicities belonging both to the light and strange sectors. Such a lower freeze-out temperature is suggested also from another analysis of lattice calculations [22] and recent LHC data [23], but further studies to solve this issue will be performed. Preliminary

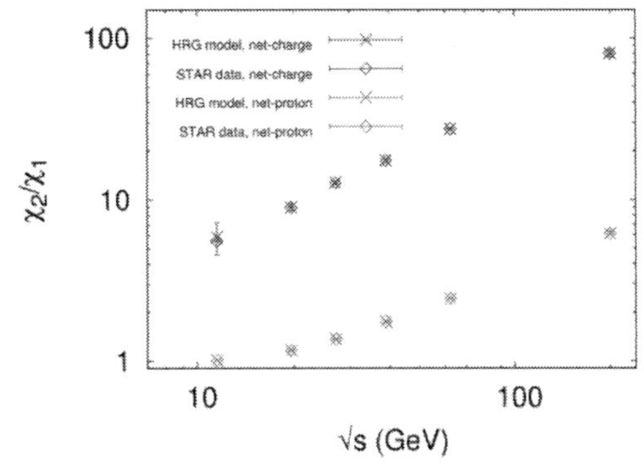

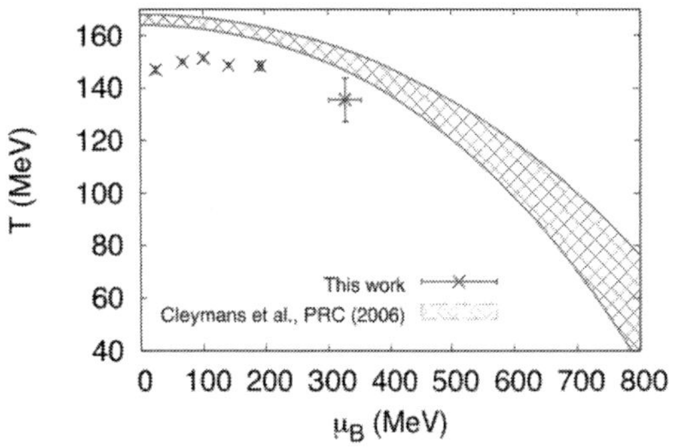

Figure 2. (Color online) Left panel: χ_2/χ_1 calculations using the HRG model with our freeze-out parameters compared to the STAR data [15]; right panel: our freeze-out parameters (blue crosses), compared to the previous result by [21] (red band).

results on the sensitivity of the higher order cumulants are available [24], and confirm our present finding that the lower order fluctuations can be a good candidate to explore the QCD phase diagram.

Acknowledgments

This work is supported by the Italian Ministry of Education, Universities and Research under the FIRB Research Grant RBFR0814TT, a fellowship within the Postdoc-Program of the German Academic Exchange Service (DAAD), and the US Department of Energy grants DE-FG02-03ER41260, DE-FG02-05ER41367 and DE-FG02-07ER41521.

References

[1] Adams J *et al.* [STAR Collaboration] 2005 *Nucl. Phys.* A **757** 102
[2] Borsanyi S, Endrodi G, Fodor Z, Jakovac A, Katz S D, Krieg S, Ratti C *et al.* 2010 *JHEP* **1011** 077
[3] Blaizot J P and Iancu E 1994 *Nucl. Phys.* B **417** 608.
[4] Bluhm M, Kampfer B and Redlich K 2012 *Phys. Lett.* B **709** 77
[5] Meisinger P N, Ogilvie M C and Miller T R 2004 *Phys. Lett.* B **585** 149.
[6] Castorina P, Greco V, Jaccarino D and Zappala D 2011 *Eur. Phys. J.* C **71** 1826.
[7] Ruggieri M, Alba P, Castorina P, Plumari S, Ratti C and Greco V 2012 *Phys. Rev.* D **86** 054007
[8] Braun-Munzinger P, Redlich K and Stachel J 2003 In **Hwa, R.C. (ed.) et al.: Quark gluon plasma** 491-599
[9] Andronic A, Braun-Munzinger P, Redlich K and Stachel J 2013 *Nucl. Phys.* A **904-905** 535c
[10] Karsch F, Redlich K and Tawfik A 2003 *Eur. Phys. J.* C **29** 549
[11] Hauer M and Vogel S 2011 *Phys. Rev.* C **83** 064906.
[12] Schuster T, CERN-THESIS-2011-318.
[13] Beringer J *et al.* [Particle Data Group Collaboration] 2012 *Phys. Rev.* D **86** 010001.
[14] Bluhm M, Alba P, Alberico W, Beraudo A and Ratti C 2014 *Nucl. Phys.* A **929** 157
[15] Adamczyk L *et al.* [STAR Collaboration] 2014 *Phys. Rev. Lett.* **112** 032302
Adamczyk L *et al.* [STAR Collaboration] 2014 *Phys. Rev. Lett.* **113** 092301
[16] Borsanyi S, Fodor Z, Katz S D, Krieg S, Ratti C and Szabo K K 2014 *Phys. Rev. Lett.* **113** 052301
[17] Garg P, Mishra D K, Netrakanti P K *et al.* 2013 *Phys. Lett.* B **726** 691
[18] Kitazawa M and Asakawa M 2012 *Phys. Rev.* C **86** 024904
[19] Nahrgang M, Bluhm M, Alba P, Bellwied R and Ratti C 2014 arXiv:1402.1238 [hep-ph].
[20] Alba P, Alberico W, Bellwied R *et al.* 2014 *Phys. Lett.* B **738** 305
[21] Cleymans J, Oeschler H, Redlich K and Wheaton S 2006 *Phys. Rev.* C **73** 034905
[22] Bellwied R, Borsanyi S, Fodor Z, Katz S D and Ratti C 2013 *Phys. Rev. Lett.* **111** 202302
[23] Preghenella R [ALICE Collaboration] 2012 *Acta Phys. Polon.* B **43** 555
[24] Forthcoming publication.

FAIRNESS 2014 IOP Publishing
Journal of Physics: Conference Series **599** (2015) 012022 doi:10.1088/1742-6596/599/1/012022

Higher order fluctuations of strangeness and flavour hierarchy

V. Mantovani Sarti[1], P. Alba[1], W. Alberico[1], R. Bellwied[2], M. Bluhm[3], M. Nahrgang[4], C. Ratti[1,2]

[1] Department of Physics, University of Turin, [2] University of Houston, [3] North Carolina State University, [4] Duke University

E-mail: mantovan@to.infn.it

Abstract. We present a preliminary analysis on the sensitivity to the chemical freeze-out temperature of higher order moments of strange particle multiplicity distributions in heavy ion collisions. Within the Hadron Resonance Gas (HRG) model we evaluate ratios of cumulants for kaons ($K^{\pm}$) and hyperons ($\Lambda, \Sigma^{\pm}, \Xi^{-,0}, \Omega^{-}$) as a function of the temperature and compare them to the sensitivity profiles obtained from ratios of particle yields. We show that ratios of higher order fluctuations of strange baryons could provide a useful tool to extract the range of freeze-out temperature, once experimental data are available. Finally, a connection to lattice data through the fourth to second cumulant ratio is made. The deconfinement transition on the lattice seems to indicate the possibility of a flavour hierarchy, namely strange quarks seem to deconfine at a higher temperature. We would like to test the possibility for the same scenario to occur at the chemical freeze-out and we show how the inclusion of multi-strange baryons in the evaluation of higher order cumulants might provide a sensitive observable to extract the freeze-out temperature.

1. Introduction

Strange particles have been a fundamental observable in the search for a possible state of deconfined matter and as a signal for the formation of a Quark Gluon Plasma (QGP) in relavitistic heavy-ion collisions (HICs). The so-called *strangeness enhancement* has been indicated as a signal of deconfinement since the early eighties [1].

The reason why strangeness is so easily produced in the presence of free quarks and gluons is that processes as gluon fusion $g + g \to s + \bar{s}$ and quark-antiquark annihilation $q + \bar{q} \to s + \bar{s}$ are energetically favoured with respect to reactions involving confined hadrons. The Q-value of gluon fusion and $q\bar{q}$ annihilation at temperatures close to the critical one, T_c, is roughly $2m_s \approx 200$ MeV, while to produce strangeness in a hadronic environment, e.g. through associated production $N + N \to N + \Lambda + K$, nearly $Q = m_\Lambda + m_K - m_N \approx 700$ MeV is needed. The abundant production of strange quark pairs in the early stages of QGP leads to the presence, in the subsequent hadronization process, of kaons and hyperons, which otherwise would be rarely produced.

Experimental observations of such an enhancement are present at both RHIC and LHC energies [2], where an increase of hyperon production ($\bar{\Lambda}$, $\bar{\Xi}^{+}$, Ω) at mid-rapidity as a function of the number of participants has been seen.

The production of particles with strangeness content is also fundamental in the description of the hadrochemistry at chemical freeze-out. The particle abundances in HICs in a very wide range of energies are well reproduced by statistical hadronization models which assume a common freeze-out surface for all particle species [3]. Nevertheless latest results on SHM fits to ratios of particle yields at LHC energies show a tension between light (protons) and strange particles (hyperons) [4][5]. These recent outcomes might represent a possible indication that in order to reproduce ratios or yields of strange hadrons a higher chemical freeze-out temperature is needed with respect to protons (see Fig.1). A different behaviour between light quarks and strange quarks is also present in lattice results [6]. Here the authors evaluate the ratio of cumulants χ_4/χ_2 for (u, d) and s quarks as a function of the temperature and compare the results to a HRG calculation. In the low temperature regime HRG results and lattice data show an agreement within the error bars in both light and strange sector. In Fig. 2 it is evident that the non-monotic behaviour of the lattice curves indicates that the system undergoes a transition, from a confined phase at low T to a deconfined one at higher T. This evolution is stressed by the separation of lattice data from HRG results as the temperature increases. Strange quarks start to show this separation at a higher temperature than light quarks: a flavour hierarchy seems to occur during the deconfinement transition.

Our analysis prepares the ground for an investigation of a similar flavour hierarchy in the chemical freeze-out process by studying higher order fluctuations of strange particles, in order to provide sizeable observables once experimental data are available [7, 8].

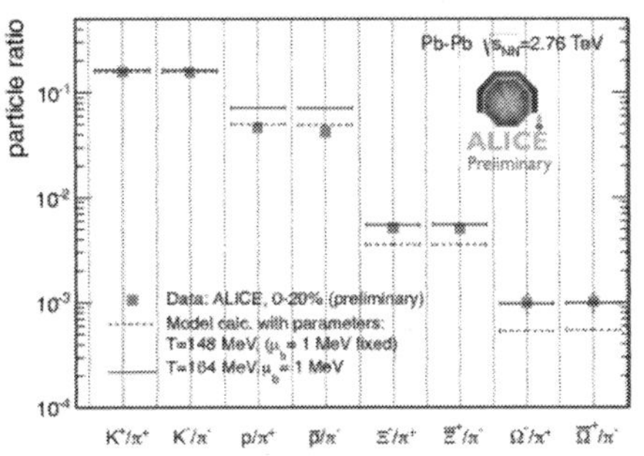

Figure 1: Particle to pion yield ratios at LHC and comparison to thermal models from [4].

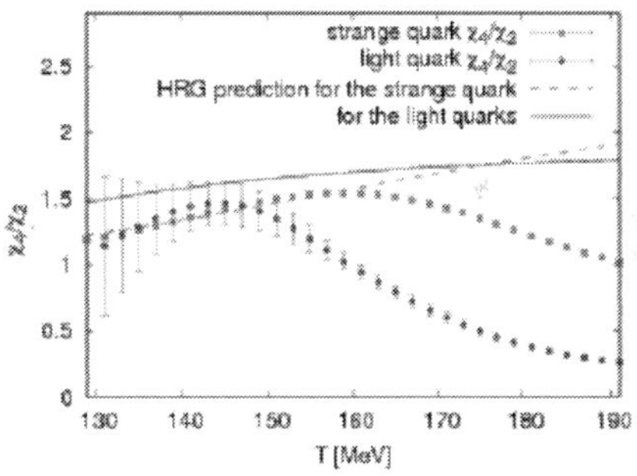

Figure 2: Comparison of lattice data in the continuum limit to HRG model calculations for χ_4/χ_2 for light and strange quarks from [6].

2. The Hadron Resonance Gas model

The results presented here are obtained using a HRG model in partial chemical equilibrium, which means that contributions from strong decays from resonances with mass up to 2 GeV/c^2 are included. Consistently with what is done in the experiment, the contributions from weak decays have not been taken into account: the HRG results have been compared to feed-down corrected experimental data. The hadron spectrum of all resonant states is taken from the PDG list [9]. The observables to be compared to experimental data are the susceptibilities of conserved charges such as baryonic number B, electric charge Q and strangeness S, defined as:

$$\chi^{lmn}_{BQS} = \frac{\partial^{l+m+n} P/T^4}{\partial(\mu_B/T)^l \partial(\mu_Q/T)^m \partial(\mu_S/T)^n}.$$

Susceptibilities are directly related to higher moments of particle multiplicity distributions through the definitions of mean (M), variance (σ^2), skewness (S) and kurtosis (κ) and the following volume-independent ratios can be extracted from experimental data and then evaluated in the HRG approach:

$$M/\sigma^2 = \chi_1/\chi_2 \,, \qquad S\sigma^3/M = \chi_3/\chi_1 \,,$$
$$S\sigma = \chi_3/\chi_2 \,, \qquad \kappa\sigma^2 = \chi_4/\chi_2 .$$

Further details on the HRG model used can be found in [10], where we obtained the freeze-out parameters (T, μ_B) from the comparison of HRG results to net-charge and net-proton distribution measurements from the STAR collaboration (see Fig. 3). We obtain a lower freeze-out curve, approximately 20 MeV less than the value given by the statistical hadronization model fits to all available particle multiplicities ($T \approx 166$ MeV). This difference in our opinion is due to the fact that net-charge and net-proton moments are mainly driven by particles containing light quarks (protons and pions), while the fit of particle multiplicities contains also contributions from strange particles such as $K, \Lambda, \Sigma, \Xi, \Omega$. As already pointed out in the introduction, lattice data for χ_4^S/χ_2^S seem to indicate a flavour hierarchy in the high T region. If an analogous mechanism occurs at the chemical freeze-out, then strange particle multiplicities and moments might yield a higher freeze-out temperature. An indication of this decoupling between light particles (π, K, p) and multi-strange baryons has already been shown in [10]. Another possibility is that higher lying strange resonances, predicted by quark models but so far not detected experimentally, might lower the freeze-out temperature for strange hadrons and bring it closer to the one we find for light particles [11].

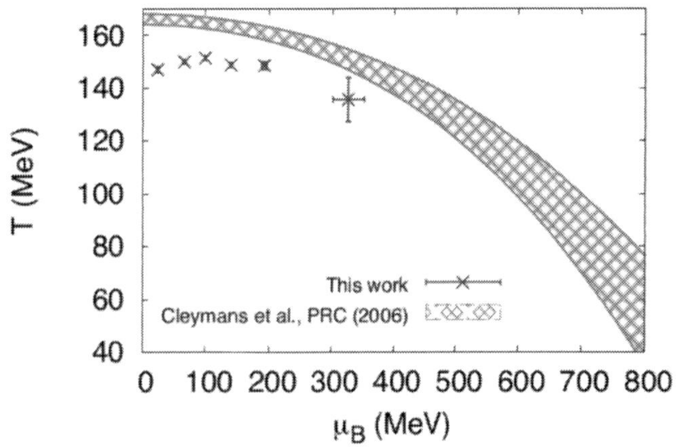

Figure 3: Freeze-out curve in the $T - \mu_B$ plane obtained from the fit of σ^2/M for net-electric charge and net-proton from [10].

3. Results

At the moment the study of higher order fluctuations in experiments is available only for light particles, namely pions, protons and kaons.

A possible way to explore the flavour hierarchy hypothesis at chemical freeze-out is to perform an analysis on lower moments of net-kaons and to compare to data once available. Our preliminary studies show that the sensitivity to the temperature of χ_2/χ_1 for net-kaons is much higher with respect to ratios such as $K^{\pm}/\pi^{\pm}$ [12]. The analysis of ratios of lower moments might already provide a sensitive tool to extract the freeze-out parameters. In Fig. 4 ratios of moments for net-kaons and net-hyperons are plotted as a function of the temperature at baryochemical potential $\mu_B = 24.3$ MeV, corresponding to the $\sqrt{s} = 200$ GeV at STAR. The finite value of μ_B leads to a non-zero amount of net-densities for multi-strange baryons which explains the opposite behaviour of the ratio χ_2/χ_1 with respect to the net-kaon result. A detailed comparison with experimental data of lower moments (χ_2/χ_1, χ_3/χ_2) could already give some

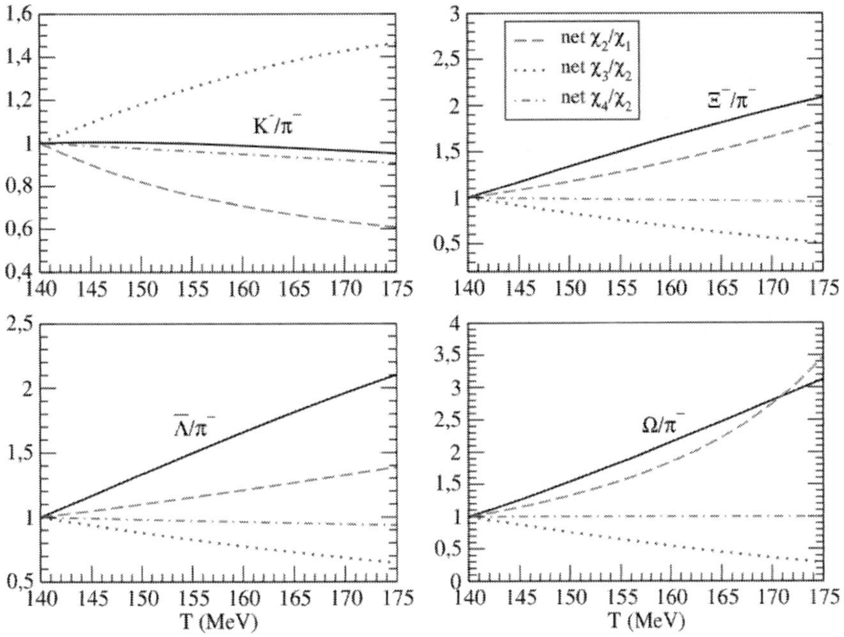

Figure 4: Ratios of moments (normalized to the value at $T = 140$ MeV) for kaons and hyperons at $\mu_B = 24.3$ MeV as a function of T.

precise estimate of the freeze-out temperature in the strange sector. In particular, once efficiency corrected data on net-kaons lower moments from STAR collaboration will be available, we could perform a similar analysis to the one used for net-charge and net-proton moments in [10]. Since lower moments of net-kaons seem to be very sensitive to variations of temperature, as shown in Fig. 4, we expect that a study of kaons alone might be sufficient to extract the freeze-out parameters.

As expected the ratio χ_4/χ_2 looks quite flat for all particle species, so a future employment as a tool to investigate the chemical freeze-out stage might be excluded. Nevertheless this ratio can be directly linked to lattice calculations, since it involves even moments, which are the only non-zero ones for strangeness at $\mu_B = 0$. In Fig. 5 results for the ratio χ_4/χ_2 in our HRG model are presented for a few specific sets of strange particles, along with results from the lattice [6]. Since this ratio is proportional to the strangeness content of the particles involved, the inclusion of hyperons in the calculation of χ_4 and χ_2 is crucial in order to achieve a sensitivity to the temperature closer to the lattice one, which will allow the determination of the freeze-out temperature once compared to data. A high accuracy of experimental data on higher moments for multi-strange baryons represents a challenging issue at the moment.

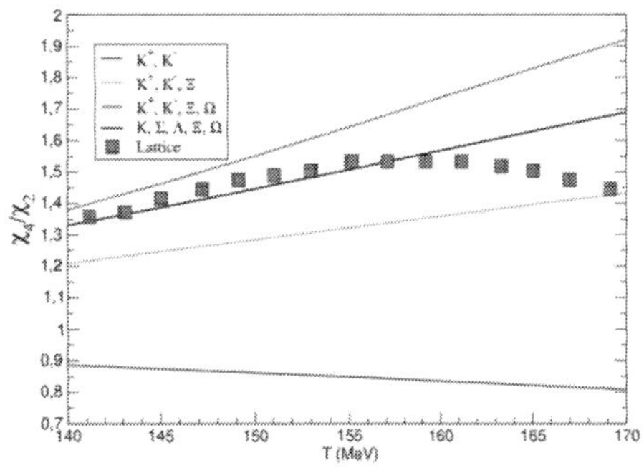

Figure 5: Comparison between lattice results for χ_4/χ_2 for net-strangeness and our HRG model curves including different sets of particles.

4. Conclusions

Several indications coming from both latest experimental data [5] and lattice results [6] suggest that particles with strangeness content might show a higher chemical freeze-out temperature with respect to light particles such as protons and pions.
Higher moments for strange mesons and baryons, within a HRG model approach, have been shown to provide a complementary tool to explore the chemical freeze-out stage and to extract the corresponding T and μ_B, once data are available.

In our study, we show that a detailed analysis of lower moments of net-kaons could already show a discrepancy on the extracted freeze-out temperature with respect to the value obtained from the study of net-charge and net-proton multiplicity distributions at STAR [10].

5. Acknowledgements

This work is supported by the FIRB Research Grant RBFR0814TT, DAAD and the DOE grants DE-FG02-03ER41260, DEFG02-05ER41367 and DE-FG02-07ER41521.

References

[1] Rafelski J and Muller B 1982 *Phys.Rev.Lett.* **48** 1066
[2] Abelev B B *et al.* (ALICE Collaboration) 2014 *Phys.Lett.* **B728** 216–227 (*Preprint* 1307.5543)
[3] Cleymans J, Oeschler H, Redlich K and Wheaton S 2006 *Phys.Rev.* **C73** 034905 (*Preprint* hep-ph/0511094)
[4] Preghenella R (ALICE Collaboration) 2012 *Acta Phys.Polon.* **B43** 555 (*Preprint* 1111.7080)
[5] Floris M 2014 (*Preprint* 1408.6403)
[6] Bellwied R, Borsanyi S, Fodor Z, Katz S D and Ratti C 2013 *Phys.Rev.Lett.* **111** 202302 (*Preprint* 1305.6297)
[7] Alba P, Alberico W, Bluhm M, Ratti C and Bellwied R 2013 *PoS* **CPOD2013** 060
[8] Bluhm M, Alba P, Alberico W, Bellwied R and Ratti C 2014 *J.Phys.Conf.Ser.* **509** 012050
[9] Beringer J *et al.* (Particle Data Group) 2012 *Phys.Rev.* **D86** 010001
[10] Alba P, Alberico W, Bellwied R, Bluhm M, Mantovani Sarti V *et al.* 2014 *Phys.Lett.* **B738** 305–310 (*Preprint* 1403.4903)
[11] Bazavov A, Ding H T, Hegde P, Kaczmarek O, Karsch F *et al.* 2014 *Phys.Rev.Lett.* **113** 072001 (*Preprint* 1404.6511)
[12] Forthcoming publication

Journal of Physics: Conference Series **599** (2015) 012023 doi:10.1088/1742-6596/599/1/012023

Fluctuations of Conserved Quantities in High Energy Nuclear Collisions at RHIC

Xiaofeng Luo

Institute of Particle Physics and Key Laboratory of Quark & Lepton Physics (MOE), Central China Normal University, Wuhan, 430079, China.

E-mail: `xfluo@mail.ccnu.edu.cn`

Abstract. Fluctuations of conserved quantities in heavy-ion collisions are used to probe the phase transition and the QCD critical point for the strongly interacting hot and dense nuclear matter. The STAR experiment has carried out moment analysis of net-proton (proxy for net-baryon (B)), net-kaon (proxy for net-strangeness (S)), and net-charge (Q). These measurements are important for understanding the quantum chromodynamics phase diagram. We present the analysis techniques used in the moment analysis by the STAR experiment and discuss the moments of net-proton and net-charge distributions from the first phase of the Beam Energy Scan program at the Relativistic Heavy Ion Collider.

1. Introduction

One of the main goal of heavy-ion collisions is to explore the phase structure of the hot and dense nuclear matter. The phase structure can be displayed in the two dimensional Quantum Chromodynamics (QCD) phase diagram (temperature, T vs. baryon chemical potential, μ_B). It is conjectured on the basis of theoretical calculation, at finite temperature and vanishing μ_B ($\mu_B = 0$) region, there is a crossover transition between hadronic phase and partonic phase, while at large μ_B region, the phase transition is of the first order. Thus, there should be the so called QCD Critical Point as the endpoint of the first order phase boundary towards the crossover region, which is a second order phase transition point [1]. Since the *ab initio* Lattice QCD calculation encounters a serious so called sign problem at finite μ_B region, there are large uncertainties in determining the location of the CP or even its existence [2, 3]. Experimental confirmation of the existence of the CP will be an excellent verification of QCD theory in the non-perturbative region and a milestone of exploring the QCD phase structure. It is one of the main goals of the Beam Energy Scan (BES) program at the Relativistic Heavy Ion Collider (RHIC). By tuning the colliding energy of the gold nuclei from high to low values, we can vary the T and μ_B of the nuclear matter created in heavy-ion collisions [4]. This enables us to probe a broad region of the QCD phase diagram. On the other hand, fluctuations of conserved quantities, such as net-baryon (B), net-charge (Q) and net-strangeness (S), are predicted to be sensitive to the correlation length of the system [5–7] and directly connected to the susceptibilities computed in the theoretical calculations [8, 9]. Thus, it can serve as a powerful tool to probe the phase transition and CP signal in heavy-ion collisions. Experimentally, the STAR experiment has measured the energy dependence of moments (up to the fourth order) of event-by-event net-proton (proton minus anti-proton number, proxy of net-baryon [10]) and net-charge multiplicity distributions in Au+Au collisions at $\sqrt{s_{NN}}$= 7.7, 11.5, 19.6, 27, 39, 62.4 and 200 GeV. Those data are collected from the first phase of the RHIC BES in the years 2010 and 2011. In the year 2014, another energy point 14.5 GeV is successfully taken and can fill in a large μ_B gap between 11.5 and 19.6 GeV. In this paper, we will present recent experimental results from RHIC/STAR experiment for fluctuations of conserved quantities as well as the analysis techniques. The physics implication of the results will then be discussed.

The paper is organized as follows: In the second section, we will give a brief discussion of the techniques used in the moment analysis. The experimental results and discussion will be presented in the third section. Finally, we will give a summary and outlook.

2. Analysis Techniques

During the last five years, a series of analysis techniques have been applied to the moments of conserved quantities distributions in heavy-ion collisions. Those include : (1) Centrality bin width correction [11,12]. (2) Novel centrality determination to account for the effects of centrality resolution and auto-correlation [12]. (3) Efficiency correction and error estimation [13–16]. Those techniques are very important to address the background effects and extract the dynamical fluctuation signals from the observables. In heavy-ion collisions, we cannot directly measure the collision centrality and/or initial collision geometry of the system of two nuclei. The centrality in heavy-ion collisions is generally determined by comparing the measured particle multiplicity with the Glauber Monte Carlo simulations. It is denoted as a percentage value (for e.g. 0-5%, 5-10%,...) for a collection of events to represent the fraction of the total cross section. This in general can cause two undesirable effects in the moment analysis of particle multiplicity distributions within finite centrality bin. One is the so called centrality bin width effect, which is caused by volume variation within a finite centrality bin size and the other one is centrality resolution effect, which is due to the initial volume fluctuations. Let's review those techniques one by one.

2.1. Centrality Bin Width Correction

The centrality bin width correction is to address the volume fluctuations effects on the higher order moments of conserved quantities distributions within a finite width centrality bin. It is also called centrality bin width effect. This effect needs to be eliminated, as an artificial centrality dependence could be introduced by it. To do this, the centrality bin width correction is applied to calculate the various moments of particle multiplicities distributions in one wide centrality bin. Experimentally, the smallest centrality bin is determined by a single value of particle multiplicity. Experimental results are usually reported for a wider centrality bin (a range of particle multiplicity), such as $0-5\%, 5-10\%$,...etc., to reduce statistical errors. To eliminate the centrality bin width effect, we calculate the various order cumulants (C_n) for each single particle multiplicity within one wider centrality bin and weighted averaged by the corresponding number of events in that multiplicity.

$$C_n = \frac{\sum_{r=N_1}^{N_2} n_r C_n^r}{\sum_{r=N_1}^{N_2} n_r} = \sum_{r=N_1}^{N_2} \omega_r C_n^r \tag{1}$$

where the n_r is the number of events for multiplicity value r and the corresponding weight for the multiplicity r, $\omega_r = n_r / \sum_{r=N_1}^{N_2} n_r$. N_1 and N_2 are the lowest and highest multiplicity values for one centrality bin. Once getting the centrality bin width corrected cumulants via Eq. (1), we can calculate the various order moments, for e.g. $\kappa\sigma^2 = C_4/C_2$ and $S\sigma = C_3/C_2$, where the

Journal of Physics: Conference Series **599** (2015) 012023 doi:10.1088/1742-6596/599/1/012023

κ and S are kurtosis and skewness, respectively. The final statistical error of moments for one centrality can be evaluated by standard error propagation based on Eq. (1). For more details, one can see [12].

2.2. Centrality Resolution and Auto-correlation Effects

Particle multiplicity are usually used in the centrality determination, as it can reflect the initial geometry of heavy-ion collision. However, the relation between measured particle multiplicities and collision geometry is not one-to-one correspondence and there are fluctuations in the particle multiplicity even for a fixed collision geometry. Thus, one could obtain a finite resolution of initial collision geometry (centrality resolution) by using particle multiplicity to determine the centrality. The more particles are used in the centrality determination, the better centrality resolution and smaller fluctuation of the initial geometry (volume fluctuation) we get. This may affect moments of the event-by-event multiplicity distributions. On the other hand, we use the multiplicity of charged kaon and pion to define the collision centrality in Au+Au collisions. This is to prevent the effect of auto-correlation between protons/antiprotons involved in our moments analysis and in the centrality definition. The auto-correlation effect will results in the suppressing values of the moments. To avoid the auto-correlation effect, we should exclude the corresponding proton/antiproton from the centrality definition. The two background effects can be well avoided by a novel centrality definition using particle multiplicities. For more details, one can see [12, 17].

2.3. Efficiency Correction and Error Estimation

Finite detecting efficiency for particle yield measurement in heavy-ion collisions can be easily corrected. However, it is not straightforward to get the efficiency corrected results for higher moments of particle multiplicity distributions. Since the fluctuation analysis is statistics hungry, it is crucial to get the correct statistical errors with limited statistics. We provide a unified description of efficiency correction and error estimation for various order moments of multiplicity distributions [16]. The basic idea is to express the moments and cumulants in terms of the factorial moments, which can be easily corrected for efficiency effects. By knowing the covariance between multivariate factorial moments, we use the standard error propagation based on Delta theorem to obtain the error formula for efficiency corrected moments. This method can also be applied to the phase space efficiency case, where the efficiency of proton or anti-proton are not constant within studied phase space. One needs to note that the efficiency correction and error estimation should be done for events with the same particle multiplicity used for centrality determination and just before the centrality bin width correction.

3. Experimental Results

Recently, the STAR experiment has published the beam energy dependence of moments (up to forth order) of event-by-event net-proton [18] and net-charge [19] multiplicity distributions in Au+Au collisions at $\sqrt{s_{\rm NN}}$= 7.7, 11.5, 19.6, 27, 39, 62.4 and 200 GeV. For net-proton analysis, the protons and anti-protons are identified with ionization energy loss in the Time Projection Chamber (TPC) of the STAR detector within the transverse momentum range $0.4 < p_T < 0.8$ GeV/c and at mid-rapidity $|y| < 0.5$. In the net-charge case, the charged particles are measured within transverse momentum range $0.2 < p_T < 2$ GeV/c and pseudo-rapidity range $|\eta| < 0.5$.

Figure 1 shows the energy dependence of σ^2/M, $S\sigma$ and $\kappa\sigma^2$ of net-proton and net-charge distributions of Au+Au collisions for two centralities (0-5% and 70%-80%) at $\sqrt{s_{\rm NN}}$= 7.7, 11.5, 19.6, 27, 39, 62.4 and 200 GeV. The Skellam (Poisson) expectations shown in the figure reflect a system of totally uncorrelated, statistically random particle production. It predicts the $\kappa\sigma^2$ and $S\sigma$/Skellam to be unity for Skellam expectations as well as in the hadron resonance gas model. For the net-proton results, the most significant deviation of $S\sigma$ and $\kappa\sigma^2$ from Skellam distribution is observed at 19.6 and 27 GeV for 0-5% Au+Au collisions. At energies above 39 GeV, the results are close to Skellam expectation. As the statistical errors are large at low energies (7.7 and 11.5 GeV), more statistics is necessary to quantitatively understand the energy dependence of $S\sigma$ and $\kappa\sigma^2$. To understand the effects of baryon number conservation etc., UrQMD model calculations (a transport model which does not include a CP) for 0-5% are presented and the results show a monotonic decrease with decreasing beam energy. For more details on baseline comparison, one can see [20]. For the net-charge results, we did not observe non-monotonic behavior for $S\sigma$ and $\kappa\sigma^2$ within current statistics. The expectations from negative binomial distribution can better describe the net-charge data than the Poisson (Skellam) distribution. More statistics is needed for net-charge moment calculation.

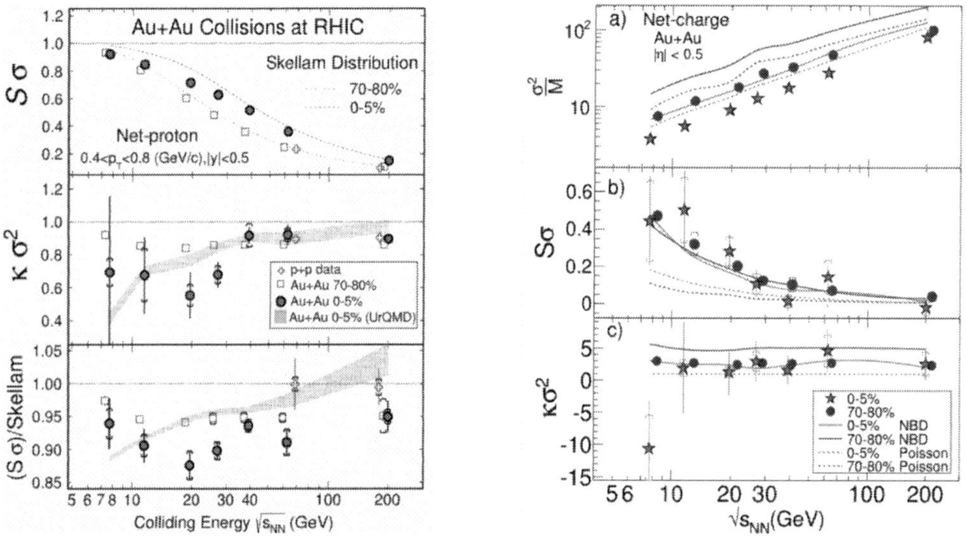

Figure 1. (Color online) Energy dependence of moments of net-proton (left) and net-charge (right) distributions for Au+Au collisions at RHIC BES energies. The statistical and systematical error are shown in bars and brackets, respectively.

We would like to point out: (1) The STAR experiment has carried out moment analysis for net-proton, net-kaon, and net-charge. Different measurements are affected by kinematic cuts, resonance decays, and other dynamical effects differently. In search for the QCD critical point, careful studies are called for. (2) So far, the resonance decay effects are contained in the experimental results of net-proton and net-charge moments. One can estimate the decay effects by theoretical calculations and/or models. Based on the hadron resonance gas model calculation [21], the decay effects for net-proton $\kappa\sigma^2$ is small and at 2% level. But for the net-charge, the decay effects are large. (3) Based on the Delta theorem, statistical error of cumulants ($\Delta(C_n)$) are related to the width of the distribution as $\Delta(C_n) \sim \mathrm{O}(\sigma^n)$ [13,16]. Thus, the wider is the distribution, the larger are statistical errors for the same number of events. That's explains why we got larger statistical errors for net-charge moments than that of net-proton, as the former has much wider distribution. (4) It is predicted that the coupling strength of pions to the critical fluctuation is smaller than protons/anti-protons. Thus, the fluctuation of net-protons will be more sensitive to the critical fluctuations than the fluctuation of net-charges (dominated by pions) [6,22].

In the year 2018, the second phase of the beam energy scan program (BES-II) at RHIC will get started. During the BES-II, we will fine tune the beam energies below 20 GeV and accumulate more events due to the increase of the luminosity of 3-10 times by using stochastic electron cooling technique. Several sub-detector upgrades for the STAR experiment are ongoing and will be ready when BES-II starts. Two of the upgrades are important for moment analysis. The inner TPC (iTPC) upgrade will improve the tracking efficiency and enlarge the TPC acceptance. The Event Plane Detector (EPD) is a forward detector used to determine the event plane for Au+Au collisions. It can provide centrality determination with particles far away from the central collision region.

4. Summary

In this paper, we discussed the analysis techniques used in the moment analysis and the energy dependence of moments of net-proton and net-charge distributions published by the STAR experiment. For the net-charge results, we didn't observe non-monotonic behavior for $S\sigma$ and $\kappa\sigma^2$ within current statistics. For the net-proton results, the most significant deviation of $S\sigma$ and $\kappa\sigma^2$ from Skellam distribution is observed at 19.6 and 27 GeV for 0-5% Au+Au collisions. More statistics are needed at low energies. In principle, one can extend the transverse momentum coverage for proton/anti-proton up to about 2~3 GeV/c with time of flight (ToF) detector at STAR for particle identification. This would allows us to have larger acceptance for protons and anti-protons in the analysis. Large acceptance is crucial for fluctuation of conserved quantities in heavy-ion collisions to probe the signals of the phase transition and the QCD critical point.

Acknowledgement

The work was supported in part by the MoST of China 973-Project No. 2015CB856901, NSFC under grant No. 11205067, 11221504 and 11228513.

References

[1] K. Rajagopal and F. Wilczek. *"At the Frontier of Particle Physics / Handbook of QCD"*, volume 3. World Scientific, 2001.
[2] Z. Fodor, S. Katz, JHEP 0203, **014** (2002); 0404, **050** (2004); R. V. Gavai, S. Gupta, *Phys. Rev. D* **71**, 114014 (2005); **78**, 114503 (2008).
[3] R. V. Gavai, arXiv:1404.6615.
[4] M. M. Aggarwal *et al.* (STAR Collaboration), arXiv: 1007.2613.
[5] M. A. Stephanov, Phys. Rev. Lett. **102**, 032301 (2009).
[6] C. Athanasiou *et al.*, *Phys. Rev. D* **82**, 074008 (2010).
[7] M. A. Stephanov, Phys. Rev. Lett. **107**, 052301 (2011).
[8] S. Gupta, X. Luo, B. Mohanty, H. G. Ritter, N. Xu, Science **332**, 1525 (2011).
[9] R. V. Gavai and S. Gupta, *Phys. Lett. B* **696**, 459 (2011).
[10] Y. Hatta and M. A. Stephanov, Phys. Rev. Lett. **91**, 102003 (2003).
[11] X. Luo (for the STAR Collaboration), J. Phys.: Conf. Ser. **316**, 012003 (2011).
[12] X. Luo, J. Xu, B. Mohanty, N. Xu, J. Phys. G **40**, 105104 (2013); X. Luo, (for the STAR Collaboration), J. Phys.: Conf. Ser. **316**, 012003 (2011).
[13] X. Luo, J. Phys. G: Nucl. Part. Phys. **39**, 025008 (2012).
[14] Adam Bzdak and Volker Koch, Phys. Rev. C **86**, 044904 (2012).
[15] Adam Bzdak and Volker Koch, arXiv:1312.4574.
[16] X. Luo, arXiv:1410.3914.
[17] X. Luo (for the STAR Collaboration), PoS(CPOD 2013)019. [arXiv:1306.3106].
[18] L. Adamczyk *et al.*, (STAR Collaboration), Phys. Rev. Lett. **112**, 032302 (2014).
[19] L. Adamczyk *et al.*, (STAR Collaboration), Phys. Rev. Lett. **113**, 092301 (2014).
[20] X. Luo, B. Mohanty, N. Xu, Nucl. Phys. A 931, 808-813 (2014).
[21] P. Garg *et al.*, Phys. Lett. B **726**, 691 (2013).
[22] Priviate communication with Misha Stephnov.

Time-based Cellular Automaton track finder for the CBM experiment

Valentina Akishina[1,2,3,4] **and Ivan Kisel**[1,2,3]

[1]Goethe University, Frankfurt am Main 60325, Germany
[2]Frankfurt Institute for Advanced Studies, Frankfurt am Main, 60438, Germany
[3]GSI Helmholtzzentrum für Schwerionenforschung, Planckstraße 1, 64291 Darmstadt, Germany
[4]Joint Institute for Nuclear Research, Joliot-Curie 6, 141980 Dubna, Russia

E-mail: v.akishina@gsi.de

Abstract. The future heavy-ion experiment CBM (FAIR/GSI, Darmstadt, Germany) will focus on the measurement of rare probes at interaction rates up to 10 MHz with data flow of up to 1 TB/s. The beam will provide free stream of particles without bunch structure. That requires full online event reconstruction and selection not only in space, but also in time, so-called 4D event building and selection. This is a task of the First-Level Event Selection (FLES) package. The FLES reconstruction and selection package consists of several modules: track finding, track fitting, short-lived particles finding, event building and event selection. The input data are distributed within the FLES farm in a form of so-called time-slices, in which time length is proportional to a compute power of a processing node. A time-slice is reconstructed in parallel between cores within a CPU, thus minimising communication between CPUs. After all tracks of the whole time-slice are found and fitted, they are collected into clusters of tracks originated from common primary vertices. After that short-lived particles are found and the full event building process is finished.

1. Introduction

The CBM experiment [1] will address fundamental questions of strong interaction physics. The experiment will investigate the properties of highly compressed baryonic matter as it is produced in relativistic nucleus-nucleus collisions. The CBM detector is designed to measure both bulk observables with large acceptance and rare diagnostic probes such as charmed particles and vector mesons decaying into lepton pairs.

Since the CBM physics program includes measurement of rare observables, the experiment is required to cope with extremely high interaction rates in order to obtain sufficient statistics. Namely, CBM is being designed to operate with collision rates up to 10 MHz, which was never a case in heavy-ion experiments before. Taking into account that rare probes are to be measured in a heavy-ion collision environment (up to 1000 charged particles per collision) one should expect a data flow rate of 1 TB/s. Such a huge data rate makes it mandatory to select interesting events online with a reduction factor of three orders of magnitude in order to meet a recordable data rate of 1 GB/s. In addition to such high input collision rate and complicated event topology, the full event reconstruction for the event selection is to be done online at the First Level Event Selection (FLES) stage, due to absence of simple hardware triggers. This poses challenges for the CBM computing and requires the algorithms to be both efficient and fast.

On top of that the FAIR accelerator beam will be a continuous stream of particles. As a result data from different collisions may overlap in time, making traditional event-by-event analysis not applicable in CBM. Instead of the event-based approach, a so called time-slice-based reconstruction will be implemented. Such a time-slice, containing not only space coordinates, but also time information from a number of possibly overlapping in time events, will be provided as an input to the reconstruction algorithm to perform a so-called four-dimensional (4D) event building and selection.

This problem is to be solved on a dedicated many-core CPU/GPU computer farm by the FLES package [2]. The package is being developed as a platform and operating system independent package, which includes several modules of the selection chain: track finding, track fitting, short-lived particles finding, event building and event selection. It has to be fast, precise and able to fully utilise potential of modern many-core architectures in order to be suitable for online data processing. The Cellular Automaton (CA) track finder [3] is used to reconstruct tracks of charged particles within a time-slice.

2. Event-based Cellular Automaton track finder

Finding trajectories of charged particles in a typical dense environment of heavy ion collisions is often considered as the most challenging and time-consuming stage of the reconstruction procedure. The reason for that usually is a very specific problem of dealing with combinatorial enumeration, which grows fast with track multiplicity, while grouping together detector measurements into tracks in the presence of noise. The CA track finder proposes a solid solution for the combinatorial search optimisation. Namely, this method benefits from drastic suppression of combinatorial enumeration by introducing a phase of building up short track segments at an early stage before going into the main search. In addition to that, the method is intrinsically local with respect to data processing and, thus, can be run in parallel on modern many-core CPU/GPU computer architectures. The CA method features made the algorithm an appropriate solution for the track reconstruction in the main tracking detector Silicon Tracking System (STS).

In the case of the CBM experiment short track segments at the first stage are triplets: all possible three measurement combinations on a neighboring STS stations are checked to be physical taking into account the multiple scattering. The hit measurement information is locally processed and stored in a new consolidation extent format, namely triplets, with no need to consider individual hits again later in the algorithm. The optimization is also achieved due to saving computational and memory access efforts by combining tasks. For example, at the stage of triplets creation the track finder also checks on a fly the neighboring relation between the triplets in order to remember potential neighbors and estimate the position of the triplet in the track. These links are stored together with triplets and used later at the next stage of connecting triplets in order to build track candidates out of them. The neighboring information allows to get a tree of candidates just by following the neighboring relation. Moving from the first triplet to the next neighboring triplet, the track finder obtains the full tree of possible tracks. In the last stage a competition between the track candidates takes place: only the longest tracks with best χ^2-value sharing no hits in common are to survive.

The task of finding tracks in CBM is split in several iterations to make the reconstruction fast and reliable in the case of a high track density: at the first iteration track finder searches for high-momentum primary tracks only, at the second one — for low-momentum primary tracks, and at the last one — for secondary tracks as well. After each iteration all hits from reconstructed tracks are marked as used and removed from further consideration, thus significantly reducing the combinatorial enumeration.

For the performance purposes we define the tracking efficiency as a ratio of reconstructed and reconstructable tracks. A track is considered as reconstructable if it has at least 4 consecutive

Monte-Carlo points. By definition, a reconstructed track is assigned to a generated particle, if at least 70% of its hits have been produced by the particle. A generated particle is regarded as found, if it has been assigned to at least one reconstructed track. If a particle is found more than once, all additionally reconstructed tracks are regarded as clones. A track is called ghost, if it is not assigned to any generated particle according to the 70% criterion.

The track reconstruction efficiency for different sets of tracks and the level of clones and ghost tracks for the case of event-based analysis are shown in the first (3D) column of Table 1. The majority of tracks of particular physics interest are particles with momentum higher than 1 GeV/c originating from the region very close to the collision point (e.g. decay products of D-mesons, charmonium, light vector mesons). The efficiency for this group is 96.1%. Since the high-momentum secondary particles (e.g. from decays of K_s^0 and Λ and cascade decays of Ξ and Ω) are created far from the primary vertex, their reconstruction is more complex and the efficiency is 76.6%. A significant multiple scattering influence on low-momentum tracks in the material of the detector system and larger curvature in the magnetic field make it a more complicated task to reconstruct them. Thus, their reconstruction efficiency is 79.8%. The levels of clones and ghost tracks are 0.4% and 0.2% respectively.

3. Track finding in the case of extreme track multiplicities

As a special study of the CA track finder stability the algorithm behavior was investigated with respect to the track multiplicity. For the study a super-event, which includes a number of minimum bias events, was reconstructed with no time information taken into account. To create such a super-event we combine space coordinates of hits from a number Au+Au minimum bias events at 25AGeV ignoring such information as event number or time. The super-event is given to the CA track finder as an input and reconstructed as a regular event with no changes in the reconstruction procedure. It is important to mention that such an approach does not precisely correspond to a pile-up simulation, since in this situation fake hits in STS are created from strips on the event level, not within the whole super-event.

Varying the number of minimum bias events in a super-event we have studied the track reconstruction efficiency with respect to the track multiplicity. The efficiency dependence is stable (see Figure 1). In particular, the efficiency of the algorithm decreases by 4% only for the extreme case of 100 minimum bias events in the super-event (see (3+1)D column of Table 1), comparing to the case of single minimum bias events (see 3D column of Table 1). The efficiencies for the reference tracks ($p > 1$ GeV/c) remains high for all track multiplicities range. The efficiencies for extra (100 MeV/c $< p <$ 1 GeV/c) and secondary tracks are also stable. The level of ghost tracks is always beyond 10% and changes slowly.

Summarising, the CA track finder reconstruction algorithm shows high stability with respect to the track multiplicities up to the extreme case of more than 10 000 reconstructed tracks.

4. In-event level parallelism of CA track finder

The newest high-performance architectures tend to have more CPUs, number of physical and logical cores per processor and longer vector registers. This tendency requires reconstruction algorithms to scale with number of cores and vector width. A vectorised sequential version of the CA track finder was taken as a starting point for developing a parallel one. Each step of the algorithm was parallelilized inside a super-event with OpenMP [4] and Pthreads [5] interfaces.

The sequential implementation of algorithms needed to undergo certain changes in order to be run in parallel. Some parts of the code, being essentially sequential, had to be significantly changed. Due to the fact that on modern architectures calculating time is negligible in comparison with data access time, memory optimisation and data structures scheme are essential for parallel programming. One should keep frequently used data in the fast cache memory and thus reduce the time for memory access, since the main memory is slower by about two orders doi:10.1088/1742-6596/599/1/012024

Efficiency, %	3D	(3+1)D	4D
All tracks	83.8	80.4	83
Primary high-p	96.1	94.3	92.8
Primary low-p	79.8	76.2	83.1
Secondary high-p	76.6	65.1	73.2
Secondary low-p	40.9	34.9	36.8
Clone level	0.4	2.5	1.7
Ghost level	0.1	8.2	0.3
Time/event/core	8.2 ms	31.5 ms	8.5 ms

Table 1: Track reconstruction performance for 3D event-by-event analysis, super-event (3+1)D and time-based 4D reconstruction for 100 minimum bias Au+Au collisions at 25 AGeV.

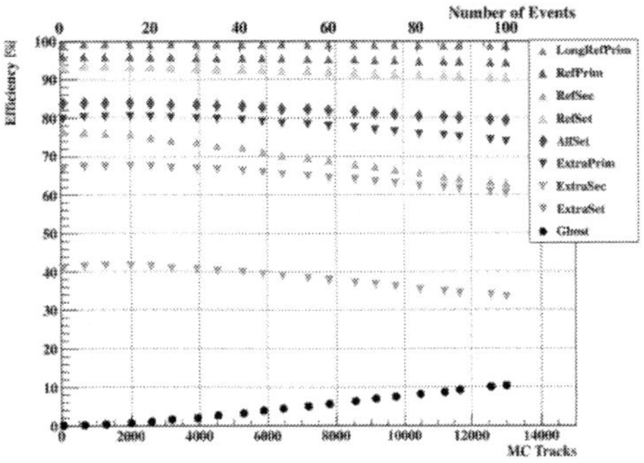

Figure 1: Track reconstruction efficiencies and ghost rate for different sets of tracks versus track multiplicity.

of magnitude than the low-level cache. A lot of optimisation efforts are required for a certain computing architecture in order to benefit from the layered structure of the CPUs caches.

The optimisation and testing of the parallel CA track finder were performed on a server with 4 Intel Xeon E7-4860 processors with 10 physical cores each. The processor supports the Intel Hyper-Threading technology: architecturally, every core consists of two logical ones. Unlike the multiple processor systems with independent processor units, in case of the Hyper-Threading the logical cores share some resources, like caches and the execution engine. As a result a logical core doesn't give 100% of a performance gain, but an estimated gain of 30% instead.

As it was mentioned above, the algorithm consists of several parts. First, a short (2% of the execution time) initialisation takes place, when we prepare the hit information for tracking. The main and the most time consuming part of triplet construction takes about 90% of the sequential execution time. Out of triplets we construct tracks, that takes about 4%, and when we prepare information for the next iteration (3.4%). All steps of the algorithm were parallelised inside a super-event using different sources of parallelism in each step: hits in the initialisation and the final stages, triplets for the major part, track candidates for the track construction step. In order to have enough sources of parallelism to fill a whole CPU, a super-event of 100 minimum bias events was processed. The resulting speed-up factors for different steps as well as for the full algorithm within one CPU (20 hyper-threaded logical cores) are presented in Figure 2.

The algorithm shows a linear scalability. Due to hyper-threading theoretically one can expect the maximum speed-up factor of about 13 on such a CPU. The achieved speed-up factor is 10.6 for the CA track finder reconstruction algorithm on a 10 hyper-threaded physical cores CPU [6].

5. Reconstruction of time-slices

Since resolving different events is a non-trivial task in CBM, an efficient time-based track finder is essential in order to define exact borders of events within a time-slice and grouping tracks into event-corresponding clusters. In order to include time measurement into the algorithm an event start time was assigned to each minimum bias event in a 100 events group during the simulation phase. The start time was simulated with the Poisson distribution, assuming the interaction rate of 10^7 Hz. A time stamp, which we assign to a certain hit, was simulated by this event start time plus a time shift due to the time of flight from the collision point to a detector plane. In order to obtain time measurement we smear a hit time stamp according to the Gaussian distribution with the σ value of the detector resolution 5 ns.

The time measurement information is to be used in the tracking. At the triplet building

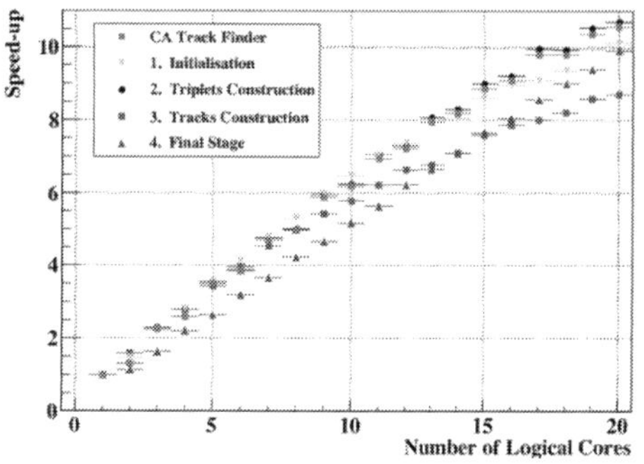

Figure 2: Speed-up factor due to parallelisation for different steps and the full algorithm on Intel Xeon E7-4860 CPU with 10 physical cores and hyper-threading.

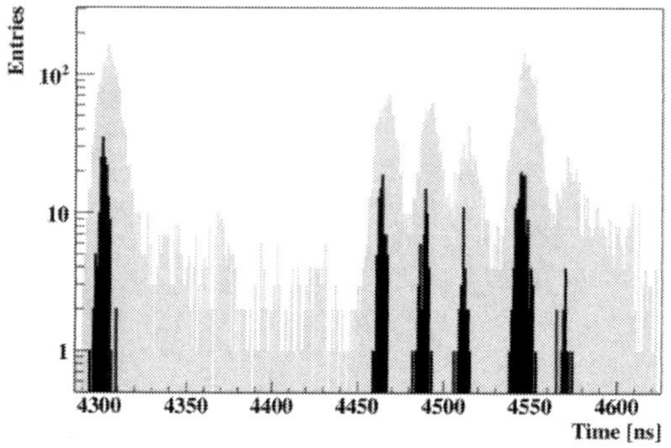

Figure 3: Distribution of time measurement in a part of a time-slice: hit time measurement (light blue), track time (black).

stage triplets are only build out of hits with the same time measurement within 3σ of the detector resolution. It is a justified assumption, since the time of flight between detector planes is negligible in comparison to the detector precision. The same rule is valid in the search for neighboring triplets. Apart from that, the reconstruction is performed in a regular way.

Including the time information in the tracking has resulted in a higher efficiency of the CA track finder (see 4D column in Table 1). In particular the time information drastically decreased ghost and made the reconstruction 3.7 times faster than without the time information ((3+1)D column of Table 1). The speed now is 8.5 ms and comparable with the event-based analysis.

The initial distribution of hits measurements representing the complexity of defining event borders in an overlapping region of a time-slice at the interaction rate of 10^7 Hz is shown Figure 3 with blue color. The resulting distribution of reconstructed track measurements is shown in black. Reconstructed tracks clearly represent event-corresponding groups.

6. Conclusions

The standalone FLES package for the CBM experiment contains all reconstruction stages: track finding, track fitting, short-lived particles finding, event building and event selection. The Cellular Automaton track finder is used to reconstruct tracks of charged particles in a time-slice. The algorithm is parallelized between cores with a speed-up factor of 10.6 on a CPU with 10 physical cores with hyper-threading. Since resolving different events is a non-trivial task in the CBM experiment, the FLES package includes an event building, the process of defining exact borders of events within a time-slice. Thus, the CA track finder is suitable for the task of event building for the CBM experiment.

References

[1] B. Friman, C. Hohne, J. Knoll, S. Leupold, J. Randrup, R. Rapp and P. Senger, Lect. Notes Phys. **814**, 1 (2011)

[2] I. Kisel, I. Kulakov and M. Zyzak, Standalone First Level Event Selection package for the CBM experiment, IEEE Trans. Nucl. Sci. (2013), vol. 60, no. 5, 3703-3708

[3] I. Kisel, "Event reconstruction in the CBM experiment," *Nucl. Instr. and Meth.*, vol. A566, pp. 85–88, 2006

[4] OpenMP API specification for parallel programming, `http://openmp.org/wp`

[5] Pthreads, IEEE. Information Technology-Portable Operating System Interface (POSIX)-Part 1: System Application: Program Interface (API) [C Language]. IEEE/ANSI Std 1003.1, 1996 Edition

[6] I. Kisel, V. Akishina *Online 4-Dimensional Event Building in the CBM Experiment*, RT-2014 - 19th Real-Time Conference (Nara, Japan, May 2014)

Overview of the Silicon Tracking System for the CBM experiment

Pavel Larionov[1], **for the CBM Collaboration**

[1] Goethe University Frankfurt, Germany

E-mail: p.larionov@gsi.de

Abstract. The Compressed Baryonic Matter (CBM) experiment will explore the QCD phase diagram in the region of high net baryonic densities and moderate temperatures. CBM will start it's physics program at the SIS-100 and later at the SIS-300 synchrothrons which will deliver beams for heavy ion collisions at energies up to 45 AGeV with hit rates up to 10 MHz. The Silicon Tracking System (STS) is the core tracking detector of the experiment. It will be used for the reconstruction of the tracks of charged particles and determination of their momenta. For this task eight tracking stations constructed with double-sided sililcon micro-strip sensors will be accomodated downstream the target inside the superconducting magnet. The required momentum resolution for particles with momenta above 1 GeV/c is of the order of $\Delta p/p = 1$ %. To aim this, the material budget should be reduced to the level of 1% X_0. The STS will provide track reconstruction efficiency for fast primary tracks on the level of 95 % and higher. In this paper, the detector concept is summarized along with the detector components and simulation of the system's performance. Also, the quality assurance tests of the silicon sensors in which the author is involved in, are presented.

1. The CBM experiment

The CBM experiment is one of the four large experiments of the new Facility for Anti-proton and Ion Research (FAIR) that will be built at GSI, Darmstadt [1]. CBM is an experiment with a comprehensive physics program about the formation of matter, quark-gluon plasma (QGP), the nature of the transition between QGP and hadronic gas, the search of the critical point, and others. To achieve this, heavy nuclei accelerated in the SIS-100 and later in the SIS-300 synchrothrons to energies from 2 to 14 GeV/nucleon and up to 45 GeV/nucleon, respectievly, will be collided with nuclear targets yielding interaction rates up to 10^7 collisions/sec. The showers of charged particles created in these collisions will flow into the CBM detector systems and will be reconstructed and identified. The experimental challenge is to measure particles and multi-differential observables including multi-strange hyperons and so called rare probes such as short-lived light vector mesons (ρ, ω, ϕ) and charmonium (J/ψ) with unprecedent precision and statistics.

2. The Silicon Tracking System

2.1. Detector concept

The Silicon Tracking System (STS) will be located inside the superconducting dipole magnet comprising 8 tracking stations between 30 cm and 1 m downstream of the taget. The task of STS is to reconstruct the tracks of all charged patricles flying through the detector after the beam-target interactions and to determine their momenta. This requires large detector acceptance to cover rapidities from center of mass to beam rapidity. This is implemented in the STS as a polar aperture from 2.5° to 25° (Fig. 1). According to UrQMD simulations of Au-Au collisions at 25 AGeV, around 1000 charged particles will be created per central interaction and around 800 of them will be captured by the detector's acceptance (Fig. 2). With the double-sided microstrip sensors having strip pitch of 58 μm, the single-hit resolution is identified to be around 25 μm, which will provide the required momentum resolution on the level of $\Delta p/p = 1$ % for particles with momenta above 1 GeV/c. For particles with the same momentum, the track reconstruction algorithms shall enable the trajectory finding with efficiencies exceeding 95 % [2].

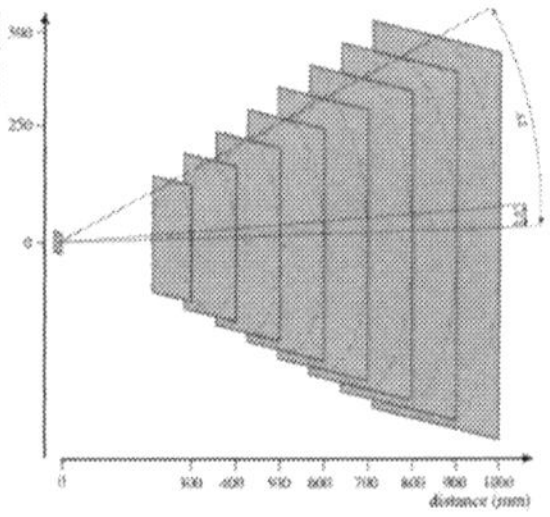

Figure 1: STS stations concept.

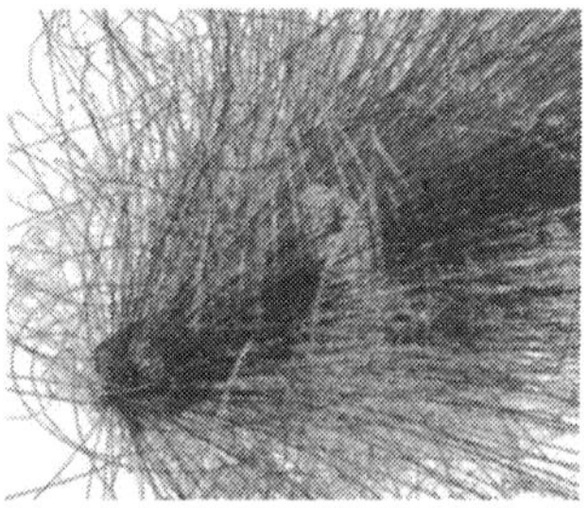

Figure 2: A simulated Au-Au collision at 25 AGeV energy with about 1000 charged particles produced.

In the inner parts of the stations hit rates up to 10 MHz/cm^2 are expected. Sensors of three different sizes have been chosen for different areas of the stations to keep the maximum strip occupancy at the level of a few percent. All sensors will be read-out via ultra-thin microcables by readout electronics outside of the physics aperture, restraining the material budget of about 1% X_0. A high level of radiation damage is expected to impact on the sensors. Beyond the maximum exposure with 1×10^{14} in 1 MeV neutron equivalent that will be reached after several years of running depending on the physics program, the replacement of the sensors is planned.

2.2. Detector components

The combination of either a single silicon sensor or a daisy-chained set of two sensors with attached micro-cables and two readout electronics boards form the *module* (Fig. 3). A module is the smallest assembled functional unit of the STS. Modules of different types will be built due to varying sizes of sensors and different lengths of readout cables. The modules will be arranged on carbon fiber supporting structures, forming ladders, and futher half-stations which will be mechanically and electrically integrated into tracking stations (Fig. 4).

Each station will be built from 300 μm thick double-sided n-type silicon microstrip sensors in three different sizes. The double-sided technology was chosen according to avdanced space-point determination in the same amount of silicon material, in contrast with two back-to-back single-sided sensors having strips implemented only on one side, obeying the low-mass requirement. The sensors have a stereo angle of ± 7.5° between the front and back side strips, providing the optimal spatial resolution in the magnet's bending plane, along the horizontal axis. The neutron equivalent fluence in the detector, expected to be lower than $1 \times 10^{14}\ n_{eq}\ cm^{-2}$ for the harshest running scenario of the experiment at SIS-300, makes double-sided sensor technology applicable for the STS.

Very thin low-mass micro cables will be used to deliver signals from the sensors to the front-end electronics. The design of the cables has been chosen such that the capacitive impact of

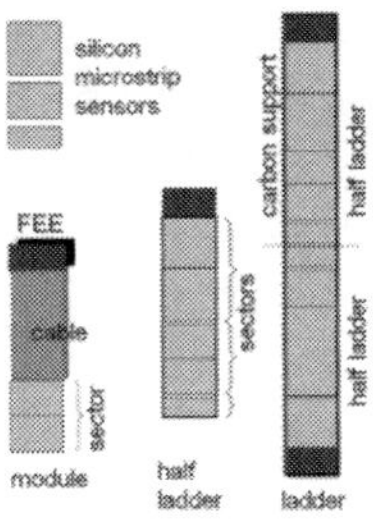

Figure 3: STS building blocks.

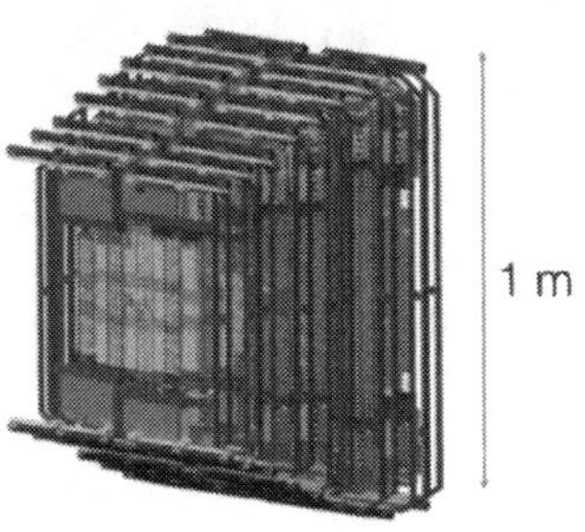

Figure 4: View of the assembled tracking stations.

the cables on the transmittable signal is on the minimum level, achieved by implementing an additional spacer layer between the conducting layers [2].

The read-out electronics of the STS will be based on the custom-designed self-triggering STS-XYTER ASIC. It will provide both timing and amplitude information for the signals transmitted from the sensors. For this, the chip will have two different signal processing chains: fast and slow. The fast path, optimized for the determination of the arrival time of the input charge comprises a fast shaper with the 30 ns peaking time, a dicriminator, a pulse stretcher and a time stamp latch. The slow path, optimized for low noise and accurate energy measurement, consists of a slow shaper with the 80 ns peaking time, a 5-bit flash ADC, and a digital peak detector [3].

2.3. Detector perfomance simulation

In the STS, the reconstruction of charged particles tracks is based on the Cellular Automaton (CA) track finder method using the registered hits.

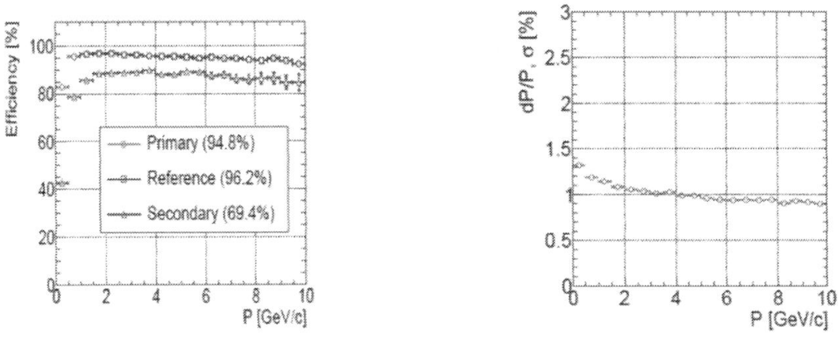

(a) (b)

Figure 5: Track reconstruction efficiency (a) and momentum resolution (b) in the STS as a function of momentum for all tracks in central Au-Au collision at 25 AGeV.

The Kalman Filter (KF) based track fit is used for precise estimation of the track parameters. The detailed information about CA and KF algorithms can be found in [4]. The track finding procedure is organized in several iterations that make the reconstruction fast and reliable in presence of a high track density: the high-momentum primary tracks are reconstructed at the first iteration, then the low-momentum primary tracks and finally all other tracks. The short-lived particles, which decay before reaching the tracking stations or inside the STS, can be reconsructed only via their decay products. The KF particle finder, which is based on the KFParticle package is used in order to find and reconstruct the parameters of short-lived particles by combining already found tracks of the long-lived charged particles. Figure 5 exhibits the track reconstruction efficiency and momentum resolution of charged particles as a function of particle momentum, calculated using the CBMROOT framework with CA and KF packages implemented. The simulated values of these parameters show good agreement with system requirements.

3. Quality Assurance for the silicon sensors of the STS

The STS will comprise 1220 silicon micro-strip sensors. Thus, three different quality assurance centers located at GSI, at the University of Tübingen (Germany), and at Dubna (Russia) are being equipped for quality assurance tests of the silicon sensors, the readout micro-cables and the detector modules after the mass production of the STS components. For the quality assurance of the sensors, a test stand has been set up in the cleanroom environment of GSI's Detector Laboratory with temperature and humidity control. The setup includes a wafer prober yielding a movement precision of 1 μm and number of devices for measurement of passive electrical characteristics of the sensors. The variety of quality tests that are being performed at the moment for STS prototype sensors include visual inspection, current-voltage and capacitive-voltage tests to check the general health of sensors, and detailed tests, in partucular the quality tests of each single strip. The latter requires automatization of the measurement procedure. For this, the author developed a dedicated LabView software which communicates with the wafer prober and allows performing single-strip measurements by stepping over all strips and extracting passive electrical parameters [5]. The manufacting process of double-sided silicon strip sensors is a complicated procedure during which several defects of the single strips are possible to be created. The developed software allows identifying defects of the strip such as

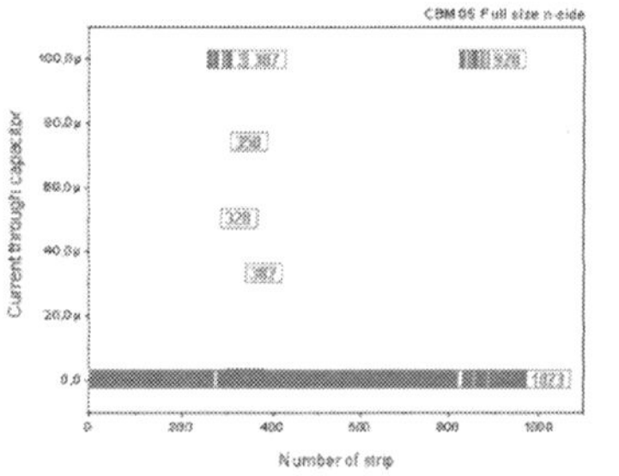

Figure 6: Pinhole test of the CBM05 sensor prototype.

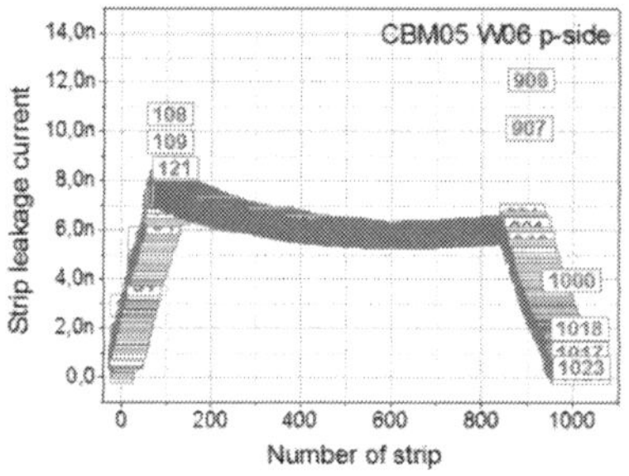

Figure 7: Measurement of the strip leakage current distribution of the CBM05 sesnor prototype.

pinholes, shorts between neighbouring metal strips, metal or implant breaks, and also measuring important parameters such as coupling capacitance and strip leakage current. Figure 6 illustates the result of a pinhole test for the junction side of the a prototype sensor pointing a large amount of pinholes found. The measurement is being done by contacting one by one the strip's DC and AC pads with two needles and measuring the current after a voltage of 20 V is applied. The origin of those defects turned out to come not from the manufacturing process but from a pronounced scratch on the sensor surface. Figure 7 depicts the measured leakage current distribution of each single strip on the junction side of the the prototype sensor. Due to the presence of the stereo angle the strip lengths decrease from the center to the corners. The leakage current is proportional to the strip lengths.

3.1. Long-term stability of sensors

The Silicon sensors of the STS will be operated during long periods of time continuously. Because of that the stability of their operation becomes important. The sensor is considered to be stable if it's leakage current does not deviate from the saturation value in time considering the temperature and humidity to be constant. Measurement of the long-term stability can be performed under laboratory conditions by monitoring the leakage current of the sensor for a substantional period of time. The dedicated LabView software has been developed for leakage current stability tests of STS sensors. It allows monitoring of the leakage currents and correcting them for their temperature and humidity dependence. Temperature and humidity monitoring was implemented in the software as well due to the fact that the leakage current has strong temperature dependence. Results of the long-term stability tests for different sensor prototypes are shown in Fig 8. The behavior of the curves with slight deviation from flatness can be explained by temperature change impact which is shown in detail in Fig 9.

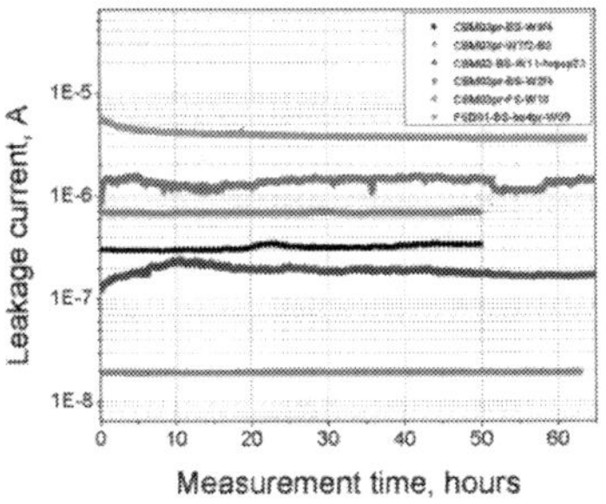

Figure 8: Long-term stabililty test of the CBM sensor prototypes.

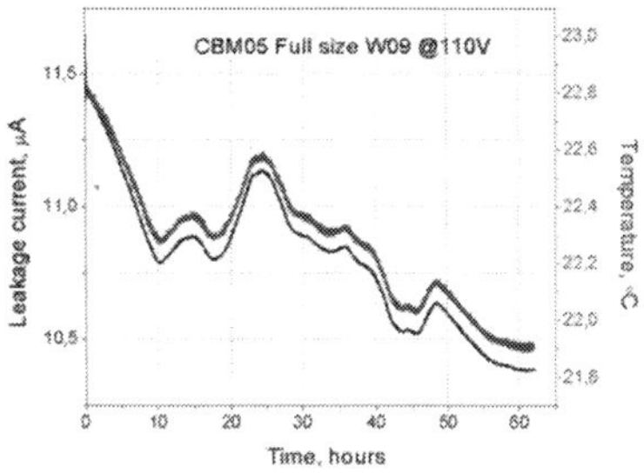

Figure 9: Leakage current and temperature monitoring.

4. Conclusions

This overview of the Silicon Tracking System of the CBM experiment introduced the detector concept and the detector components. Simulation studies of the system perfomance show that the STS is able to meet the requirements for momentum resolution and track reconstruction efficiency. The results of the developed Quality Assurance testing procedures that are being performed to current prototype sensors has been shown. The Technical Design Report of STS has been approved in 2013. The production of STS components starts in 2015.

Acknowledgments

This work is supported by HGS-HIRe, H-QM and HIC for FAIR.

References

[1] FAIR GmbH, http://fair-center.eu.
[2] CBM collaboration, Technical Design report for the CBM Silicon Tracking System, GSI Report 2013-4.
[3] CBM collaboration, CBM Progress Report 2013, http://repository.gsi.de/record/54798.
[4] I. Kisel *et al.*, Standalone First Level Event Selection Package for the CBM Experiment, *IEEE Transactions on Nuclear Science*, vol. **60**, No. 5, October 2013, p. 3703.
[5] P. Larionov, An automated testing procedure for miscrostrip sensors for the CBM Silicon Tracking System, DPG 2014, Frankfurt.

FAIRNESS 2014 IOP Publishing
Journal of Physics: Conference Series **599** (2015) 012026 doi:10.1088/1742-6596/599/1/012026

Verification of Electromagnetic Calorimeter Concept for the HADES spectrometer

O Svoboda[1], C Blume[2], W Czyżycki[3], E Epple[4], L Fabbietti[4], T Galatyuk[5], M Golubeva[6], F Guber[6], S Hlaváč[7], A Ivashkin[6], M Kajetanowic[8], B Kardan[2], W Koenig[9], A Kugler[1], K Lapidus[4], S Linev[9], E Lisowski[3], P Ott[11], P Otte[11], O Petukhov[6], J Pietraszko[9], A Reshetin[6], P Rodríguez-Ramos[1,10], A Rost[5], P Salabura[8], P Skott[11], Y G Sobolev[1], O Steffen[11], A Thomas[11], P Tlustý[1], and M Traxler[9]

[1] Nuclear Physics Institute, Academy of Sciences of the Czech Republic, Řež, 250 68, Czech Republic
[2] Institut für Kernphysik, Goethe-Universität, 60438 Frankfurt, Germany
[3] Institute of Technology Cracow, 31-155 Cracow, Poland
[4] Excellence Cluster 'Origin and Structure of the Universe', Technische Universität München, 85748 Garching, Germany
[5] Technische Universität Darmstadt, 64289 Darmstadt, Germany
[6] Institute for Nuclear Research, Russian Academy of Sciences, 117312 Moscow, Russia
[7] Institute of Physics, Slovak Academy of Sciences, 845 11 Bratislava, Slovakia
[8] Smoluchowski Institute of Physics, Jagiellonian University of Kraków, 30-059 Cracow, Poland
[9] GSI Helmholtzzentrum für Schwerionenforschung GmbH, 64291 Darmstadt, Germany
[10] Faculty of Nuclear Sciences and Physical Engineering, Czech Technical University in Prague, 115 19 Prague, Czech Republic
[11] Institut für Kernphysik, Johannes Gutenberg-Universität, 55128 Mainz, Germany

E-mail: svoboda@ujf.cas.cz

Abstract. The HADES spectrometer currently operating on the beam of SIS18 accelerator in GSI will be moved to a new position in the CBM cave of the future FAIR complex. Electromagnetic calorimeter (ECAL) will enable the HADES@FAIR experiment to measure data on neutral meson production in heavy ion collisions at the energy range of 2-10 A GeV on the beam of the new accelerator SIS100. Calorimeter will be based on 978 massive lead glass modules read out by photomultipliers and a novel front-end electronics. Secondary gamma beam with energies ranging from 81 MeV up to 1399 MeV from MAMI-C Mainz facility was used to verify selected technical solutions. Relative energy resolution was measured using modules with three different types of photomultipliers. Two types of developed front-end electronics as well as energy leakage between neighbouring modules under parallel and declined gamma beams were studied in detail.

1. Introduction – HADES experiment

The HADES detector (High Acceptance DiElectron Spectrometer) is focused on precise measurement of electron-positron pairs coming from the high energy interactions. A wide range of reactions ranging from proton – proton, pion – proton, pion - nucleus up to gold on gold [1] collisions were examined in the past decade. HADES is currently operating on the beamline of SIS18 accelerator at GSI Darmstadt, Germany. The current setup consists of a diamond start detector (START), Ring Imaging Cherenkov detector (RICH), four layers of multiwire drift chambers (MDC), a superconducting toroidal magnet (ILSE), time of flight walls from resistive plate chambers (RPC) and plastic scintillators (TOF), pre-shower detector (SHOWER) and Forward Wall detector [2].

Within the international project FAIR being built next to GSI, a brand new accelerator SIS100 will start to deliver heavy ion beams after 2018. The Hades spectrometer will be moved to a newly build cave shared with the CBM experiment and plans to continue in di-electron mass spectra measurements with beams of accelerated protons up to 10 GeV and heavier ions up to 8 AGeV.

2. Motivation and limits for a build-up of calorimeter

The electromagnetic calorimeter (ECAL) will extend the HADES experimental possibilities in several directions. First of all, the ECAL will enable measurement of gamma photons from various decays of mesons or neutral $\Lambda(1405)$ and $\Sigma(1385)$ resonances. The ECAL will also improve electron/pion separation at large momenta over 400 MeV/c (at lower momenta sufficient electron/hadron identification is provided by the RICH, RPC and TOF detectors).

Basic design of the ECAL was given by the HADES geometry – six separate sectors covering almost full azimuthal angle and polar angle 18° to 45°, more detailed description in [3]. Lead glass was loaned from the OPAL collaboration from the former endcup detector [4], so the size of the modules is fixed. More than 600 pieces of 1.5" EMI 9903KB photomultipliers were obtained from former MIRAC experiment [5].

3. ECAL module

Each module will consist of 92x92x420 mm^3 lead glass crystal. The glass is of CEREN 25 type with density of 4.06 g/cm^3 (weight of a single glass crystal is 14.4 kg). More details on glass properties can be found in [4]. Lead glass will be wrapped in white TYVEK paper 1060B, Cherenkov light will be read out by a photomultiplier (PMT). Beside the available 1.5" EMI PMTs a new 3" Hamamatsu R6091 PMTs are being bought to cover the missing number of pieces. The high price of new 3" PMTs invoked a study of possible use of 1" Hamamatsu R8619 PMTs . Photomultipliers will be shielded against residual magnetic field using a magnetic shielding. Each module will be also equipped with an optical fiber of CERAMOPTEC multimode type, which will guide well defined light pulses for stability and monitoring purposes directly to the glass. The module will be held together by a brass can with a wall thickness of 0.45 mm and an aluminum closing at the end. More information about the module layout can be found in [3].

4. Beam tests

Two beam tests of ECAL modules were already performed in the past. In September 2009, a gamma beam at MAMI Mainz was used to measure energy resolution of ECAL modules equipped with different light collector and reflector configurations. Modules with silver Mylar foil, white paint and Tyvek paper were tested. Light collector from the same lead glass placed between the glass and 1.5" EMI photomultiplier was also tested, but does not resulted in significantly better results. An electron/pion separation power was measured using a secondary e^-/π^- beam from T10 beam line at CERN in May 2010. More details about both two tests can be found in [3] and [6].

4.1. Motivation for a new beam test

Only measurements with cosmics muons and LED monitoring pulses were done since the two beam tests of ECAL modules. To conclude all laboratory tests and verify selected technical solutions, a new gamma beam test at MAMI-C Mainz was proposed. Main tasks of this test were measurement of relative energy resolution of modules with different PMT sizes, test the novel front-end boards, which were not tested with real beam data before, and to study energy leakage between neighboring modules.

4.2. Experiment layout

A new beam test of calorimeter modules and read-out concepts was performed at Institute für Kernphysik, Johannes Gutenberg Universität Mainz. Primary electron beam of 1508.4 MeV was directed to a copper target, energy of produced photons was determined using the method of tagged electrons in dedicated TAGGER [7] detector in MAMI A2 hall. Recoiled electrons from the copper target were bended in a magnetic field of 1.83204T inside the TAGGER and detected by a set of scintillation detectors. The Tagger has 352 detector channels representing photon energies from 1401 MeV down to 81 MeV. For the ECAL measurements only eight channels were selected according to the table 1.

Table 1. Selected TAGGER channels and corresponding photon energies.

Tagger channel	E_γ Low (MeV)	E_γ High (MeV)	E_γ Mean (MeV)	E_γ Bite (MeV)
2	1398.322	1400.327	1399.325	2.005
66	1216.127	1219.411	1217.769	3.284
121	1030.452	1034.414	1032.433	3.962
170	841.208	845.549	843.379	4.341
210	675.710	680.453	678.081	4.743
261	458.898	463.720	461.309	4.821
306	268.440	273.348	270.894	4.907
352	78.992	83.746	81.369	4.754

Four ECAL modules were placed on a movable platform 1m behind the tagger shielding. Gamma beam was shaped using collimators of 2mm diameter placed inside the tagger shielding. The beam spot on the module front surface was approximately 6 mm.

One ECAL module was equipped with 1.5" PMT (same module as in previous beam tests at MAMI and CERN was used, so a direct comparison of results is possible). Two modules were equipped with 3" PMTs and last module contained 1" PMT. High voltage for the photomultipliers was tuned to get the same output amplitude on cosmic muons for all three types of photomultipliers.

4.3. Comparison of different read-out boards

To have an independent ADC measurement, a standard CAEN DT5742 digitizer was used. This CAEN ADC stores 1024 values in a 1 μs window for each of the 16 input channels. Output from the four ECAL modules were read by the first four ADC channels. The 8 trigger signals were put to the last eight ADC channels and also stored to the data. Common trigger signal for starting the read-out was produced as an OR of the individual eight trigger signals, the lowest (and the most frequent) energy was downscaled four times, the second lowest energy was downscaled two times. Active trigger signal representing corresponding gamma energy in each event was determined in the off-line analysis of the last eight ADC channels. Both direct pulses from PMTs and pulses shaped in MA8000 shaper with shaping time of 1 μs (production of GSI Darmstadt) were collected. The Rhode&Schwarz oscilloscope RTO 1044 with histogramming function was used for high resolution pulse storage and amplitude measurement (analog to CAEN ADC device). Stored pulse shapes were compared to pulses collected in measurements with cosmic muons and LED monitoring systems and they were found to be identical. Relative energy resolution of "Cracow" and PaDiWa Amps readout boards was measured and it is comparable in the full energy range, see Figure 1. Energy resolution is slightly better in comparison with the MA8000 + CAEN setup. This was caused by former optimizing of both two readout boards on the PMT pulses.

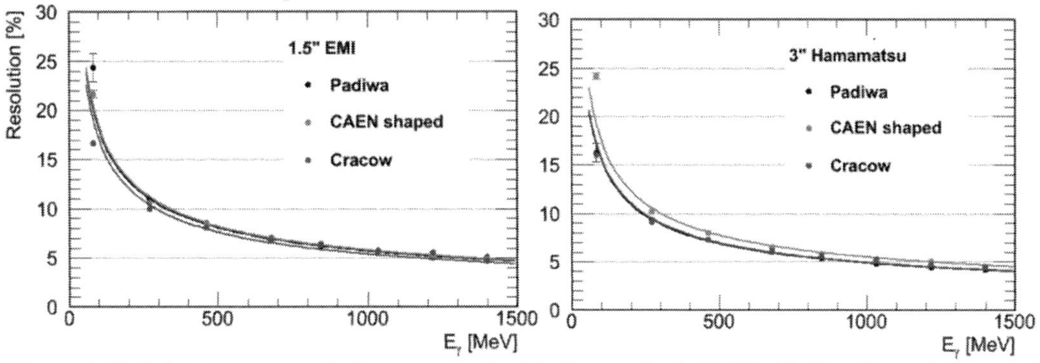

Figure 1. Relative energy resolution measured with the standard CAEN ADC and novel front-end boards "Cracow" and PaDiWa Amps. Data are measured using the module with 1.5" photomultiplier (left) and 3" photomultiplier (right).

4.4. *Relative energy resolution of modules with different photomultipliers*

As already mentioned in the section 3, three types of PMTs with different size were under considerations. Energy resolution with all three types of modules was intensively measured using cosmic muons and LED induced signals in the past ([3] or [6]), but real beam data were missing to make the final decision. Using different gamma beam energies at MAMI Mainz and three different types of pulse processing systems we got a decisive data set.

Module with 1.5" as well as with 3" PMT showed comparable energy resolution (5.8% at 1 GeV photon with 1.5" EMI and 5.5% with 3" Hamamatsu), please see Figure 3 - left. Module with 1" PMT had worse relative energy resolution by ~ 1.6%. Moreover, the 1" PMT has to work at higher input voltage to compensate for lower output amplitudes. At this voltage, the response of the PMT starts to be nonlinear, see Figure 2 left. Dependence of relative energy resolution on high voltage supplied to 1" PMT and measured with shaper MA8000 and CAEN ADC are in Figure 2 - right.

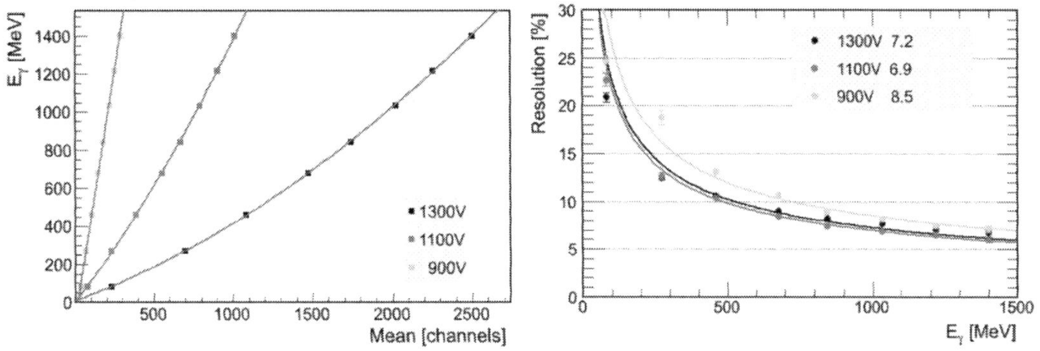

Figure 2. The (non)linear response of 1" Hamamatsu R8619 photomultiplier as a function of high voltage (left). Relative energy resolution of ECAL module equipped with 1" Hamamatsu R8619 photomultiplier as a function of high voltage (right). Numbers behind the names stand for resolution with 1 GeV photons.

4.5. *Energy leakage between the modules*

Two identical modules equipped with 3" PMTs were irradiated with gamma beam parallel with the module longitudinal axis. Three measurements were done with the beam position in the center of one module, 2 cm from the center closer to the second module and 4 cm from the module center closer to the other module (only 0.7 cm from the module border). The same measurement was repeated also with a gamma beam declined by 6° and 12° from the longitudinal module axis. The Figure 3 - right shows only a modest deterioration of the relative energy resolution in dependence on the beam angle.

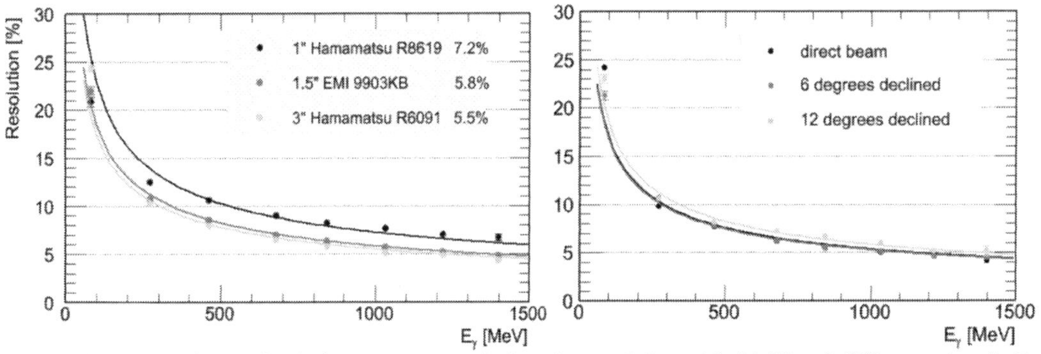

Figure 3. Comparison of relative energy resolution for modules with PMTs of different size (left). Relative energy resolution of the ECAL module equipped with 3" PMT with respect to the various declination of the incident photon (combination of measurements with different angle) (right).

5. Summary

Electromagnetic calorimeter ECAL is being built to enhance experimental possibilities and physics program of the HADES experiment. The ECAL will enable to measure gamma photons coming from various neutral particle decays.

Gamma beam test at MAMI-C Mainz showed that both two novel front-end boards have comparable relative energy resolution and are able to operate under real beam conditions. The ECAL modules equipped with 1.5" and 3" photomultipliers deliver similar relative energy resolution, whereas the module with 1" photomultiplier gives by 1.6% worse energy resolution and non-linear response.

Energy leakage between the neighbour modules was tested with parallel and declined beams. Energy of original photon was successfully recovered as a sum of energies deposited in each of the modules. Some part of the original energy was lost only in the case of parallel gamma beam hitting close to module boarder.

Design of single ECAL module was settled and successfully verified in beam tests. Mass production of modules and final setup of the calorimeter depends on the completion of the CBM cave, which is planned beyond the year 2018.

Acknowledgements

Special thanks belong to the staff of the MAMI facility for providing excellent beam. The ECAL development has been supported by the European Community FP7 – Capacities, contract HadronPhysics3 n° 283286, Czech MSMT LG 12007, GACR 13-067595 and AS CR M100481202.

References

[1] Agakishiev G et al (HADES collaboration), J. Phys.: Conf. Ser. 420 (2013) 012013A
[2] Agakishiev G et al (HADES collaboration), Eur. Phys. J. A 41 (2009) 243
[3] Svoboda O et. al., JINST 9 (2014) C05002
[4] Akrawy M et al, Nucl. Instrum. Meth. A 290 (1990) 76
[5] Awes T C et al., Nucl. Instrum. Meth. A 279 (1989) 479.
[6] Czyzycki W et al, Technical Design Report, arXiv:1109.5550
[7] McGeorge J C et.al., Eur.Phys.J.A37(2008)129-137

Backtracking algorithm for lepton reconstruction with HADES

P. Sellheim[1] for the HADES Collaboration

[1]Goethe-Universität Frankfurt, Max-von-Laue-Str. 1, 60438 Frankfurt am Main, Germany

E-mail: p.sellheim@gsi.de

Abstract. The High Acceptance Di-Electron Spectrometer (HADES) at the GSI Helmholtzzentrum für Schwerionenforschung investigates dilepton and strangeness production in elementary and heavy-ion collisions. In April - May 2012 HADES recorded 7 billion Au+Au events at a beam energy of 1.23 GeV/u with the highest multiplicities measured so far. The track reconstruction and particle identification in the high track density environment are challenging. The most important detector component for lepton identification is the Ring Imaging Cherenkov detector. Its main purpose is the separation of electrons and positrons from large background of charged hadrons produced in heavy-ion collisions. In order to improve lepton identification this backtracking algorithm was developed. In this contribution we will show the results of the algorithm compared to the currently applied method for $e^{+/-}$identification. Efficiency and purity of a reconstructed $e^{+/-}$ sample will be discussed as well.

1. Introduction

Modifications of hadrons at high densities and high temperatures might be accessed in the laboratory by colliding heavy-ions at (ultra-)relativistic energies. A detailed analysis of particle production offers insights to new phases of strongly interacting matter. Dileptons are considered to be an ideal probe of such phases, since the invariant mass of lepton pairs monitors directly an in-medium modified spectral function of hadrons.

In the energy range of SIS18 long-lived fireballs with high net-baryon densities and moderate temperatures can be created [1]. Nuclear resonances like Δ and N^* represent the main source for particle production at SIS18 energies [2]. For description of dilepton production, resonances have to be taken into account since they can decay into a nucleon and a dilepton via Dalitz decays. Because of the unknown form factor of this process, additional theoretical descriptions but also experimental data are needed. One explanation of the particle production mechanism is given by the vector meson dominance model [3]. It states that a virtual photon decay can acquire a hadronic character by decaying into an $e^{+/-}$ pair. This is possible since vector mesons carry the same isopin and parity ($J^P = 1^-$) like photons do. Photons and leptons do not interact strongly with produced matter, therefore they are a good probe for investigation of the dense and hot collision phase without any modification by interaction with the surrounding strongly interacting hadronic matter. Especially the ρ meson is the best probe since it has a short lifetime ($\tau_\rho \approx 1.3fm/c$), for this reason it is more likely to decay into leptons within the dense collision phase ($\tau_{Fireball} \approx 10fm/c$).

Unfortunately, electromagnetic probes are produced during the whole collision evolution, in the first chance collisions (Δ, N^*), in the hot and dense fireball (ρ) and in the freeze out stage

(π^0, η, ω). Furthermore they are very rare probes, e.g. $BR(\rho \to e^+e^-) = 4.72{\cdot}10^{-5}$. Moreover at SIS18 energies vector mesons are produced far below their nucleon-nucleon production threshold. For these reasons all contributions have to be known precisely in order to obtain a reliable measurement of possible in-medium modifications of the ρ meson. Significant excess radiation of dileptons, beyond final state hadron decays was established at the CERN-SPS [4],[5], at the RHIC[6],[7], at the Bevalac/SIS18[8]. From these measurements one could extract a total lifetime of the fireball and an average temperature. To complete the systematics at low beam energies HADES measured dilepton production in Au+Au collisions at 1.23 GeV/u.

2. HADES

The High Acceptance DiElectron Spectrometer (HADES) is located at the GSI Helmholtzzentrum für Schwerionenforschung in Darmstadt [9]. The data analyzed in this work are from the Au+Au campaign measured in 2012. These are the highest multiplicities which where measured with HADES so far, with this aim high granularity detectors with a precise lepton to hadron separation are required.

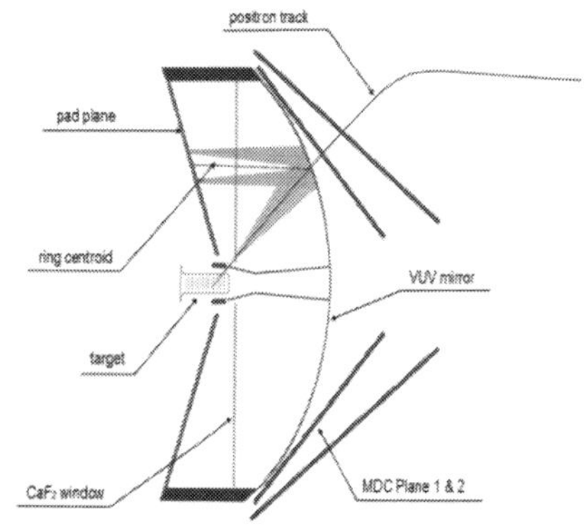

Figure 1. Schematic side view of the RICH detector. A positron track shows usage of angular information to determine a ring center.

The RICH detector [10] (see Fig. 1) is the most important HADES detector component for lepton identification. It consists of a volume filled with C_4F_{10} as radiator gas, in which particles in case they move faster than the speed of light emit Cherenkov radiation in a cone shaped pattern. Photons are reflected by a UV-mirror towards the photon detection plane, that consists of a CsI photon converter and a multiwire proportional chamber operated with pure CH_4. Both gas volumes are separated by a UV transparent CaF_2 window. Due to reflection by the mirror, ring shaped patterns are detected on the pad plane and are identified as good rings by a pattern recognition algorithm. In case their position matches with a reconstructed particle track, hits are combined and defined as lepton candidate.

3. Backtracking

3.1. Motivation

Dileptons are rare probes and their measurement is dominated by hadronic background. On one hand, due to the large number of hadrons, already a small fraction of wrongly identified particles decreases the purity of a lepton sample significantly. On the other hand high lepton reconstruction efficiency is important. The dominant background sources originate from Dalitz decays of π^0 and γ conversion on detector material in the target region. The latter are characterized by small opening angles ($\alpha < 2°$). Another problem arises in the case of overlapping rings where their geometrical shape is distorted and identification of both rings by geometrical means is complicated.

In order to improve lepton identification, compared to the currently applied ring finder, an algorithm with improved efficiency but maintaining a high purity sample is necessary. Identification of tracks with a small opening angle below the current ring resolution helps to identify correlated signal pairs instead of creating background by identifying only one single lepton.

Backtracking is a new approach which uses additional tracking information to improve ring identification efficiency and to enhance capability of rejecting incomplete conversion pairs. In

Journal of Physics: Conference Series **599** (2015) 012027 doi:10.1088/1742-6596/599/1/012027

case a particle candidate fulfills velocity and energy loss criteria that correspond to leptons, the angular information is used to point back at the RICH detector and fix the possible ring center on the pad plane (see Fig. 1). By fixing the ring center position, requirements for good rings can be reduced to enhance identification efficiency. Even stronger improvements can be obtained by applying this method for very close tracks. Since the angular track resolution is better than 2° while the ring resolution is about 4°, tracking information offers the possibility to use the expectation values of close pairs to search for overlapping rings.

3.2. Method

Backtracking procedures from assigning the ring expectation to interpreting the measurement are explained below. In the beginning of the procedure, the angular information from track candidates has to be transformed by geometrical mapping in order to match with the photon positions on the RICH pad plane. Afterwards the ring has to be parametrized to define the radius and the width of its photon distribution. Fig. 2 shows that ring attributes are not constant for different ring sections. Moreover they are not constant over the polar and azimuthal angles of the detector coordinates. Additionally, rings are correlated with the position of the particle emission vertex. Therefore functions depending on these four parameters are necessary to describe the ring shape and to transform it to discrete RICH detectors pads. The number of measured photons per ring is the main indicator of a good ring. However the amount of measured clusters do not provide a direct measurement of the photon number because close photons can merge to a single cluster. Additionally, the noise disturbs the signal, and it can also form clusters. An identification of this noise would lead to a more pure lepton identification. For the case of close pairs it is also necessary to determine how much rings are overlapping. As a basic observable, number of clusters contributing to a ring is used. Additionally, local maxima of clusters are searched for and counted. Furthermore, pads and charge of clusters or of pads directly matching with the ring prediction area are summed up. The fit of the two dimensional charge distribution with a Gaussian function allows a more precise maximum determination. By calculating the distances between the predicted point of the most probable hit and the maximum position, a ring quality can be extracted. The comparison of the fired pads matching with prediction area and fired pads in a 16x16 matrix around the ring center gives a ring matrix observable. For close pairs, shared maxima information is provided.

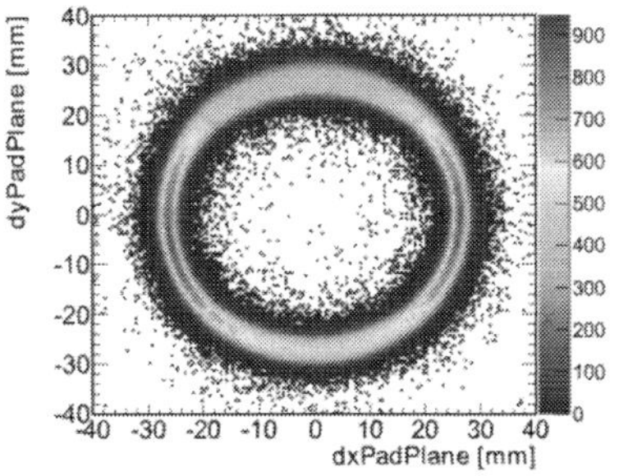

Figure 2. Photon hit distribution in the RICH pad plane for fixed ring center at (0,0).

Searching for local maxima is the best approach for a direct measurement of the photon number and it is a good estimation of signal quality. Using pads and charge information of rings a large difference between sum of directly matching pads and sum of clusters can be used as an indicator for noise or overlapping rings. Random noise hits do not match well with the ring expectation area, for which reason rings created by noise hits have worse ring quality values. Using the information of the number of shared maxima allows identification of close pairs and exposure of hadronic tracks picking up charge from leptonic tracks. Finally a set of observables is provided for ring identification, noise rejection and close pair analysis.

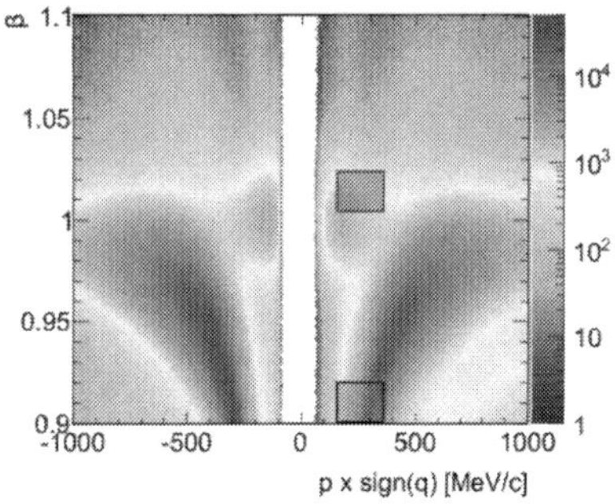

Figure 3. Momentum versus velocity distribution shows lepton separation by usage of time-of-flight information. Training sample selection is indicated by shaded areas (upper one is for signal, lower one is for background).

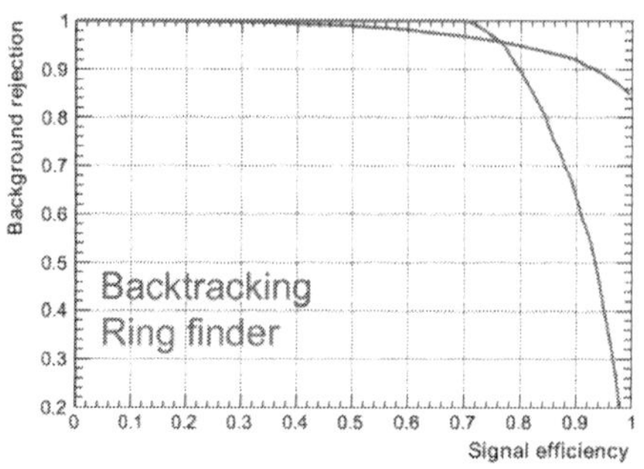

Figure 4. Signal efficiency as function of background rejection for backtracking and currently applied ring finder algorithm.

4. Application

4.1. Single lepton identification

The HADES sub detectors provide a set of observables for particle identification. Information provided by MDC, ToF and PreShower detectors are velocity, momentum, specific energy loss, track quality and matching quality between MDC and RPC or TOF detectors. Velocity measured with ToF detectors offer a high purity lepton identification up to momenta of 350 MeV/c. But above this momentum threshold it is essential to apply additional backtracking information for efficient hadron suppression (see Fig. 3).

From the large set of particle observables of sub detectors and especially of backtracking only one- or two-dimensional correlations can be considered. To include multidimensional correlations a multivariate data analysis is performed [11]. Before this method is applicable to data the algorithm has to be trained with a signal and a background sample. Therefore the momentum and velocity selection, as shown in Fig. 3 is used to define a signal sample with $e^{+/-}$ and background sample with $\pi^{+/-}$. The resulting efficiencies for given background rejection are shown in Fig. 4. A comparison of the backtracking algorithm with a similar analysis using RICH ring finder information shows that backtracking achieves similar background rejection but is able to achieve much higher efficiencies by maintaining slightly smaller background rejection.

4.2. Lepton pairs

To obtain the dilepton invariant mass spectrum, single leptons have to be combined into pairs. However, decays from the hot and dense stage are not the only leptonic particle source. Additionally, electrons and positrons decaying from π^0 or η contribute. Since the origin of the decay particles can not be identified a priori all possible pair combinations have to be introduced. The uncorrelated lepton pairs represent combinatorial background (CB). The CB can be determined by using the same-event like-sign method. To do so like-sign pairs (N_{++}, N_{--}) from the same event are combined and used for calculation of the geometrical mean $CB_{geom} = 2\sqrt{N_{++}N_{--}}$. The signal of correlated pairs results from subtraction of the calculated background distribution from the unlike-sign spectrum.

Higher lepton reconstruction efficiencies of backtracking increase the probability of detecting incomplete lepton partners. Additional CB rejection power is achieved when applying an opening angle cut. Changes after application of the close pair rejection are shown in Fig. 5. Here we can see that the combinatorial background from the analysis with backtracking is reduced by up to a factor of 4.

5. Conclusion

In this work we presented a new method for lepton identification with HADES. Tracking information from drift chambers is used to reduce the degrees of freedom for the ring search. The general strategy as well as output observables of the algorithm were presented. A comparison with the currently used RICH ring finder showed an enhanced efficiency for lepton reconstruction by similar purity of the lepton sample. Furthermore, pair observables of backtracking were used and showed improvements for rejection of close pairs. Finally, a successful application of backtracking algorithm was shown. The CB can be reduced by up to an additional factor of 4. It will allow for detailed multi-differential (p_T, angular distribution) analysis of the dilepton invariant mass spectrum.

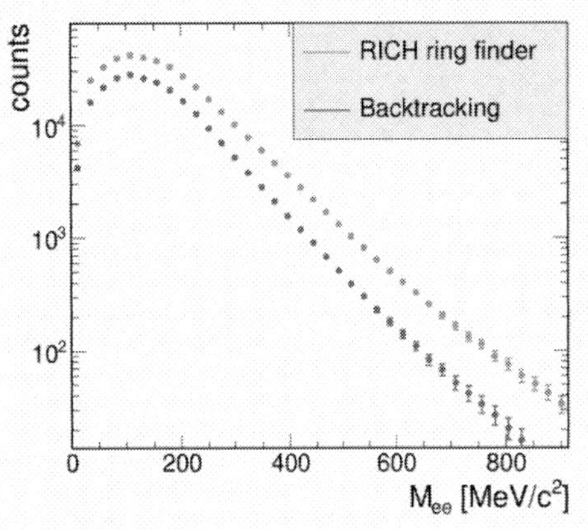

Figure 5. CB distribution for currently applied ring finder (yellow) and backtracking (blue). Close pair rejection is applied.

Acknowledgements

The collaboration gratefully acknowledges the following funding: INFN-LNS Catania (Italy); LIP Coimbra (Portugal): PTDC/FIS/113339/2009; SIP JUC Cracow (Poland): 2013/10/M/ST2/00042 and NN202198639; GSI Darmstadt (Germany): Helmholtz Alliance HA216/EMMI; TU Darmstadt (Germany): VH-NG-823, Helmholtz Alliance HA216/EMMI; HZDR, Dresden (Germany): 283286, 05P12CRGHE; Goethe-University, Frankfurt (Germany): Helmholtz Alliance HA216/EMMI, HIC for FAIR (LOEWE), GSI F&E, BMBF 06FY9100I; TU Muenchen, Garching (Germany): BMBF 06MT7180; JLU Giessen (Germany): BMBF:05P12RGGHM; University Cyprus, Nicosia (Cyprus): UCY/3411-23100; IPN Orsay, Orsay Cedex (France): CNRS/IN2P3; NPI AS CR, Rez, (Czech Republic): MSMT LG 12007, GACR 13-06759S.

References

[1] Y. B. Ivanov, V. N. Russkikh, V. D. Toneev, Phys. Rev. C 73 (2006) 044904.
[2] R. Rapp, J. Wambach, Adv.Nucl.Phys. 25 (2000) 1–205.
[3] J. J. Sakurai, Annals Phys. 11 11 (1960) 1–48.
[4] I. Tserruya, in: R. Stock (Ed.), Relativistic Heavy Ion Physics, Vol. 23 of Landolt-Boernstein - Group I Elementary Particles, Nuclei and Atoms, Springer Berlin Heidelberg, 2010, pp. 176–207.
[5] H. J. Specht, AIP Conf.Proc. 1322 (2010) 1–10.
[6] F. Geurts, Nuclear Physics A 904905 (0) (2013) 217c – 224c.
[7] P. Huck, arXiv:1409.5675 [nucl-ex].
[8] G. Agakishiev, et al., Phys.Rev. C84 (2011) 014902.
[9] G. Agakishiev, et al., Eur.Phys.J.A 41 (2009) 243–277.
[10] K. Zeitelhack, et al., Instr. Meth. A 433 (1999) 201–206.
[11] A. Hoecker, P. Speckmayer, J. Stelzer, J. Therhaag, E. von Toerne, H. Voss, PoS ACAT (2007) 040.

Protons and light fragments in Ar+KCl at 1.76 AGeV measured with HADES

Heidi Schuldes[1], **Manuel Lorenz**[1,2] **(for the HADES-Collaboration)**
[1]Goethe-University Frankfurt, [2]Utrecht University
E-mail: h.schuldes@gsi.de

Abstract. We present transverse momentum spectra, rapidity distributions and multiplicities of protons, deuterons and tritons measured with the High Acceptance DiElectron Spectrometer HADES in the reaction Ar(1.76A GeV)+KCl. This completes the HADES data set measured in this reaction, comprising dielectrons and various lighter hadrons.

1. Introduction

The investigation of rare and penetrating probes in heavy-ion collisions (HIC) has attracted much interest in the recent years over all accessible collision energies. The ultimate hope is to find key observables sensitive to the onset of deconfinement [1, 2, 3], the critical point [4, 5] and chiral symmetry restoration [6, 7]. In practice it turned out that in order to extract quantitative and most of the time also qualitative results the observed production rates or phase-space distributions have to be compared to phenomenological models. To do so these models have to describe the global event characteristics first. Unfortunately, data on these global characteristics like rapidity distributions and transverse momentum spectra of e.g. protons can be rare at some energies/systems. At SIS energies valuable systematics have been collected and published by the FOPI collaboration in [8] mostly for central Au+Au collisions. Although this data covers the same energy range and include some variations of the size of the collision system, the extrapolation to semi-central collisions of Ar+KCl measured by HADES at a kinetic beam energy of 1.76 AGeV exhibit large uncertainties. Recently, the HADES collaboration had a Au+Au beamtime at 1.23 AGeV incident kinetic energy. To understand the data from the heavier system, it is important to investigate as many observables as possible in the lighter systems.

Apart from the dielectron yield [9] HADES has published hadron yields from the Ar+KCl data for this energy regime [9]. The aim of this paper is to complete the published data set by adding transverse momentum spectra, rapidity distributions and multiplicities of protons, deuterons and tritons.

2. Experiment and Analysis

HADES is a charged-particle detector consisting of a 6-coil toroidal magnet centered on the beam axis and six identical detection sections located between the coils. A detailed description of HADES is given in [10]. For the presented data an Argon beam of $\sim 10^6$ particles/s was incident with a beam energy of 1.76 AGeV on a four-fold segmented KCl target with a total thickness corresponding to 3.3 % interaction probability. A fast diamond start detector located upstream of the target was intercepting the beam and was used to determine the time-zero information. The data readout was started by a first-level trigger (LVL1) requiring a charged-particle multiplicity ≥ 16, in the TOF/TOFino detectors. Based on a full GEANT simulation, employing realistic detector responses of Ar+KCl events generated with the UrQMD transport model [11], we found that the event ensemble selected by this (LVL1) trigger condition has a mean number of participating nucleons ($\langle A_{part} \rangle$) equal to 38.5 ± 3.9.

The particle identification was done by a velocity vs. momentum × polarity correlation, where the velocity was determined by the time-of-flight measurement in the TOF and TOFino scintillators with respect to the time-zero information and the tracked flight path. If needed, additional particle discrimination power was gained from the energy loss (dE/dx) information in the MDC and scintillators. This is illustrated in Fig. 1 where the mass distribution in the TOF-region is plotted after application of track quality selections (dark blue) and after an additional selection on e.g. deuterons making use of the energy loss information (light blue).

The yields of the different particles are assessed as a function of rapidity y and transverse

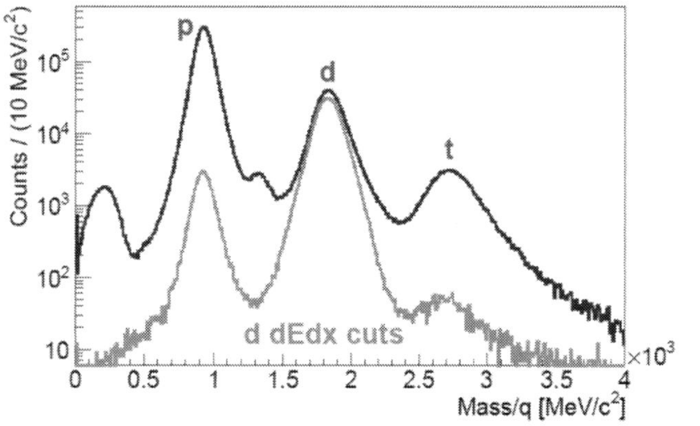

Figure 1: Mass distribution in the TOF-region after application of track quality selections (dark blue) and after additional selection on deuterons making use of the energy loss information (light blue).

mass $m_t - m_0$. Efficiency and acceptance losses are corrected for a given y and $m_t - m_0$ region iteratively using simulated data embedded into real events. In a first step the momentum distribution of the simulated particles was assumed with a white distribution, while finally with a distribution very similar to the experimentally observed assuming a thermal source of a temperature T=80 MeV and a constant radial expansion velocity β_r of 0.4. We find however only little sensitivity of the input on the correction factor (at the order of 1% in the overall correction per given region). The combined correction factor for losses due to efficiency and acceptance as a function of $m_t - m_0$ for the y region $0.4 \leq y \leq 0.45$ is displayed in Fig. 2 separately for p, d and t. For further details on particle identification and the data analysis of charged hadrons we refer to [12, 13, 14].

3. Phase-space Distributions

The resulting transverse mass spectra are displayed in Fig. 4. In order to extrapolate to the unmeasured regions the spectra are described using the following function according to Siemens-Rasmussen:

$$\frac{dN}{2\pi m_t dm_t dy} = c \cdot E \cdot \exp\left(-\frac{\gamma \cdot E}{T}\right) \cdot \frac{\sinh\alpha}{\alpha} - \frac{T}{E} \cdot \cosh\alpha, \tag{1}$$

with $\alpha = (\gamma \cdot \beta_r \cdot p)/T$ and $\gamma = 1/\sqrt{1-\beta_r^2}$. T, β_r and the constant C are free parameters. T can be associated with the temperature of a thermal source and β_r with its radial expansion

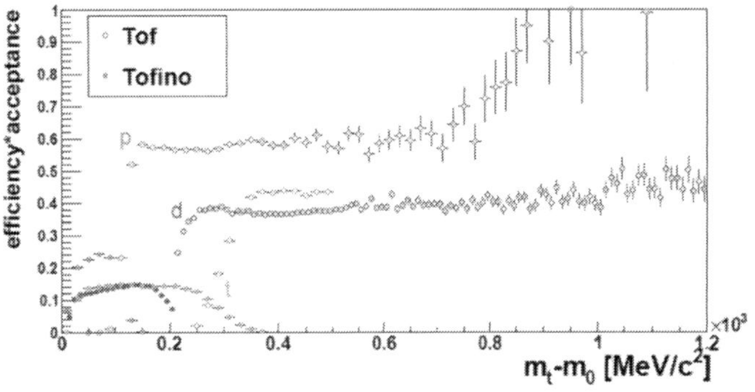

Figure 2: Combined correction factor for losses due to efficiency and acceptance as a function of $m_t - m_0$ for $0.4 < y < 0.45$ for both Tof-systems TOFino and TOF and the different particles p, d and t.

velocity. In this picture T and β_r should not depend on y. Hence a χ^2 scan in the $T - \beta_r$-plane is done globally for all three particles. The χ^2/dof distribution for the global scan is displayed in Fig. 3. As the values of χ^2/dof indicate, some spectra cannot be reproduced based on the scenario of a radially flowing thermal source.

To obtain rapidity distributions we integrate the measured data points and use the parameters of the global scan (the corresponding Siemens-Rasmussen fits are displayed together with the spectra in Fig. 4) in order to extrapolate down to small m_t and up to infinity. The resulting yields as a function of y are displayed in Fig. 5, together with the yields from integrating the distribution within the measured limits. This allows to judge on the systematic error introduced by the extrapolation down to small transverse mass.

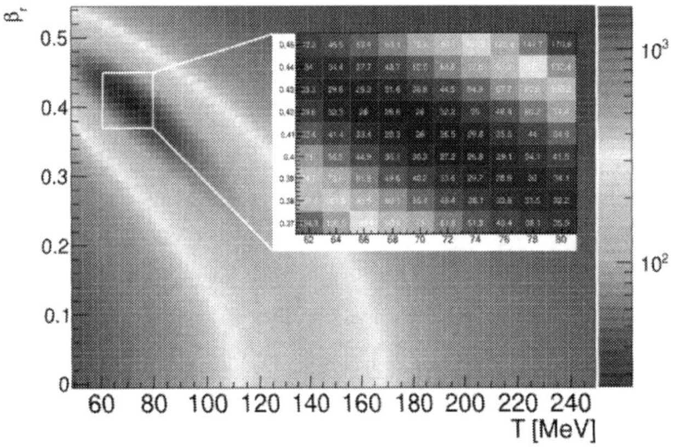

Figure 3: χ^2/dof distribution in the $T-\beta_r$-plane from fitting p, d and t transverse mass spectra simultaneously with Eq. 1 (see text).

4. Discussion and Conclusion

In semi-central Ar+KCl collisions at 1.76 AGeV incident kinetic energy, the rapidity distributions of p, d and t exhibit distinct local maxima close to target/projectile rapidity. The location of the maximum differential count rate dN/dy moves closer to target/projectile rapidity, respectively, the heavier the mass of the particle is, e.g. comparing protons with tritons. Also, heavier particles are more abundant close to target/projectile rapidity than at midrapidity.

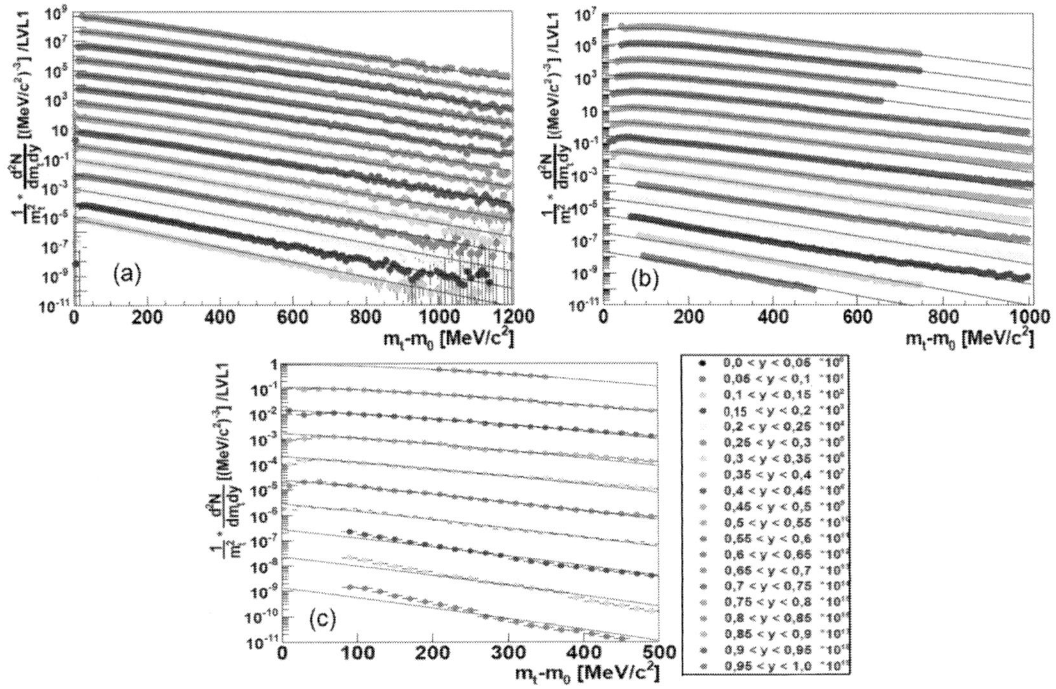

Figure 4: Efficiency and acceptance corrected transverse mass spectra of protons (a) (rapidity range $0.1 \leq y \leq 1$), deuterons (b) (rapidity range $0.05 \leq y \leq 0.8$) and tritons (c) (rapidity range $0.05 \leq y \leq 0.55$) scaled with exponentials. Black lines correspond to Siemens-Rasmussen-functions with fixed parameters for temperature and radial flow (see text).

This experimental observation is consistent with a scenario, where heavier particles do mainly origin from target/projectile fragmentation in this reaction system. However, the position of the two peaks are slightly shifted away from target/projectile rapidity which might be explained by a convolution of a thermal like source sitting at midrapidity and sources at target/projectile rapidity. In addition, this effect might be enhanced due to the low p_t-acceptance-cutoff acting stronger on the target/projectile-source then on the thermal one. It is interesting to note that indeed, the spectral distribution can be simultaneously parametrized assuming a thermal source at midrapidity with T=70±7 MeV, which radially expands with β_r=0.41±0.03. these numbers are compatible with results from other analyses in the SIS18 energy range and emphasize the role of (radial) flow on the emission of particles in this energy regime.

The analysis, together with the broad set of published particle yields demonstrate the outstanding particle identification capabilities of the HADES detector, even for heavy fragments. After a detector upgrade, where for example the TOFino szintillator wall was replaced by a resistive plate chamber and a forward wall was installed, the collaboration recorded 7.3 x 10^9 events of Au+Au collisions at 1.23 AGeV. With this improved setup refined data of particle production at SIS18 energies, covering fragment emission as well, will be soon available [15, 16].
We gratefully acknowledge the HADES-Collaboration, GSI, HIC for FAIR and HGS-HIRe.

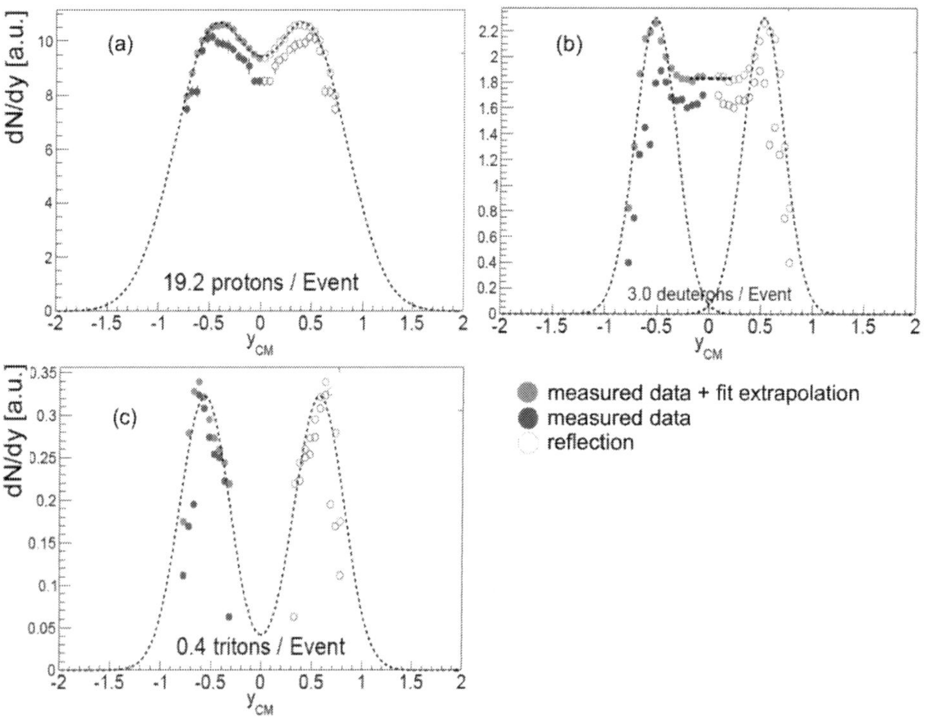

Figure 5: dN/dy-distribution of protons (a), deuterons (b) and tritons (c) from integrating the distribution within the measured limits (blue) and extrapolated with Siemens-Rasmussen-functions (red). Dashed lines correspond to two gaussian (plus polynomial for d) parametrizations to estimate the total yield of the particles.

References

[1] E.V.Shuryak, Phys.Lett.B**78** (1978) 150 [Sov.J.Nucl.Phys.**28** (1978) 408] [Yad.Fiz.**28** (1978) 796].
[2] T. Matsui and H. Satz, Phys. Lett. B **178** (1986) 416.
[3] J. Rafelski and B. Muller, Phys. Rev. Lett. **48** (1982) 1066 [Erratum-ibid. **56** (1986) 2334].
[4] Z. Fodor and S. D. Katz, JHEP **0404** (2004) 050 [hep-lat/0402006].
[5] M. A. Stephanov, K. Rajagopal and E. V. Shuryak, Phys. Rev. D **60** (1999) 114028 [hep-ph/9903292].
[6] S. Weinberg, Phys. Rev. **166** (1968) 1568.
[7] R. Rapp and J. Wambach, Adv. Nucl. Phys. **25** (2000) 1 [hep-ph/9909229].
[8] W. Reisdorf *et al.* (FOPI Collaboration), Nucl. Phys. A **876** 1 (2012).
[9] G. Agakishiev *et al.* (HADES Collaboration), Phys. Rev. C **84**, 014902 (2011); Phys. Rev. C **80**, 025209 (2009); Phys. Rev. C **82**, 044907 (2010); Phys. Rev. Lett. **103**, 132301 (2009); Eur. Phys. J. A **47** 21 (2011).
[10] G. Agakishiev *et al.* (HADES Collaboration), Eur. Phys. J. A **41**, 243 (2009).
[11] S.A. Bass *et al.*, Prog. Part. Nucl. Phys. **41** 225 (1998).
[12] A. Schmah, PhD thesis, Technical University Darmstadt, Darmstadt (2008).
[13] M. Lorenz, PhD thesis, Goethe-Universität, Frankfurt (2012).
[14] H. Schuldes, Master thesis, Goethe-Universität, Frankfurt (2012).
[15] M. Lorenz (HADES Collaboration), NUPHA 20167 (2014).
[16] T. Galatyuk (HADES Collaboration), NUPHA 20289 (2014).

FAIRNESS 2014 IOP Publishing
Journal of Physics: Conference Series **599** (2015) 012029 doi:10.1088/1742-6596/599/1/012029

Reconstruction of π^0 and η mesons via conversion method in Au+Au at 1.23 GeV/u with HADES

C Behnke[1] for the HADES Collaboration

[1]Goethe Universität Frankfurt, Max-von-Laue-Str. 1, 60438 Frankfurt am Main

E-mail: C.Behnke@gsi.de

Abstract. Lepton pairs emerging from decays of virtual photons represent promising probes of matter under extreme conditions. In the energy domain of 1 - 2 GeV per nucleon, the HADES experiment at GSI Helmholtzzentrum fuer Schwerionenforschung in Darmstadt studies di-electrons and strangeness production in various reactions, i.e. collisions of pions, protons, deuterons, heavy-ions, and nuceous with nuclei. An accurate determination of the medium radiation depends on a precise knowledge of the underlying hadronic cocktail composed of various sources contributing to the net spectra. Therefore, a measurement of the neutral meson yields together with the dileptons is crucial. In this contribution, the capability of HADES to detect e^+e^- pairs from conversion of real photons will be demonstrated. We will present results from a two-photon analysis of Au+Au collisions at 1.23 GeV/u providing information on neutral π^0 and η mesons.

1. Introduction

Nuclear matter under extreme conditions can be created via heavy-ion collisions in the laboratory. At SIS 18 energies three possible stages of such a collision (see Fig. (1)), the first chance collisions, the hot and dense stage and the freeze-out, are assumed. Direct photons (γ) and dileptons (e^+e^-) are penetrating all these stages without strong interactions. Therefore lepton pairs, are an ideal probe to test the properties of the hot and dense stage of the collision. Furthermore it is necessary to understand the lepton pair production mechanisms in the other stages. Measurements of reference spectrum from elementary reactions, i.e. pp and np, are used to estimate the contribution of the first chance collisions [1, 2, 3]. Mesons with a long lifetime (long compared to the lifetime of the fireball), i.e. the neutral π^0 and η, are the dominant contribution to the dilepton invariant mass spectra from the freeze-out stage. For the normalization of dilepton invariant mass spectra, the understanding the π^0 is crucial. The cross section of the η is essential to determine the non trivial enhancement of low-mass ($M_{e^+e^-}$ between 0.15 GeV/c^2 and 0.55 GeV/c^2) lepton pairs, that was found by DLS [4] and HADES [5] at Bevalac/SIS18, CERES [6] and NA60 [7] at CERN, and STAR[8] and Phenix [9] at RHIC. This contribution will focus on the reconstruction of neutral π^0 and η with the HADES detector in Au+Au at 1.23 GeV/u.

Figure 1: Scematical view of a 3 possible stages of a heavy-ion collision. Dileptons are indicated in all stages.

Journal of Physics: Conference Series **599** (2015) 012029 doi:10.1088/1742-6596/599/1/012029

2. Analysis strategy

The contribution of the neutral mesons to the dilepton spectra is given by their Dalitz decays (meson $\rightarrow \gamma e^+e^-$). The detection of the γ from Dalitz decay would give opportunity to fully reconstruct the meson yield. Besides the Dalitz decay, π^0 and η mesons have dominant decay channel into two photons (meson $\rightarrow \gamma\gamma$), that also could be studied.

2.1. HADES spectrometer

The High Acceptance DiElectron Spectrometer (HADES) is installed at the SIS 18 (GSI Helmholtzzentrum für Schwerionenforschung, Darmstadt)[10] . It consists of 6 identical sectors that cover a full azimuthal angle and polar angle from 18° to 85°. The tracking is performed with 4 x 6 Multiwire Drift Chambers and a superconducting toroidal magnet. For the time-of-flight measurement the ToF detector and Resistive Plate Chamber walls are used. HADES measured the collision system Au+Au at the highest (achievable at SIS 18) beam energies, E_{kin} = 1.23 GeV/u in April/May 2012. In total 7.3 x 10^9 events, that corresponds to 140 TByte of data have been collected. The trigger on hit multiplicity in ToF_{Mult} =>20 (PT3) corresponds an impact parameter $b_{max} \approx 10$ fm.

2.2. Conversion Probability

Since HADES has no photon detector yet, the measurement of the electromagnetic decays of π^0 and η is only possible via external conversion of photons in detector material. The conversion probability can be extracted with the help of the simulation tool GEANT3. In this work the conversion probability is defined as a reconstructible (which means traversing the HADES acceptance) dilepton coming from a γ which originate from a π^0. The differential cross section for such process depends on the atomic number Z of the material in which the interaction occurs. In a compound material the element i in which the interaction occurs is chosen randomly according to the probability:

$$Prob(Z_i, E_\gamma) = \frac{n_{ati}\sigma(Z_i, E_\gamma)}{\sum_i[n_{ati} \cdot \sigma_i(E_\gamma)]}, \quad (1)$$

where Z_i stands for the atomic number of the material, n_{ati} is the number of atoms per volume of the ith element and E_γ the energy of the photon. In Fig. 2a the conversion probability inside the HADES as a function of the photon energy and the polar Θ angle is shown. Since the target is segmented into 15 vertically aligned gold discs (r =1.2 mm, thickness = 0.25mm), a larger polar angle corresponds to a longer flightpath trough target material. The dependence of Eq. 1 is reflected here. In Fig. 2b the conversion probability for the main inner detector parts is presented. The systematic errors are in the order of 5% due to uncertainties in the material budget, for details see [11].

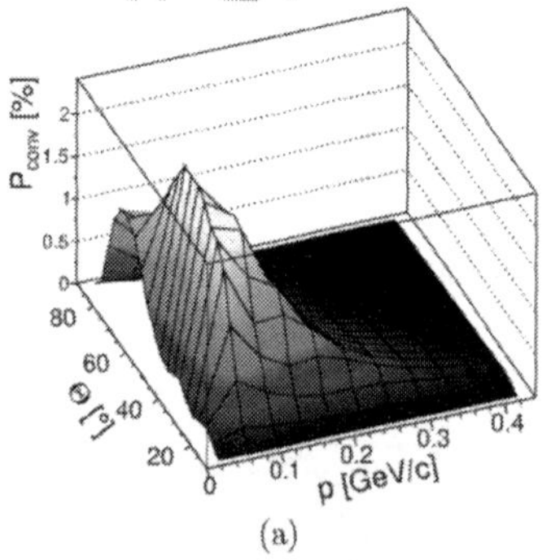

(a)

Detector component	Material	% (π^0)
Target	Gold	0.05
δ-shield	PE	0.05
Beam pipe	Carbon	0.04
Radiator gas	C4F10	0.11
Mirror	Carbon	0.08
Sum		0.36

(b)

Figure 2: (a) Conversion probability (in %) as function of the γ energy and the Θ angle. (b) Conversion probability (in %) for different material in the center of the spectrometer for γ coming from π^0 decays

2.3. Reconstruction of neutral mesons

Leptons can be identified within the HADES spectrometer by various observables, i.e. RICH ring properties, particle velocity, energy loss in ToF detector and MDC chambers, etc. Lepton pairs coming from conversion are characterized by a very small opening angle and a low momentum. Pairs with this small opening angles will be identified as a single ring in the RICH detector. Algorithms to reconstruct close pairs are under development. Furthermore conversion in radiator gas and the mirror will not be taken into account if a ring would be required. Therefore the identification of leptons is realized using momentum versus velocity information in the RPC and ToF detectors. This leptons are combined into opposite charge pairs. At least two pairs are reqired in one event. To identify neutral mesons, topological cuts on the opening angles between the leptons and the reconstructed γs are applied. In Fig. 3a the opening angle α distribution for dilepton-pairs coming from γ, π^0-Dalitz and η-Dalitz decays is shown. For identification with the full conversion method at least one of the reconstructed pairs need to come from a real photon. Therefore one of the lepton pairs need to have an opening angle $\alpha_1 < 2.5°$. Since the second pair could origin from a virtual photon of a Dalitz decay the cut is less strict : $\alpha_2 < 20°$. Another cut is applied on an opening angle between the two reconstructed photons $\Theta_{\gamma\gamma}$. In Fig. 3b the $\Theta_{\gamma\gamma}$ from different sources can be seen. Uncorrelated leptons (black dashed curve, labled with Lep4) and photons (black solid curve, labled with Fake2Conv) reslut in a wide distribution, where photons coming from π^0 and η decays appear in certain opening angle regions.
For π^0 the used topological cut is $10° < \Theta_{\gamma\gamma} < 40°$, for η $40° < \Theta_{\gamma\gamma} < 140°$.

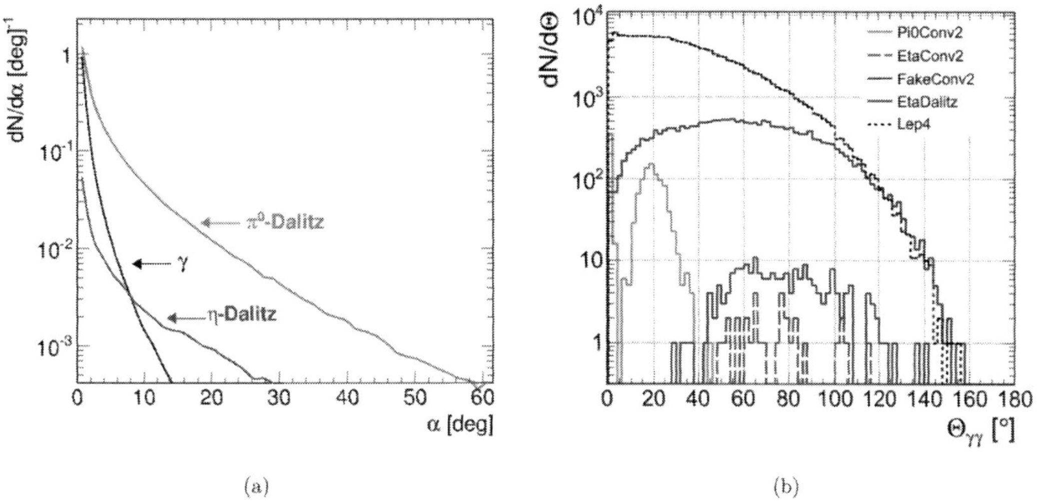

Figure 3: (a) Opening angle (α) distribution for dilepton-pairs coming from γ, π^0-Dalitz and η-Dalitz decays. (b) Opening angle ($\Theta_{\gamma\gamma}$) distribution for γ originating from π^0 and η decays in comparison to uncorrelated γ.

3. Results

After applying topological cuts to identify π^0 the resulting four lepton invariant mass spectrum is shown in Fig. 4a (black circles) together with the event-mixed background (red curve). At 130 MeV/c^2 a clear π^0 peak is visible. In Fig. 4b UrQMD simulations show the different contributions to the invariant mass spectrum. Real π^0 are shown with the orange curve. Uncorrelated real ($\gamma\gamma$ black curve) and virtual photons ($\gamma\gamma^*$ black dashed line) are the main background sources. Contribution from misidentified particles (blue curve) is minor. To subtract the combinatorial background the event-mixing technique was used. Here photons and virtual photons from different events where mixed.

The event-mixing background was normalized to the integral of events used for the like sign analysis. In Fig. (5a) and Fig. (5b) the four lepton mass spectra after subtraction of the event-mixed background are shown. In both cases the event-mixed technique was able to describe the background. Integration of the spectrum in the 2σ region around the peak gives ≈ 9600 counts for π^0 and ≈ 450 counts for η. This amount of signal will allow for a multi differential analysis.

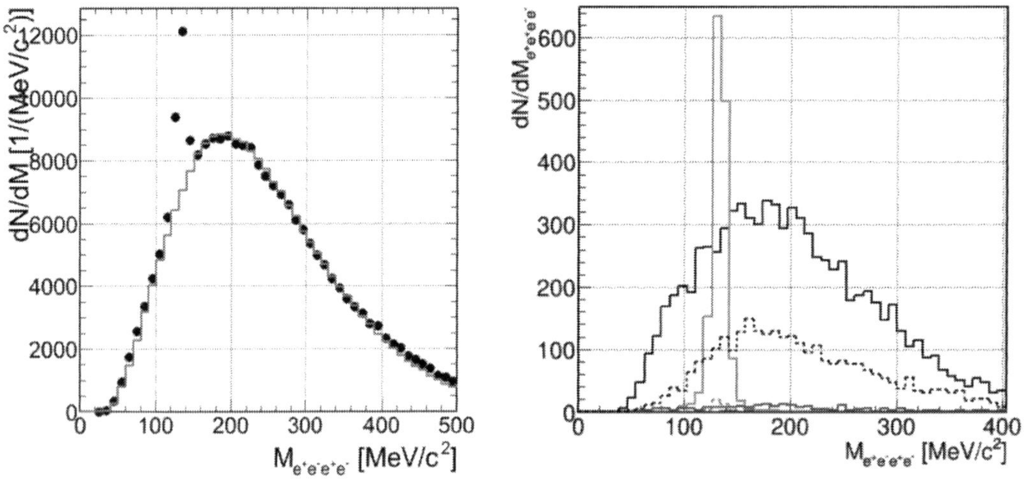

(a) (b)

Figure 4: (a) Four lepton invariant mass spectrum (black circles) after applying topological cuts for π^0 reconstruction, together with the event-mixed background (red curve). (b) Four lepton invariant mass spectrum from UrQMD simulation. For details see text.

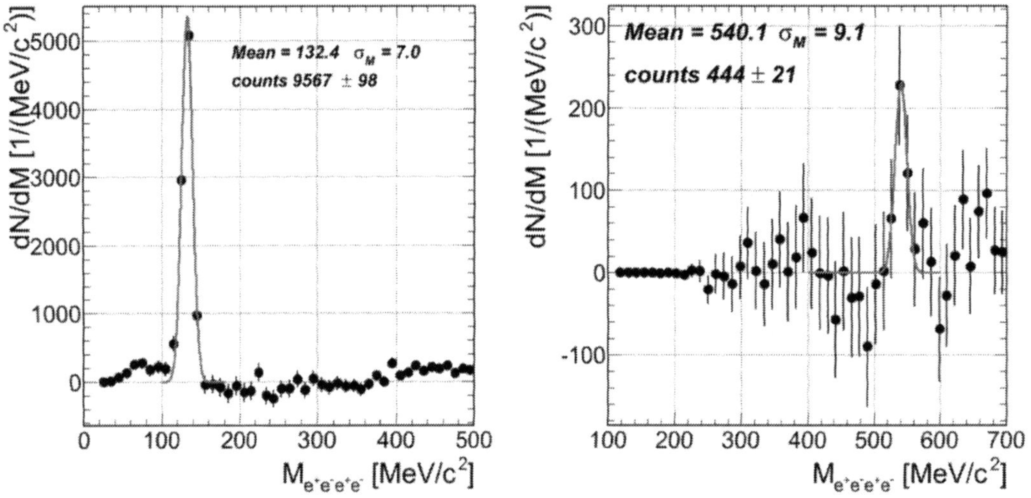

(a) (b)

Figure 5: Four lepton invariant mass spectrum (black circles) after subtraction of event-mixed background. Gaussian functions (red curve) are fitted to estimate the yields. Topological cuts are optimised for (a) π^0 and (b) η

4. Summary and Outlook

In this work we study the reconstruction of neutral mesons π^0 and η via photon conversion. The conversion probability in the inner parts of the HADES detector was calculated using simulated photons coming from π^0 decay. Lepton candidates were identified using velocity and momentum and combined to 4 lepton invariant mass spectrum. Both mesons were found and the yield was estimated to be $N_{\pi^0} \approx 9600$ counts and $N_\eta \approx 450$. Acceptance and efficiency corrections will be extracted in the next step. The multiplicities, y and $p_\perp$ for both mesons will be reconstructed for Au+Au at 1.23 GeV/u and a cross section will be extracted. In future an electromagnetic calorimeter will be installed at HADES and it will support the reconstruction of photons and neutral mesons.

Acknowledgments

The collaboration gratefully acknowledges the following funding: INFN-LNS Catania (Italy); LIP Coimbra (Portugal): PTDC/FIS/113339/2009; SIP JUC Cracow (Poland): NN202198639; GSI Darmstadt (Germany): Helmholtz Alliance HA216/EMMI; TU Darmstadt (Germany): VH-NG-823, Helmholtz Alliance HA216/EMMI; HZDR, Dresden (Germany): 283286, 05P12CRGHE; Goethe-University, Frankfurt (Germany): Helmholtz Alliance HA216/EMMI, HIC for FAIR (LOEWE), GSI F&E, BMBF 06FY9100I, HGS-Hire, H-QM research school TU München, Garching (Germany): BMBF 06MT7180; JLU Giessen (Germany): BMBF:05P12RGGHM; University Cyprus, Nicosia (Cyprus): UCY/3411-23100; IPN Orsay, Orsay Cedex (France): CNRS/IN2P3; NPI AS CR, Rez, (Czech Republic): MSMT LG 12007, GACR 13-06759S.

References

[1] G. Agakishiev, et al., Origin of the low-mass electron pair excess in light nucleus-nucleus collisions, Phys.Lett. B690 (2010) 118–122. arXiv:0910.5875, doi:10.1016/j.physletb.2010.05.010.

[2] G. Agakishiev, et al., Inclusive dielectron production in proton-proton collisions at 2.2 GeV beam energy, Phys.Rev. C85 (2012) 054005. arXiv:1203.2549, doi:10.1103/PhysRevC.85.054005.

[3] G. Agakishiev, et al., Inclusive dielectron spectra in p+p collisions at 3.5 GeV, Eur.Phys.J. A48 (2012) 64. arXiv:1112.3607, doi:10.1140/epja/i2012-12064-y.

[4] R. J. Porter, et al., Dielectron Cross Section Measurements in Nucleus-Nucleus Reactions at 1.0A GeV, Phys. Rev. Lett. 79 (1997) 1229–1232. doi:10.1103/PhysRevLett.79.1229.

[5] G. Agakishiev, et al., Dielectron production in Ar+KCl collisions at 1.76A GeV, Phys.Rev. C84 (2011) 014902.

[6] G. Agakichiev, et al., e+ e- pair production in Pb - Au collisions at 158-GeV per nucleon, Eur.Phys.J. C41 (2005) 475–513. arXiv:nucl-ex/0506002, doi:10.1140/epjc/s2005-02272-3.

[7] R. Arnaldi, et al., First measurement of the rho spectral function in high-energy nuclear collisions, Phys.Rev.Lett. 96 (2006) 162302. arXiv:nucl-ex/0605007, doi:10.1103/PhysRevLett.96.162302.

[8] B. Huang, Dielectron differential cross section in Au + Au collisions at different beam energies at STAR, Nucl.Phys. A904-905 (2013) 565c–568c. doi:10.1016/j.nuclphysa.2013.02.077.

[9] A. Adare, et al., Detailed measurement of the e^+e^- pair continuum in $p+p$ and Au+Au collisions at $\sqrt{s_{NN}} = 200$ GeV and implications for direct photon production, Phys.Rev. C81 (2010) 034911. arXiv:0912.0244, doi:10.1103/PhysRevC.81.034911.

[10] G. Agakishiev, et al., The High-Acceptance Dielectron Spectrometer HADES, Eur. Phys. J. A41 (2009) 243–277. arXiv:0902.3478, doi:10.1140/epja/i2009-10807-5.

[11] G. Agakishiev, A. Balanda, D. Belver, A. Belyaev, J. Berger-Chen, et al., Inclusive pion and eta production in p+Nb collisions at 3.5 GeV beam energyarXiv:1305.3118.

FAIRNESS 2014 IOP Publishing
Journal of Physics: Conference Series **599** (2015) 012030 doi:10.1088/1742-6596/599/1/012030

V^0 Reconstruction of Strange Hadrons in Au+Au Collisions at 1.23 AGeV with HADES

T. Scheib[1] for the HADES Collaboration

[1]Goethe Universität Frankfurt, Max-von-Laue-Str. 1, 60438 Frankfurt am Main

E-mail: t.scheib@gsi.de

Abstract. Preliminary results on the production of weakly decaying strange hadrons are reported for collisions of Au+Au at 1.23 AGeV beam energy studied with the HADES detector at GSI in Darmstadt. At this collision energy all strange particles are created below their elementary threshold. The reconstruction of the investigated particles (i.e. Λ and K^0_s) via the topology of their charged decay products (V^0 reconstruction) is presented in detail. From the corrected yields of Λ and K^0_s the ratio K^0_S/Λ can be calculated and included into a statistical model fit.

1. Introduction

Hadrons containing strange quarks are well suited probes of the high density phase of nuclear matter created in heavy-ion collisions at SIS energies. For example, the investigations on sub or near threshold production of K^+ provided constraints on the nuclear equation of state at matter densities of about 2-3 ρ_0 [1, 2, 3], which are typical for the SIS energy regime. A rather consistent picture of in-medium K^+ potentials emerged and the resulting equation of state agrees fairly well with K^+ phase space distributions and flow patterns [4, 5, 6, 7].

Furthermore, as a consequence of strangeness conservation the kaon production is directly connected to the production of hyperons. Whereas the strangeness production is well understood in elementary collisions of nucleons, in heavy-ion collisions further mechanisms have to be taken into account i.e. effective in-medium potentials or multi-step processes involving baryon resonances or mesons. Strangeness exchange channels like $\pi\Lambda \to NK^-$ have been proposed to explain observed K^- yields [8, 9]. Additionally it was shown that the feed down from the Φ meson decaying into a $K^+ K^-$ pair gives a non-neglibile contribution to the K^- yields [10, 11]. For a deeper understanding of strangeness production and propagation in a density dominated environment more information on all produced strange hadrons at these energies needs to be collected.

In this contribution, preliminary results on the production of the weakly decaying strange hadrons Λ and K^0_s from Au+Au data at 1.23 AGeV incident beam energy collected with the HADES experiment are presented. At this energy all strange particles are produced below their free nucleon-nucleon threshold including the Λ and K^0_s observed for the first time subthreshold. A special emphasis will be put on the analysis methods used to reconstruct these particles and to correct for acceptance and efficiency effects of the detector system. Finally the preliminary ratio of K^0_s to Λ will be included into a statistical model fit.

2. Au+Au Data Sample

Au+Au data at 1.23 AGeV with a beam rate of (1.2 - 1.5) $\cdot 10^6$ ions/s was collected in April and May 2012 comprising in total 557 hours. In order to reduce the amount of data and select more central events a multiplicity trigger ($M > 20$ in the outer detector region selecting $0-40\%$ most central collisions) was used, leading to an average data rate of 8 kHz and 200 MByte/s during the spills and a mean number of participating nucleons per reaction of $<A_{part}> \approx 174$. In the selected data sample 7.3 billion events were recorded giving a total amount of 140 x 10^{12} bytes.

3. Particle Identification

The High Acceptance Di-Electron Spectrometer covers almost fully the azimuthal angle ϕ and largely the polar angle ($\theta = 18^\circ - 85^\circ$). With two planes of Multi-Wire Drift Chambers in front and two behind the superconducting toroidal magnet ILSE the tracks of charged particles can be reconstructed and their momenta determined. For this purpose a new algorithm for high track density environments has been developed. By additionaly measuring the time of flight of the tracks with a Diamond-Start-Detector plus the two time-of-flight walls RPC and TOF (covering inner and outer polar angles) the particles can be identified. More details on the HADES detector can be found in [12].

4. V^0 Reconstruction of Λ and K_s^0

Λ and K_s^0 decay via the weak interaction which leads to relatively long life-times compared to strong decays. Both decay into two charged particles:

$$\Lambda \rightarrow \pi^- + p$$

$$K_S^0 \rightarrow \pi^- + \pi^+$$

These decay particles are then detected with HADES and from their reconstructed tracks the invariant mass can be calculated. Due to the fact that most of the pions and protons are produced thermally in the collision zone the obtained invariant mass spectra are mainly populated by uncorrelated pairs. Because of their long life-times the secondary decay vertex of Λ and K_s^0 can be separated from the primary collision vertex and therefore constraints on the decay topology can be set. The topology is shown in Fig. 2a exemplarily for the Λ decay. Applied conditions are: minimum distance of primary to secondary vertex (d_v), minimum distance of the daughter tracks to the primary vertex (d_2, d_3), distance of closest approach for daughter tracks (d_t), distance of closest approach of mother track to primary vertex (d_1) and minimum opening angle of the daughter tracks ($\Delta\beta$).

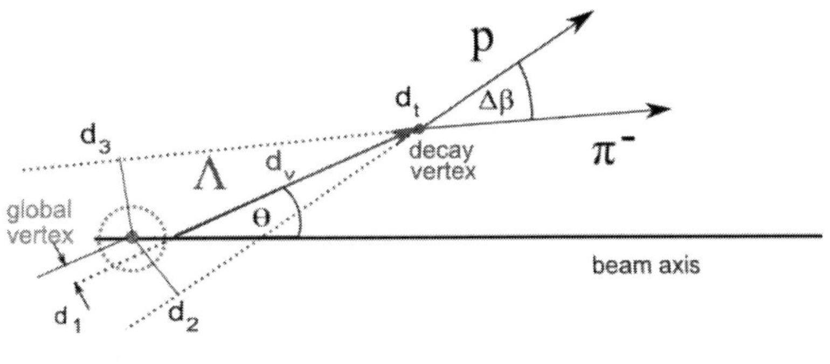

Figure 1: Topology of a Λ decay.

The background is estimated with the event-mixing technique. Here the invariant mass of the two decay particles is calculated under the condition that they come from different collisions. Hence the pairs are by definition uncorrelated and therefore their invariant mass spectrum reproduces the background of the same event spectrum. The events are mixed under the condition that they are close in centrality and have comparable detector performances as well as proximate reaction vertices.

The invariant mass spectra obtained after applying all topology cuts are shown in Fig. 2a and Fig. 2b. The corresponding cut values are listed in Table 1.

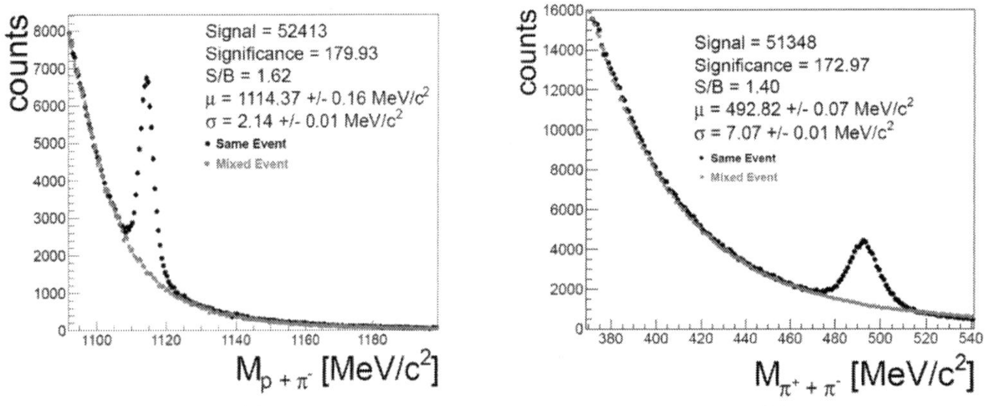

(a) (b)

Figure 2: Invariant mass spectrum of negative pions plus (a) protons and (b) positive pions, respectively, coming from the same event (black) and different events (red). A highly significant peak is visible close to the expected mass for (a) Λ and (b) K^0_S.

Cut		$d_v [mm]$	d_1 [mm]	d_2 [mm]	d_3 [mm]	d_t [mm]	$\Delta\beta$ [°]
Value	Λ	> 55	< 5	> 21	> 6	< 7	> 15
	K^0_S	> 20	< 10	> 8	> 8	< 10	> 15

Table 1: Values for the topological constraints chosen for the V^0 reconstruction

A clear peak emerging from the background close to the expected mass is visible showing a highly significant Λ and K^0_S sample respectively. The significance is comparable to that obtained in Ar+KCl, the second largest collision system investigated with HADES [10]. The high statistics allow for a double differential analysis as a function of transverse mass and rapidity.

Furthermore, the reconstructed signal has to be corrected for the non-perfect acceptance and efficiency of the detector. For this purpose the strange hadrons are generated in a full Monte-Carlo simulation within the Pluto framework [13]. Then the simulated particles are subjected to a GEANT simulation [14] giving a realistic detector response of HADES. Finally they are propagated through the same analysis chain that was used for data leading to a correction factor for acceptance and efficiency effects. This factor is then applied on the reconstructed signal giving the yield of initially produced particles.

After correcting the yields as a function of transverse mass and rapidity the transverse mass spectra can be fitted with a Boltzmann function in order to extrapolate to the unmeasured phase space. By integrating each fit and plotting each yield as a function of the center of mass rapidity one obtains the corrected rapidity distribution. This distribution can be fitted with a gaussian function and by integration the total multiplicity can be determined. For minimization of the systematic errors only the values at mid-rapidity are considered here.

Moreover, statistics are sufficient to do this analysis for different centrality classes.

5. Results and Discussion

Following this procedure a preliminary ratio for the yields of Λ and K^0_S can be calculated:

$$\frac{N(K^0_S)}{N(\Lambda)} = 0.26 \pm 0.11.$$

The error of this ratio includes statistical and systematic uncertainties. At this point of our analysis this error is yet dominated by systematics.

As a tool to describe hadron yields measured in heavy-ion collisions statistical models have been quite successful in the past [15, 16]. Combined with further determined preliminary results on multiplicities in this collision system a simultaneous statistical hadronization fit can be applied using the volume V, temperature T and baryochemical potential μ_B of the system as free parameters. In order to account for strangeness suppression a (strangeness-)canonical ensemble is used where the strangeness quantum number needs to be conserved exactly within a subvolume V_c with radius R_c. The freely available statistical model THERMUS [17] was used to simultaneously fit the K^0_S/Λ ratio together with π^-/p, K^-/K^+ and the ϕ/K^-, where all values were taken at mid-rapidity. Details for the fitting procedure and constraints on the parameters are described in [18]. The fit of the statistical model to the data points is shown in Fig. 3a. From this we find the baryochemical potential and the temperature to be $\mu_B = 799 \pm 34$ MeV and $T_{chem} = (47 \pm 5)$ MeV respectively. The ratio of the radius of the subvolume to the radius of the overall volume of the system is determined to be $R_C/R = 0.3 \pm 0.2$ whereas the χ^2/d.o.f. of the fit is equal to 1.2. In Fig. 3b the corresponding point of freeze-out in the $T - \mu_B$ plane is plotted showing a good agreement with systematics of other experiments [10, 15, 19].

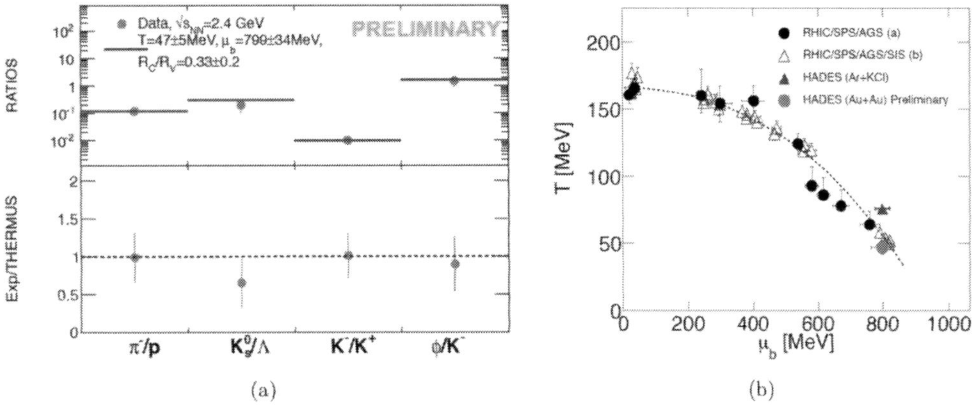

(a) (b)

Figure 3: (a) Comparison between data and statistical model for ratios of particle yields at mid-rapidity together with the parameter values obtained from the statistical model fit. (b) Corresponding freeze-out point in the $T - \mu_B$ plane together with a collection of points from other experiments [10, 15, 19].

6. Conclusion

In summary, preliminary results on the production of the uncharged strange hadrons Λ and K^0_S in the Au+Au collision system at incident beam energy of 1.23 AGeV with HADES have been presented. The particles have been reconstructed via the topology of their charged decay products. The corrected transverse mass spectra have been determined in order to calculate the total production yield. The preliminary ratio of the yields at mid-rapidity $K^0_S/\Lambda = 0.26 \pm 0.11$ – together with further hadron ratios – was compared to a statistical model resulting in an overall good agreement.

Acknowledgments

The collaboration gratefully acknowledges the following funding: INFN-LNS Catania (Italy); LIP Coimbra (Portugal): PTDC/FIS/113339/2009; SIP JUC Cracow (Poland): NN202198639; GSI Darmstadt (Germany): Helmholtz Alliance HA216/EMMI; TU Darmstadt (Germany): VH-NG-823, Helmholtz Alliance HA216/EMMI; HZDR, Dresden (Germany): 283286, 05P12CRGHE; Goethe-University, Frankfurt (Germany): Helmholtz Alliance HA216/EMMI, HIC for FAIR (LOEWE), GSI F&E, BMBF 06FY9100I, HGS-Hire, H-QM research school TU München, Garching (Germany): BMBF 06MT7180; JLU Giessen (Germany): BMBF:05P12RGGHM; University Cyprus, Nicosia (Cyprus): UCY/3411-23100; IPN Orsay, Orsay Cedex (France): CNRS/IN2P3; NPI AS CR, Rez, (Czech Republic): MSMT LG 12007, GACR 13-06759S.

References

[1] C. Sturm *et al.* [KaoS Collaboration], Phys. Rev. Lett. **86**, 39 (2001).

[2] C. Fuchs, A. Faessler, E. Zabrodin, Phys. Rev. Lett. **86**, 1974 (2001).

[3] C. Hartnack, H. Oeschler and J. Aichelin, Phys. Rev. Lett. **96**, 012302 (2006).

[4] J. Schaffner-Bielich, J. Bondorf, A. Mishustin, Nucl. Phys. A **625**, 325 (1997).

[5] W. Cassing, E. L. Bratkovskaya, U. Mosel, S. Teis, A. Sibirtsev, Nucl . Phys. A **614** 415 (1997).

[6] L. Adamczyk *et al.* [STAR Collaboration], Phys. Rev. Lett. **110** 142301 (2013).

[7] V. Zinyuk *et al.* [FOPI Collaboration], arXiv:1403.1504 [nucl-ex].

[8] A. Förster, F. Uhlig, I. Bottcher, D. Brill, M. Debowski, F. Dohrmann, E. Grosse and P. Koczon *et al.*, Phys. Rev. C **75** 024906 (2007).

[9] C. Hartnack, H. Oeschler, Y. Leifels, E. L. Bratkovskaya and J. Aichelin, Phys. Rept. **510** 119 (2012).

[10] G. Agakishiev *et al.* [HADES Collaboration], Eur. Phys. J. A **47** 21 (2011).

[11] G. Agakishiev *et al.* [HADES Collaboration], Phys. Rev. C **80** 025209 (2009).

[12] G. Agakishiev *et al.* [HADES Collaboration], Eur. Phys. J. A **41** 243 (2009).

[13] I. Froehlich, L. Cazon, T. Galatyuk, V. Hejny, R. Holzmann, M. Kagarlis and W. Kuhn *et al.*, PoS ACAT **2007** (2007) 076 [arXiv:0708.2382 [nucl-ex]].

[14] GEANT. Detector Description and Simulation Tool, 2004. http://cont.cern.ch/writeup/geant/, Online User Guide.

[15] A. Andronic *et al.*, Nucl. Phys. A **837** 65 (2010).

[16] F. Becattini, M. Gazdzicki, A. Keranen, J. Manninen and R. Stock, Phys. Rev. C **69**, 024905 (2004).

[17] S. Wheaton and J. Cleymans, Comput. Phys. Commun. **180** 84 (2009).

[18] M. Lorenz *et al.* [HADES Collaboration], Nucl. Phys. A, 931 785-789 (2014).

[19] J. Cleymans, H. Oeschler, K. Redlich and S. Wheaton, Phys. Rev. C **73**, 034905 (2006).

FAIRNESS 2014 IOP Publishing
Journal of Physics: Conference Series **599** (2015) 012031 doi:10.1088/1742-6596/599/1/012031

Variances in r-process predictions from uncertain nuclear rates

Matthew Mumpower[1]*, Rebecca Surman[1], Ani Aprahamian[1]

[1] Department of Physics, University of Notre Dame, Notre Dame, IN 46556 USA

E-mail: matt.mumpower@nd.edu

Abstract. Rapid neutron capture or 'r-process' nucleosynthesis is a critical process for creating approximately half of heavy elements above iron in the Universe. Simulations of the r-process require theoretical calculations of neutron capture and β-decay rates for thousands of short-lived nuclei for which no experimental data exists. We explore the impact of uncertainties in theoretical nuclear rates by performing global monte carlo simulations. We show that randomly varying neutron capture and β-decay rates within current theoretical model uncertainties results in large error bars for r-process predictions. We conclude that the reduction of rate uncertainties either by new measurements or by improved nuclear models will allow for more robust r-process predictions.

1. Introduction

The Facility for Anti-proton and Ion Research (FAIR) currently under construction at GSI will offer an array of insights into the frontiers of physics from heavy ion collisions to the production of rare isotopes [1]. The focus of the NUSTAR collaboration is to probe the physics of exotic nuclei with extreme neutron or proton excess [2]. Planned NUSTAR measurements will occur after the beam passes through the super-FRS (Fragment Recoil Separator) and are of particular interest to the astrophysics community as they provide insight into low energy nuclear structure, one of the biggest uncertainties that go into astrophysical models [3].

The astrophysical rapid neutron capture process or 'r-process' of nucleosynthesis is likely accountable for half of the neutron-rich nuclei found in nature [4]. Research in this area is motivated by one of the most challenging problems in theoretical physics: What is the site or sites where the r process may take place? Since most of the nuclei that participate in the r process remain unmeasured, theoretical calculations of nuclear masses and rates must be used to predict the final composition. With limited experimental data to constrain these models, predictions of the properties of short-lived nuclei can range drastically as one approaches the limits of nuclear existence.

To understand how these uncertainties impact predictions of final abundances we previously performed a global monte carlo investigation using a variation in unmeasured nuclear masses [5]. In this contribution we extend the methodology developed in our previous work to gauge the importance of uncertain neutron capture rates and β-decay rates on the final composition of the r process. We combine the set of final abundances obtained from these monte carlo studies to produce an estimation of error bars for these nuclear properties.

*This work was supported by the NSF through grant numbers PHY0822648 and PHY1419765.

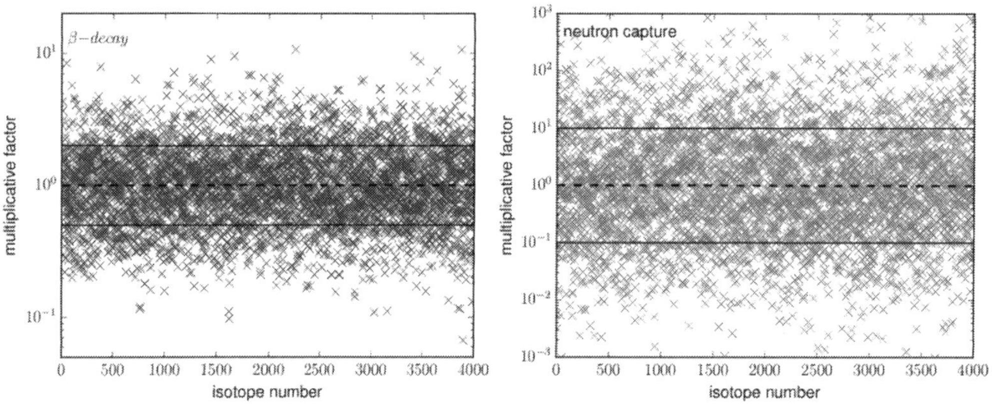

Figure 1. Multiplicative factors representing the uncertainty in β-decay rates (left panel) and neutron capture rates (right panel) sampled from the respective log-normal distributions for the first 4000 nuclei that enter into the reaction network. The dotted lines represent the mean of both distributions, $\mu = 1$. The solid lines represent the value of the log-normal σ used in each case: $\sigma = 2$ for beta-decays and $\sigma = 10$ for neutron captures.

2. Calculations

For our astrophysical r-process conditions we use a parameterized model from Ref. [6] and note that this trajectory has also been used in our previous work [7]. This trajectory represents a 'hot' wind with entropy of 200 k_B, electron fraction of $Y_e = 0.3$ and timescale of 80 ms which produces a main r-process ($A \gtrsim 130$ and $Z \gtrsim 56$) component without fission recycling. The nuclear flow in this trajectory is initially controlled by a long equilibrium phase (lasting just over a second) between neutron capture and photo-dissociation channels, the so called $(n, \gamma) \rightleftarrows (\gamma, n)$ equilibrium. The equilibrium continues until the supply of free neutrons is exhausted and the r-process path, or set of most abundant isotopes moves back towards stability. Once this equilibrium breaks down, the last phase of the r process known as 'freeze-out' begins, and individual neutron capture rates may contribute to the predicted final abundances [8, 9]. The impact of β-decay rates is more consistent however, since these rates influence not only the timescale for heavy element production but also contribute to the relative abundances of isotopic chains for the entire r-process phase [10].

We calculate evolution of the r-process with a reduced reaction network used most recently in Ref. [5]. This network only contains reaction channels most relevant to the r-process which include neutron capture, photo-dissociation, β-decay and β-delayed neutron emission. We disable fission due to our choice of astrophysical conditions. These two simplifications greatly reduces our computational time allowing for more monte carlo steps to be achieved without altering our conclusions for a hot wind outflow.

As the basis for our theoretical nuclear models we use the Finite Range Droplet Model (FRDM1995) [11], Duflo Zuker (DZ) [12] and Hartree-Fock-Bogoliubov version 17 (HFB-17) [13]. Our neutron capture rates are calculated with a recently developed statistical model code, Capture Induced Gamma-ray Reactions (CIGAR) [14]. This model is based on the Hauser Feshbach approach which implicitly assumes that the level density is sufficiently high so that averaged values of the compound system can be computed. For our weak interaction data we use the publicly available dataset from Ref. [15] which includes β-decay rates and neutron emission probabilities. This model takes into account Gamow-Teller transitions and additionally includes

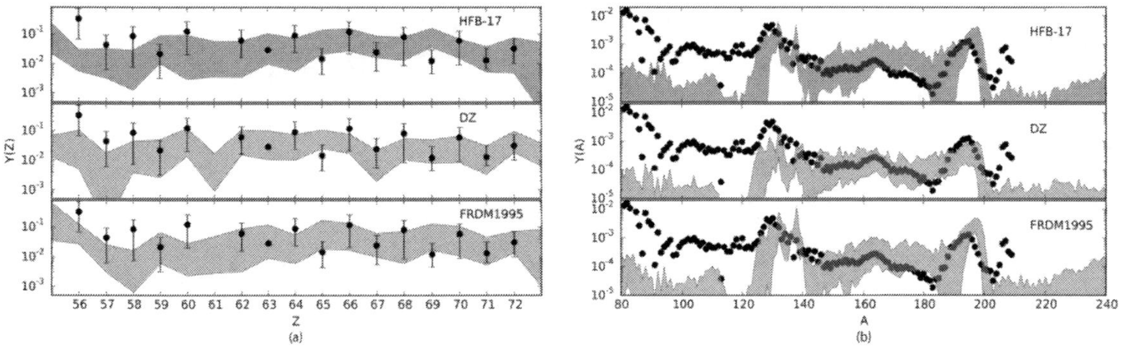

Figure 2. Variance in (a) elemental abundances and (b) isotopic abundances from current theoretical uncertainties in β-decay rates for three nuclear models (HFB-17, DZ, and FRDM1995). Elemental abundances are compared to the average of five r-process enriched metal poor stars and isotopic abundances are compared to the solar r-process residuals, both observations denoted by black dots.

a First-Forbidden component which is important for the computation of half-lives of neutron-rich nuclei near closed shells [16, 17]. We note that where available we use half-lives from the NNDC compilation [18] and that these measured values remain unvaried in our monte carlo simulations.

To estimate error bars on r-process abundances from uncertain nuclear rates we perform monte carlo variations of β-decays rates and neutron capture rates separately. We first fix our hot wind astrophysical trajectory and nuclear model. Then for a given monte carlo step, each rate that enters into the reaction network is varied by sampling multiplicative factors from a log-normal probability distribution:

$$p(x) = \frac{1}{x\sqrt{2\pi}\sigma} \exp\left[-\frac{(\mu - ln(x))^2}{2\sigma^2}\right] \tag{1}$$

where μ is the mean, and σ is the standard deviation of the distribution and x is a random variable. To approximate current theoretical uncertainties in β-decay rates we modify each half-life in the network by sampling factors from the log-normal distribution with $\mu = 1$ and $\sigma = 2$. The choice of $\sigma = 2$ for β-decay rates is motivated by Fig. 4 in Ref. [15]. To approximate current theoretical uncertainties in neutron capture rates we increase σ to 10, which is motivated by Fig. 8 in Ref. [19]. In Fig. 1 we show a representative sample distribution of rate factors for both β-decays (left panel) and neutron capture rates (right panel) for the first 4000 nuclei that enter into the reaction network. We note that with choice of $\mu = 1$, rate variations above and below theoretical values are sampled with equal weight.

Our final abundance patterns are compared to the averaged elemental abundances from five metal poor stars: CS22892-052 [20], CS31082-001 [21], HD115444 [22], HD221170 [23], and BD+17°3248 [24]. We also compare final compositions to the isotopic abundances of the solar r-process residuals [25].

3. Results

In this preliminary investigation each monte carlo study contains approximately 3000 abundance patterns. This abundance count is low relative to the size of the reaction network and consequently we plan to increase the number of monte carlo steps in a future analysis. A given

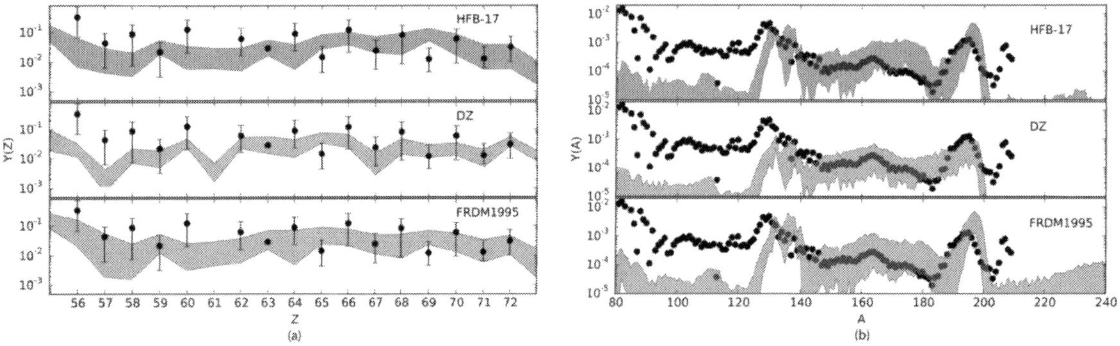

Figure 3. Variance in (a) elemental abundances and (b) isotopic abundances from current theoretical uncertainties in neutron capture rates for three nuclear models (HFB-17, DZ, and FRDM1995). Observational data (black dots) are the same as in Fig. 2.

monte carlo study contains only abundance patterns generated with a particular nuclear model and choice of rate variation, either neutron capture or β-decay. In the latter case, β-delayed neutron emission branching ratios are held fixed with values from Ref. [15]. Therefore, the impact of uncertain neutron capture rates and β-decay rates are analyzed independently. We estimate the current variability in r-process predictions by computing the standard deviation of a particular ensemble of isotopic and elemental abundances respectively.

We show the results of our β-decay monte carlo studies in Fig. 2 for both elemental, $Y(Z)$, and isotopic abundance predictions, $Y(A)$. These monte carlo studies were performed with the three aforementioned nuclear mass models, however all β-decay rate variations were based off the dataset from Ref. [15]. When plotting the elemental abundances we scale each study to the averaged halo star value of Europium ($Z = 63$) and note that the isotopic abundances are scaled to the rare earth region as outlined in Ref. [9]. The shaded color band in each panel of this figure represents the variance of the abundances from the monte carlo study performed with current theoretical uncertainties ($\sigma = 2$) for β-decay rates. We see that for the main r-process component ($A \gtrsim 130$ and $Z \gtrsim 56$) astrophysical observations (black dots) mostly lie within the error bars from our monte carlo study. The results of this study show that our ability to predict features in r-process abundances are washed out by the current uncertainties in theoretical weak decay properties.

In Fig. 3 we show the results of our neutron capture monte carlo studies with the same scaling and shading as in the previous figure. Again we find that for a main r-process component the observational constraints lie within the theoretical error bars. At first glance the results may come as a surprise since the chosen trajectory has a long duration $(n, \gamma) \rightleftarrows (\gamma, n)$ equilibrium, during which modifications to neutron capture rates are expected to play a minimal role. However, this result shows clearly that the final neutron captures during the last or 'freeze-out' stage of the r-process (once equilibrium has ended) are critical in determining final abundances, which reaffirms the conclusions drawn in our previous work [8, 9, 26]. If other astrophysical conditions are used, for example a 'cold' wind outflow, then we expect the dependence on neutron capture rates to increase resulting in a potentially larger variability in predicted abundances. This is because the r-process path in a cold outflow is controlled by a balance between β-decays and neutron captures.

4. Conclusions

We have used a monte carlo approach to estimate the variability of r-process predictions from uncertain nuclear β-decay and neutron capture rates in the context of a hot wind r-process. The current uncertainties in the thousands of nuclear rates that enter into network calculations were approximated by sampling multiplicative factors from a log-normal distribution which boosted or diminished individual rates. At each step in our monte carlo study the dataset of randomized rates was used as input into r-process network calculation to produce an abundance pattern. To estimate the error bars on r-process predictions from these nuclear model uncertainties we took the variance of the ensemble of abundances.

We have shown that the variance of abundance predictions with current theoretical calculations of β-decays and neutron capture rates span an order of magnitude (or more) for most of the main r-process component ($A \gtrsim 130$ and $Z \gtrsim 56$). It is therefore the uncertainties in nuclear models that limits the ability to resolve features found in elemental or isotopic abundance data. All astrophysical simulations of the r process use theoretical nuclear models and thus are subject to a similar resolving power. Ergo, in order to improve r-process predictions a concerted effort is required to achieve measurements on exotic nuclei. The role of FAIR and other facilities found worldwide play in this endeavor is two fold: (1) the inclusion of measured data directly into astrophysical simulations and (2) the reduction of uncertainties by providing more experimental data with which to constrain theoretical nuclear models. In addition to pursuing these difficult measurements, our group plans to continue to increase the efficacy of the methodology developed here by combining monte carlo studies of nuclear masses and rates. This extension will allow for the analysis of nuclear correlations and may reveal a new path towards understanding the reliability of nuclear models towards the neutron dripline.

References

[1] Fair users website. `http://www.fair-center.eu/`.
[2] Nustar collaboration website. `http://www.fair-center.eu/for-users/experiments/nustar.html`.
[3] M. Mumpower, R. Surman, D.L. Fang, M. Beard, and A. Aprahamian. *J. Phys. G*, 2014.
[4] E. M. Burbidge, G. R. Burbidge, W. A. Fowler, and F. Hoyle. *Reviews of Modern Physics*, 29:547–650, 1957.
[5] M. Mumpower, R. Surman, and A. Aprahamian. *arXiv:1411.3973*, 2014.
[6] B. S. Meyer. *Physical Review Letters*, 89(23):231101–+, November 2002.
[7] M. Mumpower, D.L. Fang, R. Surman, M. Beard, and A. Aprahamian. *arXiv:1410.7275*, 2014.
[8] R. Surman, J. Beun, G. C. McLaughlin, and W. R. Hix. Phys. Rev. C, 79(4):045809, 2009.
[9] M. Mumpower, G. C. McLaughlin, and R. Surman. ApJ, 752:117, 2012.
[10] D. D. Clayton. *Principles of stellar evolution and nucleosynthesis*. 1968.
[11] P. Möller, J. R. Nix, W. D. Myers, and W. J. Swiatecki. *Atomic Data & Nuclear Data Tables*, 59:185, 1995.
[12] J. Duflo and A. P. Zuker. Phys. Rev. C, 52:23, July 1995.
[13] S. Goriely, N. Chamel, and J. M. Pearson. *Physical Review Letters*, 102(15):152503, April 2009.
[14] M. Beard, E. Uberseder, R. Crowter, and M. Wiescher. Phys. Rev. C, 90:034619, 2014.
[15] P. Möller, B. Pfeiffer, and K.-L. Kratz. Phys. Rev. C, 67(5):055802–+, May 2003.
[16] T. Suzuki, T. Yoshida, T. Kajino, and T. Otsuka. Phys. Rev. C, 85(1):015802, January 2012.
[17] N. Nishimura, T. Kajino, G. J. Mathews, S. Nishimura, and T. Suzuki. Phys. Rev. C, 85(4):048801, 2012.
[18] National nuclear data center website. `http://www.fair-center.eu/for-users/experiments/nustar.html`.
[19] R. Surman and J. Engel. Phys. Rev. C, 64(3):035801, September 2001.
[20] C. Sneden, A. McWilliam, G. W. Preston, J. J. Cowan, D. L. Burris, and B. J. Armosky. ApJ, 467, 1996.
[21] V. Hill, B. Plez, R. Cayrel, T. C. Beers, B. Nordström, J. Andersen, M. Spite, F. Spite, B. Barbuy, P. Bonifacio, E. Depagne, P. François, and F. Primas. A&A, 387:560–579, May 2002.
[22] J. Westin, C. Sneden, B. Gustafsson, and J. J. Cowan. ApJ, 530:783–799, February 2000.
[23] I. I. Ivans, J. Simmerer, C. Sneden, J. E. Lawler, J. J. Cowan, R. Gallino, and S. Bisterzo. ApJ, 645, 2006.
[24] J. J. Cowan, C. Sneden, S. Burles, I. I. Ivans, T. C. Beers, J. W. Truran, J. E. Lawler, F. Primas, G. M. Fuller, B. Pfeiffer, and K.-L. Kratz. ApJ, 572:861–879, June 2002.
[25] C. Arlandini, F. Käppeler, K. Wisshak, R. Gallino, M. Lugaro, M. Busso, and O. Straniero. ApJ, 525, 1999.
[26] M. R. Mumpower, G. C. McLaughlin, and R. Surman. Phys. Rev. C, 86(3):035803, 2012.

FAIRNESS 2014 IOP Publishing
Journal of Physics: Conference Series **599** (2015) 012032 doi:10.1088/1742-6596/599/1/012032

Studies of exotic nuclei: state-of-the-art experimental tools and techniques

Stefanos Paschalis

Institut für Kernphysik, Technische Universität Darmstadt, D64289 Darmstadt, Germany

E-mail: spaschalis@ikp.tu-darmstadt.de

Abstract. As new radioactive-ion beam facilities are coming online there is an even growing need for advanced experimental apparatuses that offer unprecedented resolution and efficiency and can fully exploit the physics opportunities that open up in this new for the nuclear physics community era. In this contribution state-of-the-art equipment and techniques for nuclear physics experiments are presented.

1. Introduction

The study of the atomic nucleus dates more than a hundred years back, when early experiments indicated that an atom has a small, central, dense and positively charged core. Since then, there has been a tremendous progress in understanding the structure and properties of the atomic nucleus. The vast majority of our knowledge originates from studies of a few hundred nuclei at or near the valley of β stability. There is, however, an even larger number of nuclei that are short lived (radioactive or exotic) and do not naturally occur on earth. It is only in the last decades that the nuclear physics community gained access to this unknown territory of the nuclear chart, which became available with the advent of radioactive-ion beam (RIB) facilities. Recent major upgrades in the world's most powerful facilities opens up the way for studies of even more exotic systems and in much greater detail.

Some of the strongest motivations to extend our studies to the most exotic nuclear systems include the fact that although they are not found on earth, they play an important role in the cosmos as they are constantly created in stars. Furthermore, exotic nuclei provide a fertile ground to study quantum phenomena such as clustering, quantum dots and phenomena unique to weakly bound systems. Such phenomena are also important in other fields of physics today and in many cases a universal description can be applied. Finally, major advances in many-body theory result from investigations of nuclei with very different proton-to-neutron admixtures, since certain parts of the nucleon-nucleon interaction are particularly sensitive to the proton-to-neutron asymmetry.

2. Production of exotic nuclei

The two most popular methods to produce exotic nuclei are:

- the production at rest followed by Isotope-Separation-On-Line (ISOL-technique)
- the production and separation in flight using heavier ions (fragmentation technique)

In the fragmentation technique, a high-energy primary beam impinges on a production target and gets fragmented. In this process many different nuclei are produced including some which are very exotic – typically the cross-section for producing exotic nuclei falls rapidly as a function of their distance from the stability line. These fragments are then flying in forward angles, due to the high beam energy, and can be separated and identified through a magnetic spectrometer (commonly referred to as fragment separator). The selected and identified species (constituting the secondary beam) are then transported to the experimental halls where they are probed using a secondary target. The primary beam intensity and the acceptance of the fragment separator are two of the most important factors that determine the secondary beam intensity and thus define how far exotic systems can be probed.

Some of the largest fragmentation facilities worldwide include the GSI/FAIR facility in Germany, the NSCL/FRIB facility in the United States and the RIKEN/RIBF facility in Japan. Recent or ongoing upgrades in these facilities enable studies of exotic systems which were not feasible before. Besides the major upgrades in the production of the secondary beams, the experimental apparatuses are also being significantly improved in terms of resolution, detection efficiency and rate capability.

3. State-of-the-art experimental tools and techniques for nuclear structure studies

3.1. NUSTAR (Nuclear Structure and Astrophysics Research) at FAIR (Facility for Antiproton and Ion Research)

The NUSTAR collaboration aims at exploring the structure and properties of the atomic nuclei by exploiting the high-intensity radioactive-ion beams that will become available at FAIR. The collaboration utilises different and complimentary ways to probe the nucleus. The main experiments planned within the NUSTAR collaboration at FAIR are the following:

- Super-FRS (RIB production, identification and high-resolution spectroscopy),
- DESPEC (γ-, β-, α-, proton- and neutron-decay spectroscopy),
- HISPEC (in-beam γ-ray spectroscopy at low and intermediate energy),
- ILIMA (masses and lifetimes of nuclei in ground and isomeric states),
- LaSpec (laser spectroscopy),
- MATS (in-trap mass measurements and decay studies) and
- R^3B (kinematically complete reactions at high beam energy),

which are currently being built in the first phase of the project and the experiments:

- ELISe (elastic, inelastic, and quasi-free e^-A scattering) and
- EXL (light-ion scattering reactions in inverse kinematics),

which are foreseen in the second phase.

Some of the highlight physics questions that these experiments are aiming to address include:

- the changes in the nuclear structure as one moves away from the valley of stability; the evolution of the single-particle structure and the onset of deformations will be probed as a function of isospin,
- quantum phenomena that occur at extreme isospin values, such as neutron halos and neutron skins,
- pin down components of the nuclear force that become more prominent as the neutron-to-proton asymmetry increases,
- systematical studies of the collective response of exotic nuclei.

In the following section the R^3B experiment is discussed in more detail.

3.2. The R^3B (Reactions with Relativistic Radioactive Ion Beams) experiment

The strength of the R^3B experiment is the kinematically complete measurement of reactions with relativistic short-lived ions with energies of up to 1 *A*GeV. These rare radioactive isotopes are produced at the beginning of the Superconducting Fragment Separator (Super-FRS) when the high-energy primary beam is fragmented on the production target. The separator selects and identifies on an event-by-event basis the ions of interest and measures their momentum to a 10^{-4} precision. The maximum beam rigidity that it can accept is 20 Tm. The R^3B experiment is planned at the high-energy branch of the Super-FRS.

The key constituents of the R^3B setup, shown in Fig. 1, are:

- the large-acceptance superconductiong dipole magnet (GLAD). Its construction is completed at CEA Saclay and will be installed in cave C at GSI with all its cryogenic components beginning of 2015.
- The New Large Area Neutron Detector (NeuLAND). It is currently being built and 20% is completed already.
- The silicon tracker (R^3B-Si-Tracker). It has all components developed and is expected to be fully completed by the end of 2015.
- The photon and particle calorimeter and spectrometer (CALIFA). It is under construction with the goal to reach 20% in the year 2015.
- The in-beam tracking detectors for the heavy fragments and evaporated protons. It is under construction with the goal to be completed in 2017.

For more details on these systems the reader is referred to the corresponding Technical Design Report of each system [1, 2] and also to the R^3B Technical Proposal [3]. Once the individual

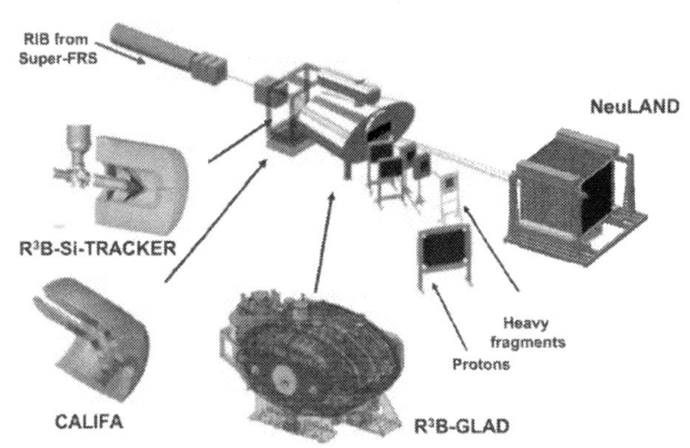

Figure 1. The R^3B setup in its startup version with its main components: the silicon tracker R^3B-Si-TRACKER, the calorimeter CALIFA, the dipole magnet R^3B-GLAD and the neutron time-of-flight spectrometer NeuLAND.

detection systems are complete, the apparatus will be able to measure all particles emerging from a reaction with high efficiency and resolution. A typical scenario would be as follows: the high-energy (1 *A*GeV) radioactive-ion beam, which is selected and identified through the fragment separator (Super-FRS), impinges on the secondary target and the beam-like fragments are moving at small forward angles and fly through the large acceptance dipole magnet GLAD. The trajectory and time of flight of the fragments is measured by the in-beam tracking detectors before and after the dipole magnet to a precision that corresponds to about 10^{-3} momentum resolution. The light charged particles that scatter at large angles are tracked precisely with the finely segmented three-layer Si-R^{3}B tracker. It is designed to offer, for example, precise tracking and vertex reconstruction by detecting the two protons originating from a (p,2p) quasifree scattering reaction (QFS). A total energy measurement of the target recoil charged particles is performed in the CALIFA calorimeter. The γ rays emerging from the excited nuclei are also measured by the CALIFA detector with very high efficiency due to its large geometrical coverage and minimum dead layers between the crystals. Its high crystal granularity allows for a precise Doppler correction; Doppler broadening is the dominant factor of the energy resolution at these relativistic energies. An unbound daughter nucleus – either excited in the continuum or lie beyond the neutron driplines – is likely to decay via neutron decay or evaporation. In this case, the in-flight emitted neutrons fly at forward angles through the dipole magnet and hit a large neutron time-of-flight wall (NeuLAND), where they are measured with very high efficiency. The high granularity of the neutron detector allows also for high neutron multiplicity measurements, which enables the study of very exotic systems and the poorly known multi-neutron correlations. In particular, once fully built it will provide an unprecedented detection efficiency of more than 50% for four neutrons evaporated from a fast moving fragment.

These kinematically complete measurements enable a rich and versatile physics program:

- QFS reactions, in which a bound nucleon is knockout out at high momentum transfer, offer a sensitive probe to study the single particle structure of exotic nuclei and also constitute an effective way to populate unbound nuclei. For example, when starting from very neutron rich nuclei as beam particles and knocking out a proton.
- Systematic studies of collective degrees of freedom in nuclei and their evolution towards more exotic systems. For instance, the low-lying strength below the Giant Dipole Resonance (GDR) identified as Pygmy Dipole Resonance (PDR), carries information about the neutron skin thickness of nuclei, which in turn is used in calculations of the symmetry term coefficient in the equation of state.
- The R^{3}B setup will also be well suited to study total reaction cross section, as well as total charge-changing and total neutron-removal cross sections. This information can be associated to the nuclear and charge radii and the thickness of the neutron skin.

3.3. The high-resolution γ-ray tracking arrays

High resolution γ-ray spectroscopy has been and still is one of the most powerful ways to probe the structure of the atomic nucleus. This is reflected in the tremendous improvement of the γ-ray spectrometers over the last decades, in terms of resolution, efficiency, peak-to-total (P/T) ratio and γ-ray multiplicity.

The Ge detectors with their unprecedented resolution are at the forefront of γ-ray spectroscopy. One of the main limitations of the Ge detectors has been the relatively low P/T ratio for a typical coaxial geometry Ge crystal, e.g. only about 1 out of 5 γ-rays of 1 MeV energy that enter the crystal will deposit their full energy in this crystal. The rest of the γ rays will deposit only part of their energy - mainly through single or multiple Compton scattering - in the crystal, before escaping. These events generate a significant background at lower energies in the crystal's energy spectrum. The identification of other smaller energy γ rays is hindered by this background. For low γ-ray multiplicity events a solution to increase the P/T ratio is to pack several crystals next to each other in a clover geometry. However, for high multiplicity events this solution is not ideal. A very effective solution to increase the P/T ratio, which has been used in the past decades, is to Compton-shield the Ge crystal with fast scintillator detectors that veto the escaping γ rays. Combining many of such Compton-suppressed modules full four-π arrays have been built. These arrays have been used very successfully over the past two decades and have served a broad physics program with an important impact for nuclear physics.

Journal of Physics: Conference Series **599** (2015) 012032 doi:10.1088/1742-6596/599/1/012032

The use of the suppression shields, however, limits the solid angle that can be covered by Ge detectors and thus limits the efficiency of the array. A pioneering solution to maintain the good P/T ratio offered by the Compton suppression and at the same time increase the efficiency is γ-ray tracking (see for example Ref. [4]). Instead of tagging on the escaping γ rays using Compton shields, one can fill the full four-π solid angle with Ge detectors and track the γ rays that scatter into the neighbouring crystal. In this way, the full γ-ray energy can be reconstructed and both the efficiency and the P/T ratio of the array improve drastically. Two such arrays have been specifically designed and built as γ-ray energy tracking arrays over the past decade: the GRETINA [5] and the AGATA [6] arrays. Both arrays cover a one π of the full solid angle and have been successfully completed. In the last couple of years they have started serving the nuclear physics community around the world.

Gamma-ray tracking uses the Compton relation to group and order the individual γ ray interaction positions inside the crystals for each γ ray. In order to perform γ ray tracking, it is required to know precisely the coordinates of each γ ray interaction position and the deposited energy in this position. In the 3D volume of large coaxial detectors with high electrical segmentation this can be achieved by recording the pulse shapes of the signals that are generated as the charge drifts towards the electrodes. The rise time of the signal depends on the drift distance between the segment (or central contact) electrode and the interaction position. In addition, the neighbouring segments see an image charge of opposite polarity during the drift of the charge in the hit segment. The magnitude of the image charge is related to the distance between this segment and the hit position inside the segment where the interaction occurred. In reality, the final signal shapes that the electrodes record for each event become rather complicated as they are folded with cross-talk effects, multiple interactions inside one segment, electronic shaping and noise. The event-by-event on-line reconstruction of the interaction positions is performed by comparison of the measured pulse shapes with a basis of simulated pulses. This is a computationally intense task and such arrays are coupled to dedicated computer farms that perform the reconstruction in real time.

4. Summary
The high-intensity radioactive-ion beams from the upgraded accelerator facilities coupled to the state-of-the-art experimental tools and techniques that are currently being developed open up the way for the nuclear physics community to explore the structure and properties of exotic nuclei that were not accessible so far. The FAIR facility and the NUSTAR experiments will play a leading role in these investigations.

References
[1] Technical Design Report of CALIFA Barrel: The R3B CALorimeter for In Flight detection of γ rays and high energy charged pArticles (2011)
`http://www.fair-center.eu/fileadmin/fair/publications\_exp/CALIFA\_BARREL\_TDR\_web.pdf`

[2] Technical Design Report of NeuLAND: The High-Resolution Neutron Time-of-Flight Spectrometer for R^3B (2011)
`http://www.fair-center.de/fileadmin/fair/experiments/NUSTAR/Pdf/TDRs/NeuLAND-TDR-Web.pdf`

[3] Technical Proposal for the Design, Construction, Commissioning and Operation of R^3B (2005)
`http://www.gsi.de/onTEAM/grafik/1130224398/13-PROP-R3B-TP-07Dez2005.pdf`

[4] M. A. Deleplanque, et al., Nucl. Instr. Meth. A **430** (1999) 292 – 310.

[5] S.Paschalis, et al., Nucl. Instr. Meth. A **709** (2013) 44 – 55.

[6] S. Akkoyun, et al., Nucl. Instr. Meth. A **668** (2012) 26 – 58.

TRIGA-SPEC: the prototype of MATS and LaSpec

S. Kaufmann[1,2], T. Beyer[3], K. Blaum[3], M. Block[4,5], Ch. E. Düllmann[4,5,6], K. Eberhardt[1,5], M. Eibach[1,3], C. Geppert[1], C. Gorges[1,2], J. Grund[6], M. Hammen[1], J. Krämer[1,2], Sz. Nagy[3,4], W. Nörtershäuser[1,2], D. Renisch[1], F. Schneider[1,7], K. Wendt[7]

[1]Institut für Kernchemie, Johannes Gutenberg-Universität, 55128 Mainz; [2]Institut für Kernphysik, Technische Universität Darmstadt, 64289 Darmstadt; [3]Max-Planck-Institut für Kernphysik, 69117 Heidelberg; [4]GSI Helmholtzzentrum für Schwerionenforschung GmbH, 64291 Darmstadt; [5]Helmholtz-Institut Mainz, 55099 Mainz; [6]PRISMA Cluster of Excellence and Institut für Kernchemie, JGU Mainz, 55128 Mainz; [7]Institut für Physik, Johannes Gutenberg-Universität, 55128 Mainz

E-mail: s.kaufmann@uni-mainz.de

Abstract. Investigation of short-lived nuclei is a challenging task that MATS and LaSpec will handle at the low energy branch of Super-FRS at FAIR. The groundwork for those experiments is laid-out already today at the TRIGA-SPEC facility as a powerful development platform located at the research reactor TRIGA Mainz. The latest status, new developments and first results of commissioning runs are presented here.

1. Introduction

LaSpec (**La**ser **Spec**troscopy of short-lived nuclei at FAIR) [1] and **MATS** (Precision **M**easurement of very short-lived nuclei using an **A**dvanced **T**rapping **S**ystem) [2] are both experiments for investigating exotic short-lived nuclei which will be produced at the low energy branch at FAIR using atomic physics techniques of laser spectroscopy and mass spectrometry. LaSpec will use collinear laser spectroscopy (CLS) to obtain charge radii from isotope shifts, nuclear moments and spins from hyperfine structure measurements in ions and atoms. MATS will perform precise mass measurements and in-trap decay spectroscopy. While FAIR just started construction, the development platforms already now provide the opportunity to optimize the experiments for their future application.

2. TRIGA-SPEC

Situated at the research reactor TRIGA Mainz the TRIGA-SPEC experiment benefits from the possibility to create short-lived nuclides by neutron-induced fission of a heavy actinide target, e.g. ^{249}Cf. A gas chamber containing the target is placed close to the core of the reactor. Fission fragments are stopped inside the chamber and transported through the biological shield of the reactor by an aerosol gas-jet and guided into the surface ion source. The source is operated on a high voltage platform and the ions are accelerated into the beamline towards a 90° mass separator on ground potential. The mass separation is followed by an RFQ cooler and buncher used to reduce the emittance and transform it into short ion bunches, which is critical for both experiments [3]. The cooling leads to a higher transport efficiency and a reduced

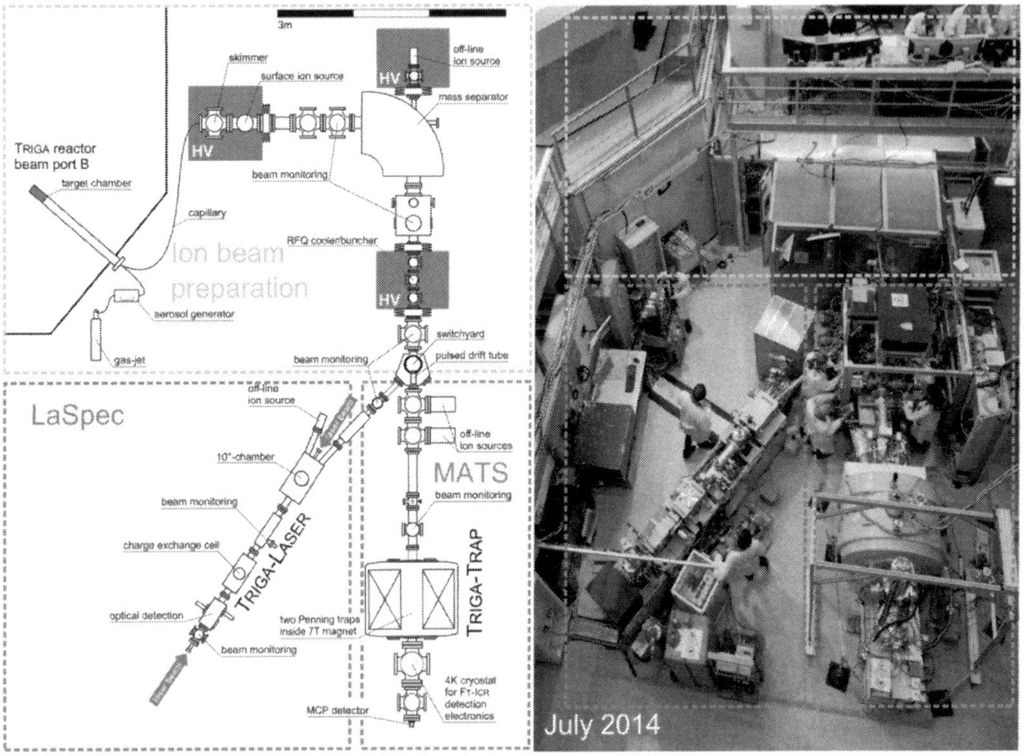

Figure 1: left: Floorplan of the TRIGA-SPEC experiment, detailed description in the text. right: picture inside the TRIGA hall with the TRIGA-SPEC experiment connected to beamport B.

broadening of spectral lines thanks to the reduced emittance. The CLS experiment benefits from the pulsed beam structure since the deterministic passing time of the ion bunch through the optical detection region (ODR) allows restricting the photon detection to the temporal length of the ion bunch (typically a few μs). This leads to a tremendous (up to 10^4) suppression of the constant background by the continuous laser light. Thus CLS on ions at rates down to 100 ions/s is feasible [4]. The Penning trap experiment benefits from the bunch structure by an increased injection efficiency into the trap. Both branches can be supplied with ion bunches in turns by use of fast switching electrodes in the switchyard.

2.1. *TRIGA-TRAP*

The Penning-trap mass spectrometer TRIGA-TRAP is used routinely to determine the masses of various nuclides with a relative precision $\Delta m/m$ of down to $5 \cdot 10^{-9}$ [5, 6, 7]. These measurements are performed so far with the well-established time-of-flight ion-cyclotron-resonance detection technique (TOF-ICR) [8]. The implementation of advanced measurement techniques such as Fourier-transform ICR [9] and phase-imaging ICR [10] is currently being prepared. The Penning trap is a superposition of a strong homogeneous magnetic field and a weak electric quadrupole field, which confines charged particles in three dimensions. Ion motion inside the trap is a superposition of an axial motion and two radial motions: a slow magnetron and a rapid modified cyclotron motion. The mass m of a charged particle orbiting around the field lines of a magnetic field B is linked to its cyclotron frequency $\nu_c = qB/(2\pi m)$, where q denotes the charge of

the particle. The magnetic field B is calibrated by frequency measurements of a reference ion. TRIGA-TRAP exclusively uses carbon clusters $^{12}C_x$ as reference ions because the mass of carbon defines the current mass standard ($m(^{12}C)$=12 u).
TRIGA-TRAP was constructed as a test bench for the MATS experiment. However, its location at the TRIGA reactor in Mainz offers access to neutron-rich fission products with mass numbers from 80 to 140. Mass measurements will reveal information about the evolution of shell structure, collectivity and shell closures [11] as well as second-order differentials such as odd-even staggering of nuclear masses [12]. TRIGA-TRAP also features a powerful off-line laser ablation ion source [13] for the measurement of stable nuclides all over the nuclear chart. Recent highlights include mass measurements around the deformed $N = 152$ neutron shell closure, which were performed on very small sample sizes of 10^{15} atoms by coupling the ion source to a miniature RFQ [6], and Q-value measurements for neutrino physics applications [14].

2.1.1. Electrostatic deceleration stage
As the TRIGA-SPEC ion source and RFQ cooler and buncher are biased on 30 kV potential, the produced ion bunch must be decelerated prior to the injection into the Penning-trap mass spectrometer. Therefore, at TRIGA-TRAP, an electrostatic deceleration stage has been built as described in, e.g. [15], to slow down incoming ion bunches prior to injection. As the deceleration of an ion bunch from the kinetic energy E_{high} to E_{low} increases its emittance by a factor of $\sqrt{E_{\text{high}}/E_{\text{low}}}$, the reduction of the ions' motional amplitudes inside the RFQ is critical. The ion bunch is then transported at a low energy (1 keV) to the Penning-trap setup.

2.1.2. The double Penning-trap setup
TRIGA-TRAP consists of two Penning traps. In the first one, the purification trap, the mass-selective buffer-gas cooling technique [16] is applied. The typical mass resolving power $m/\Delta m$ is around 10^4 and sufficient to separate isobaric contaminants from the wanted ion species. The purified ion ensemble is then transported through a small aperture, which also acts as a differential pumping barrier, to the second Penning trap, the so-called precision trap. The precision trap features an ultra-high vacuum and a hyperbolic shape to minimize electric-field inhomogeneities. In the TOF-ICR procedure, the eigenfrequency of the ion ensemble is probed by the application of a quadrupolar RF field with frequency ν_{q} around ν_{c}. Then, the ion ensemble is ejected towards an MCP detector outside the magnetic field and the time of flight is recorded. A resonant quadrupole excitation results in a minimum time of flight, and the corresponding frequency is used to extract the mass. The control and data acquisition system at TRIGA-TRAP is based on the CS framework developed at GSI [17].
In order to demonstrate the successful transport of ions from the on-line ion source into the Penning traps, a TOF-ICR of $^{85}Rb^+$ has been recorded (see Fig. 2).

2.2. TRIGA-LASER
The ions reach this branch by a right turn in the switchyard followed by a transfer tube. In the following 10°-chamber the ions are deflected by a pair of electrodes to overlap with the laser light that is coupled into the beamline through a window either in collinear or in anticollinear geometry with respect to the ion motion. The ions will then pass the charge exchange cell (CEC), where they can be neutralized if an optical excitation scheme in the atom is favored. This is followed by the ODR, where two photomultiplier tubes supported by a two stage mirror system detect the fluorescence light. Resonance spectra are recorded by changing the ion/atom velocity by applying a voltage of up to 10 keV to the ODR. The change of the ion velocity leads to a corresponding change of the laser frequency in the ion's restframe according to $\partial\nu/\partial U = \nu_{\text{Laser}} e/\sqrt{2eUmc^2}$. This is an advantage since the laser frequency can be kept fixed in the laboratory frame.

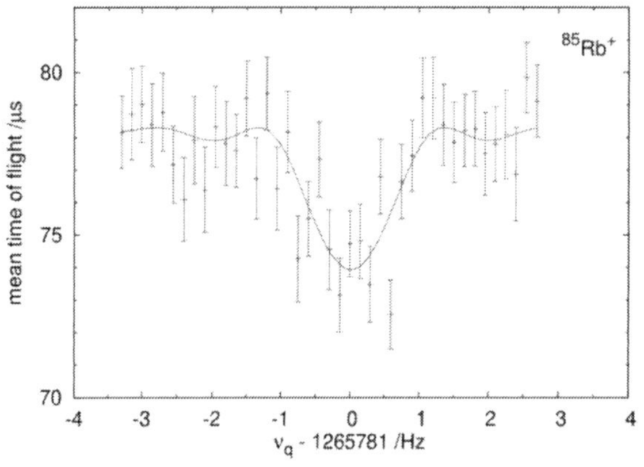

Figure 2: First TOF-ICR spectrum at TRIGA-TRAP of an ion species produced in the online surface ionization ion source, in this case $^{85}Rb^+$. The black line is the fit of the theoretical curve to the data points.

2.2.1. TRIGA pyThon cONtrol system - TRITON

In order to efficiently operate and control the setup, we have developed a control system which is custom-built for the slow controls of our experiment. Python was chosen as the programming language. TRITON is based on a modular distributed structure and the communication between different computers is realized using network sockets. This offers the opportunity to control all devices which are needed to prepare and monitor the experiment from any PC inside the network. The modularity of TRITON thereby makes it as simple as possible for the user to add a new device and reuse as much code as possible while programming a new device.

2.2.2. TRIGA Laser Data Acquisition - TILDA

TILDA is the completion to TRITON and will be responsible for the fast experiment control and the data acquisition (DAQ). CLS on bunched beams requires a separation between the photons detected during the bunch passing time and the background photons. This can be realized by recording each photon with a time stamp and therefore recording the complete timing information for off-line analysis. We developed a prototype for this DAQ with a small (Virtex-II) field programmable gate array (FPGA) and tested it successfully while commissioning the RFQ (see 2.2.4). TILDA is the follow up of this prototype and will use two independent FPGAs (Virtex 5) located in one PXI rack. One is foreseen for the fast control, e.g. loading and extracting the RFQ, the other one is acquiring data in parallel. Synchronisation is realized through the back plane of the PXI.

2.3. Commissioning of the RFQ cooler and buncher using CLS

The commissioning of bunched beam CLS with the RFQ had the goal to optimize the RFQ settings in order to create bunches with a small longitudinal emittance and hence a low energy spread and a short bunch length to achieve high background suppression and high resolution. With the time resolved DAQ, both parameters could be monitored online while optimizing the RFQ settings. The well-known D1 transition in stable Ca^+ ions was chosen for CLS with the highest abundance in $^{40}Ca^+$. The bunch length after optimization for CLS is about 0.5 μs with an energy spread of approximately 9 eV at full width at half maximum. The corresponding linewidth is about 150 MHz which is compared to a natural linewidth of $\approx$ 20 MHz still subject for further improvement. The typical TRIGA-SPEC RFQ operating parameters obtained during commissioning are listed in Tab. 1. With these parameters we measured the D1 line of the isotopes $^{42,43,44,48}Ca^+$. The least abundant one of these is $^{43}Ca^+$ with a natural abundance of 0.135% and the resonance spectra splits up into four peaks due to hyperfine splitting. The density plot of the photon signal is shown in Fig. 3 (top) with the Doppler tuning voltage on the x-axis and the time relative to the RFQ extraction pulse on the y-axis. The fluorescence rate is color-coded. Besides the resonance positions it is also observable that the photon rate is increasing while the bunch is passing by, i.e. in the time period between 21 and 21.6 μs. This is due to inelastic residual gas collisions that lead to the excitation of the residual-gas and subsequent photon emission. In the lower part of Fig. 3, the projection on the x-axis is shown with a multiple Voigt profile fitted to the data.

Parameter	Symbol	Value
radio frequency	ν_{RF}	1.0 MHz
RF amplitude	V_{RF}	100 V
axial well depth	U_{DC}	6–10 V
capture voltage	$U_{15\mathrm{hi}}$	25 V
buffer-gas pressure	p_{RFQ}	$\sim$4.6·10^{-3} mbar
accumulation time	t_{acc}	0.1–100 ms
cooling time	t_{cool}	10–20 ms
ejection frequency	f_{eject}	40 Hz
ejection voltage	$U_{15\mathrm{lo}}$	40 V
kicking voltage	$U_{13\mathrm{hi}}$	40 V

Table 1: Standard parameters for operation of the TRIGA-SPEC RFQ in bunching mode.

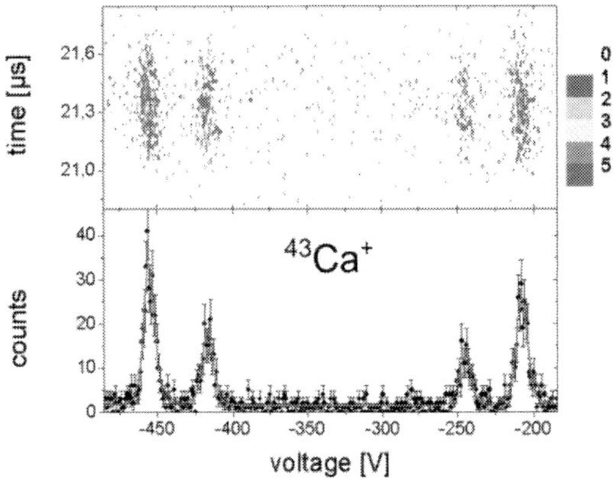

Figure 3: CLS spectra of $^{43}Ca^+$ bunches, obtained during commissioning. For details, see text.

Acknowledgments
The TRIGA-SPEC collaboration would like to thank the Bundesministerium für Bildung und Forschung (BMBF), the Max-Planck society, the Helmholtz Graduate School for Hadron and Ion Research (HGS-HIRE), the Precision Physics, Fundamental Interactions and Structure of Matter (PRISMA) cluster of excellence and the Helmholtz-Institut Mainz (HIM) for financial support.

References
[1] Nörtershäuser W, Campbell P and the LaSpec collaboration 2006 *Hyperfine Interact.* **171** 149-56
[2] Rodriguez D, Blaum K, Nörtershäuser W et al. 2010 *Eur. Phys. J. Spec. Top.* **183** 1-123
[3] Beyer T et al. 2014 *Appl. Phys* B **114** 129–36
[4] Nieminen A, Campbell P et al. 2002 *Phys. Rev. Lett.* **88** 094801
[5] Ketelaer J, Krämer J et al. 2008 *Nucl. Instr. Meth.* A **594** 162–77
[6] Eibach M et al. 2014 *Phys. Rev.* C **89** 064318
[7] Smorra C et al. 2012 *Phys. Rev.* C **86** 044064
[8] König M, Bollen G, Kluge H J, Otto T and Szerypo J 1995 *Int. J. Mass Spectrom. Ion Process.* **142** 95
[9] Ketelaer J et al. 2009 *Eur. Phys. J.* A **42** 311–17
[10] Eliseev S et al. 2014 *Appl. Phys.* B **114** 107–28
[11] Lunney D, Pearson J M and Thibault C 2003 *Rev. Mod. Phys.* **75** 1021–82
[12] Satula W, Dobaczewski J and Nazarewicz W 1998 *Phys. Rev. Lett.* **81** 3599–602
[13] Smorra C et al. 2009 *J. Phys.* B **42** 154028
[14] Smorra C et al. 2012 *Phys. Rev.* C **85** 027601
[15] Herfurth F et al. 2001 *Nucl. Instr. Meth.* A **469** 254–75
[16] Savard G et al. 1991 *Phys. Lett.* A **158** 247–52
[17] Beck D et al. 2004 *Nucl. Instr. Meth.* A **527** 567–79

An idea for the future proton detection of (p,2p) reactions with the R^3B set-up at FAIR

G Ribeiro[1], M Mårtensson[2] and A Perea[1]

[1]Instituto de Estructura de la Materia-CSIC, E-28006 ,Madrid, Spain
[2]Chalmers University of Technology, S-41296, Göteborg, Sweden

E-mail: g.ribeiro@csic.es

Abstract. The R^3B Collaboration has a long experience in probing exotic nuclei via quasi-free scattering reactions. To continue these studies a new array capable of detecting protons and gamma rays of high energy is currently being developed, the CALIFA (R^3B CALorimeter for In Flight γ arrays and high energy charged pArticles). This contribution reports on the current solution for the forward Endcap of the CALIFA detector and on the latest test results.

1. Quasi-Free Scattering reactions from GSI to FAIR

The R^3B Collaboration studies exotic nuclei using RIBs by means of quasi-free scattering (QFS) reactions as a tool to access the single particle properties in the nuclei. In this kind of reactions an incoming proton with medium-high energy (a few hundreds of MeV) knocks out a bound nucleon. Since the cross-section of the nucleon-nucleon interaction is small at these energies, the reaction can be interpreted with the impulse approximation (quasi-free). As both outgoing particles are nucleons of same mass, they are knocked out with an opening angle between 80° and 90° in the laboratory system. For a general review of QFS see [1][2][3].

In order to perform a complete kinematics analysis using QFS reaction, it is mandatory to detect the momentum vector of the outgoing nucleons and gammas from the de-excitation of the recoil nucleus. A new experimental setup (R3B) for this purpose is currently under design and construction for the future FAIR facility. One of the new systems is the gamma-ray and proton calorimeter-array CALIFA that will be surrounding the target position. This detector is composed of two sections: Barrel and EndCap (see figure 1). A Technical Design Report of its Barrel section has already been approved by the FAIR management board[4]. The inner section of the endcap will receive the highest energy gammas and protons, which typically means protons up to 400-700 MeV and gamma rays up to 30 MeV, this part will be covered by a $LaBr_3/LaCl_3$ Phoswich detector CEPA(CALIFA EndCap Phoswich Array).

2. The Phoswich approach

The collaboration sought for a solution able to fulfill the requirements and a phoswich array in the endcap section is the current answer. The first prototype was presented in [5]: two scintillators were optically coupled together as it is shown in the figure 2, using a common PMT as a readout. In order to choose the scintillator material several properties have to be taken into account; the first condition is that they must be optically compatible, meaning that the second crystal has to be transparent to the scintillation light produced by the first crystal. The

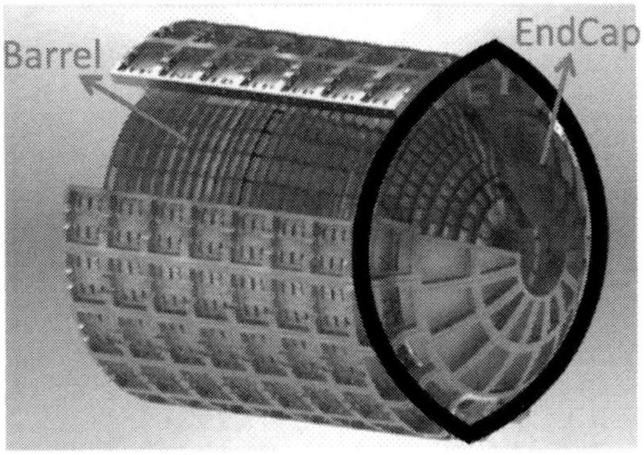

Figure 1. Model of the CALIFA detector, the barrel and Endcap section are visualised. The Barrel section will be surrounding the target position while the endcap corresponds to the most forward part of the detector, devoted to detect the higher-energy gamma rays and protons that continue in the beam direction.

second condition is to have low internal radioactivity. The materials $LaBr_3$ and $LaCl_3$ both fulfill previous conditions and have a set of good properties (see table 1): a resolution between 3-4% and a very good light output production of 63 and 49 photons/keV respectively. The difference in the decay time is a key parameter which opens the possibility of the pulse-shape analysis for doing Particle Identification and reconstructing the incident energy from the energy-loss measurement.

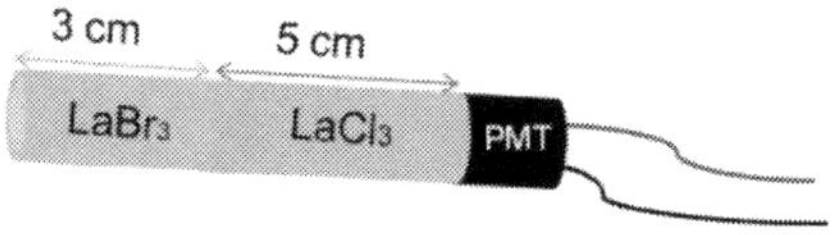

Figure 2. The first prototype was composed of a 3 cm $LaBr_3$ optically coupled to a 5 cm $LaCl_3$ crystal.

Table 1. Summary of the relevant properties of the scintillators for the phoswich array.

Material	Resolution at 662 keV	Light yield (photons/keV)	Decay Time (ns)	$\lambda_{emision}$(nm)
$LaBr_3$	2.9	63	16	380
$LaCl_3$	3.8	49	28	350

Coupling two different scintillator crystals together will produce two different light signals that are detected together in a common readout. The pulse can be carefully analysed by comparison of the integral of the whole signal against the tail. This tail depends on two parameters: the decay-time of the light emission in the material and the kind of interacting particle, thus it will allow us to separate the different light outputs from both crystals (see figure 3).

Due to the linear response of the material, it is possible to correlate the total integral directly with the tail integral as follows:

$$I_{Cl}^{tail} = a_{Cl} I_{Cl}^{total} \quad (1)$$

$$I_{Br}^{tail} = a_{Br} I_{Br}^{total} \quad (2)$$

$$I = I_{Cl} + I_{Br} \quad (3)$$

Using these three equations it is easy to solve the system of equations:

$$I_{Br}^{total} = \frac{I^{tail} - a_{Cl} I^{total}}{a_{Br} - a_{Cl}} \quad (4)$$

$$I_{Cl}^{total} = \frac{a_{Br} I^{total} - I^{tail}}{a_{Br} - a_{Cl}} \quad (5)$$

This allows the full separation of the two different contributions, and an energy calibration can be performed. The energy is directly proportional to the signal integral $I = \beta E$ where the constant is dependent on the material, thus it can be calibrated using data from previously known energies.

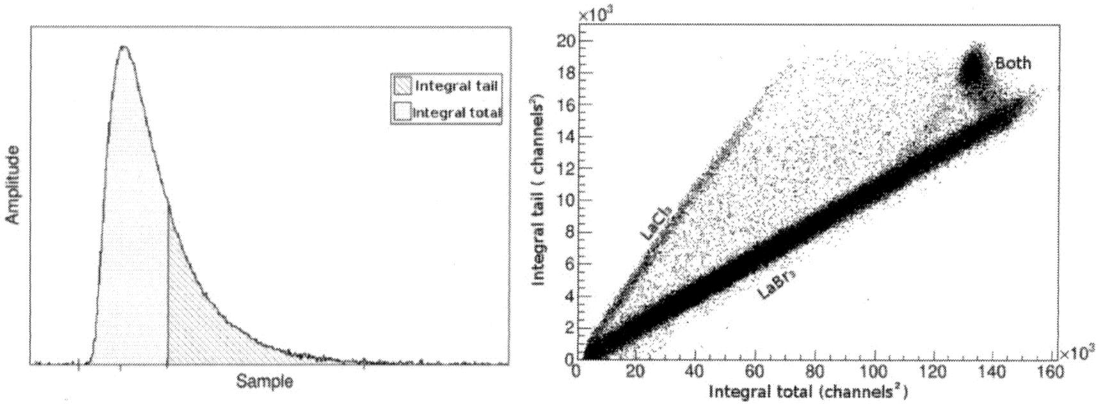

Figure 3.
Left: Typical sampled pulse signal from the photomultiplier, the total integral and the integral of the tail are indicated. The tail threshold must be calibrated in every experiment in order to optimise the separation of the signals.
Right: Integral Total *vs* Integral tail for 150 MeV protons. The two straight lines corresponds to the signals coming from only one of the scintillators, the slope difference can be easily appreciated, thus fitting to the lines gives us the needed parameters to solve the equations 4 and 5. The blob corresponds to the 150 MeV protons that loss the full energy within the phoswich, punching through the $LaBr_3$ crystal and producing the Bragg peak inside the $LaCl_3$.

3. Current prototype: CEPA4

The next step towards the final CALIFA design was to produce a small phoswich array called CEPA4, composed of four phoswich units of 4 cm of $LaBr_3$ and 6 cm of $LaCl_3$ with a common readout. The crystals are optically insulated from each other by a 1 mm of teflon reflector and packed together in an aluminium case 0.2 mm thick (see figure 4). The characterisation of the prototype and its response to high-energy protons was recently published [6].

Journal of Physics: Conference Series **599** (2015) 012034 doi:10.1088/1742-6596/599/1/012034

Figure 4.
Left: Scheme of CEPA4 with the four phoswich units (grey-$LaBr_3$ and red-$LaCl_3$) the PMTs are painted in green. The $LaBr_3$ faced the beam direction in the test beam.
Right: A picture of CEPA4 in a test experiment.

The CEPA4 proton response has been tested at the Bronowice Cyclotron Centre(CCB) in Krakow which provided a proton beam with energies from 70 to 235 MeV, allowing us to test the prototype in a broad energy range. At this beam test we were seeking protons with high energy that punch-through the whole phoswich, this was expected to happen around 200 MeV. Thus following the same procedure explained in the previous section the two different signals where separated and their contributions calibrated to produce the figure 5.

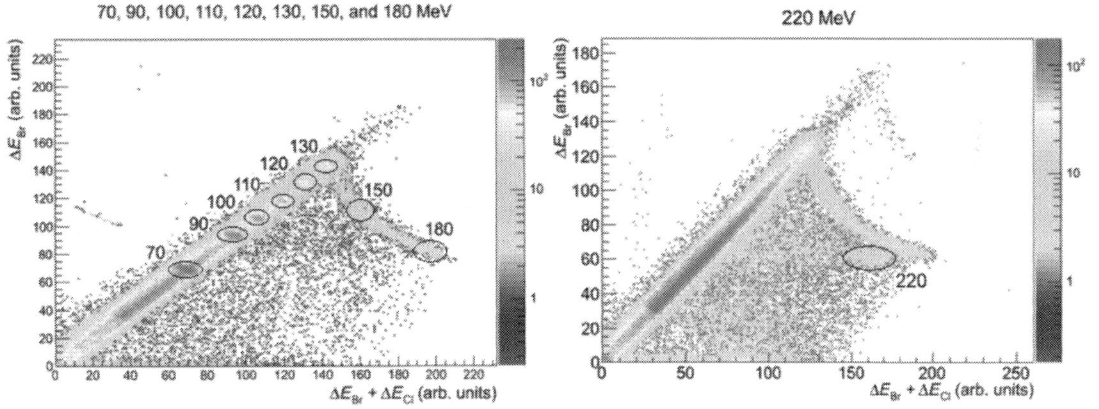

Figure 5. Calibrated Total Energy vs Energy Loss in the first crystal ($LaBr_3$).
Left (a): From 70 to 180 MeV the different blobs along the line corresponds to the Bragg peaks of the different beam energies. Up to 130 MeV are protons stopped in the first crystal($LaBr_3$) and from 140 MeV up to around 200 MeV are stopped in the second crystal ($LaCl_3$).
Right (b): The 220 MeV protons punch-through the whole phoswich, thus leaving energy in both crystals without any Bragg peak.

Protons that do not leave the Bragg peak within our detector are losing less energy on both crystals, thus the signal produced is smaller and as we are plotting two contributions in a two dimensional plot that difference can be observed. As a further proof of the goodness of the method, the higher energy protons blob has been displayed in figure 6 in order to check the good separation between the different contributions. The measured energy resolution is 7% [6] for 220 MeV protons.

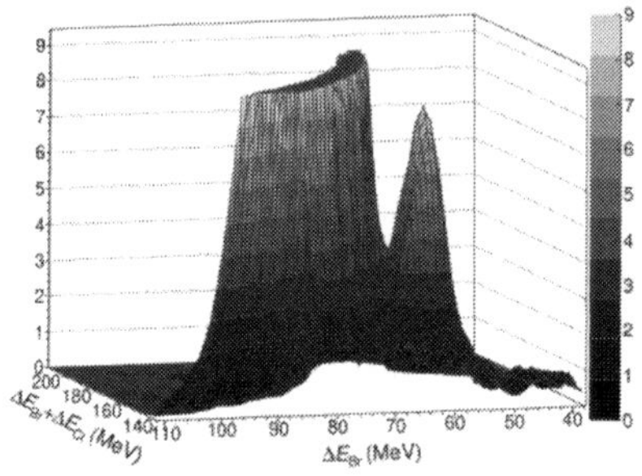

Figure 6. Three dimensional representation of the right graph of the figure 5b and zoomed close to the 220 MeV spot.

4. Conclusions

It has been described a solution that is being tested for the CALIFA EndCap. The results using the prototype CEPA4 were presented while introducing the potential of this system to separate protons that do not loose full energy within the detector.

Acknowledgments

This work was partly financed by the Spanish Research Funding Agency CICYT under Project FPA2012-32443. Further Guillermo Ribeiro acknowledges his FPI-MICINN grant.

References

[1] P.G. Hansen: Annu. Rev. Nucl. Part. Sci. **53** (2003) 219
[2] P. Kitching et al: Adv. Phys. Part. Nuclei **15** (1985) 43
[3] T. Aumann et al: Phys. Rev. C **88** (2013) 064610
[4] The CALIFA Collaboration, CALIFA Barrel Technical Design Report, 2012, URL: http://www.fair-center.eu/for-users/experiments/nustar/nustar-documents/technical-design-reports.html
[5] O. Tengblad et al: Nuclear Instruments and Methods in Physics Research A **704** (2013) 19
[6] E. Nacher et al: Nuclear Instruments and Methods in Physics Research A **769** (2015) 105

Adaptive triggering for scintillation signals

J. Vesic[1], M. Vencelj[1], K. Strnisa[2], D. Savran[3]

[1] Jozef Stefan Institute, Ljubljana, Slovenia
[2] Cosylab d.d., Ljubljana, Slovenia
[3] EMMI, GSI, Darmstadt, Germany; FIAS, Frankfurt am Main, Germany

E-mail: jelena.vesic@ijs.si

Abstract. Due to the stochastic nature of the pulse creation in a scintillation detector the output pulses are not all of the same shape but rather 'noised' with statistical fluctuations on the pulse tails, which may induce false triggers. The current state of the art in solving this kind of problems is either introducing a deadtime after each pulse which makes the detector inefficient at higher count rates or raising the trigger threshold above the fluctuactions level, which on the other side, lowers the dynamic range of the detector from the low energy side. In order to meet the ever growing demand for high precision/efficient experiments the solutions to these limitations are highly desirable. We propose a new method, the adaptive triggering for scintillation signals.

1. Introduction

Scintillation detectors will be an important part of detection systems in FAIR (Facility for Antiproton and Ion Research). In order to extract maximum information from the scintillation pulses, improved signal processing is required.

Precision of signal magnitude is limited by fluctuations. We can differ between two types of fluctuations. The first one are the fluctuations due to the stohastic nature of the pulse creation and the second are the baseline fluctuations due to the electronics. Main contribution to the fluctuations in the scintillator signals originates from the statistical fluctuations. Namely, the output pulses are not all the same shape but noised with statistical fluctuations on the pulse tails, which give rise to the afterpulsing. However, we have to note that some non-statistical processes also can give rise to the afterpulsing, namely the emission of light from the latter stages of the multiplier structure which finds its way back to the photocatode [1]. Also, another cause of afterpulsing can be an imperfect vacuum within the tube. Residual gas can be ionized by the passage of electrons through the multiplier structure. Formed positive ions will drift in the reverse direction and eventually some may find a path back to the photocatode [1].

The pulses from the scintillation detectors have exponential shapes. The main role of the signal processing that happens afterwards is to create a pulse which amplitude is proportional to the deposited energy. In the traditional analog pulse shaping, a combination of two basic filters has been used, namely CR and RC filters. The CR filter is a high pass filter and it allows only high frequency components of the pulse to pass, blocking the DC component of the signal. An integrator is a low pass filter, it only passes the low frequency components of the signal and modifies the differentiated pulse by widening its maximum. However if the original exponential pulse is noised with statistical fluctuations, the semigaussian processed signal will have the statistical fluctuations as well, which might cause detection of false pulses or the information on amplitude value not correctly deduced. One of the digital processing methods (DSP) that is commonly used is a moving window deconvolution method (MWD) [2, 3]. We chose DSP over the analog processing mainly due to its flexibility (simple to change filter parameters etc.).

In our experiment, signal from a photomultiplier tube (PMT) was firstly sampled with a fast digitizer to reconstruct the pulse and the MWD method was used consequently for the offline DSP. The concept of the MWD method is similar to the analog pulse shaping. The output of a PMT is a fast rising step followed by an exponential relaxation. In order to be able to measure its amplitude, it is favorable to reshape signal to a more readable form. One of the algorithms for shaping it is the MWD algorithm. The single exponential decay with an initial time t=0 and decay time τ is written as:

$$f(t) = \begin{cases} Ae^{-t/\tau}, & t \geq 0 \\ 0, & t < 0 \end{cases} \tag{1}$$

Knowing the value $f(t_n)$ at time t_n, the initial amplitude can be found as:

$$A = f(t_n) + A - f(t_n) = f(t_n) + A\left(1 - e^{-\frac{t_n}{\tau}}\right) = f(t_n)\frac{\int_0^{t_n} f(t)dt}{\tau}. \tag{2}$$

In order to use the previous equation in digital domain, it should be written in a discrete form:

$$A[n] = x[n] + \frac{\sum_{k=0}^{n-1} x[k]}{\tau} = \mathrm{x[n]} - \left(1 - \frac{1}{\tau}\right)\mathrm{x[n-1]} + \mathrm{A[n-1]}. \tag{3}$$

The deconvolution equation reshapes the exponential pulse into a staircase signal. To measure the height of the step signal one should differentiate the deconvulated signal. By applying the numerical differentiation, the MWD equation with a window M is obtained [3]:

$$MWD_M[n] = A[n] - A[n-M] = x[n] - x[n-M] + \frac{1}{\tau}\sum_{k=n-M}^{n-1} x[k]. \tag{4}$$

The MWD algorithm converts an exponentially decaying signal into a step signal with length M. However, at this point the signal to noise ratio is not yet improven. To impove the signal to noise ratio, a low pass filter is applied after the MWD module, implemented as a moving window averager. In the MWD algorithm the shaped pulse is a trapezoid which emerges as a convolution sum of the digitized signal with a kernel function. After the MWD processing the signals with good statistics and well-defined exponential shape transform into trapezoids. However, the signals with bad statistics, 'noised exponential shape' convert into the 'noised' trapezoids with the statistical fluctuations on the trapezoid tails (see figure 1). These statistical fluctuations may produce false triggers.

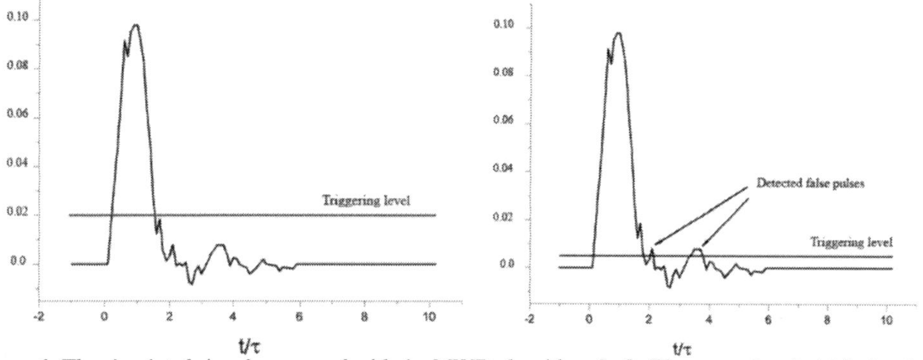

Figure 1. The simulated signal processed with the MWD algorithm. Left :The example of a high threshold. Right: The example of a low threshold.

2. The adaptive triggering method

The problem of the 'noised' pulse tails is solved at the time with the two approaches. The first one is to raise the trigger threshold above the fluctuations level. The second one is to introduce a deadtime after each pulse. However, both methods have their shortcomings. The low triggering level causes the detection of false pulses (see figure 1). On the other hand, with increasing triggering level above the fluctuation level, the false pulses will not be detected. However, the pulses which amplitudes are below the triggering level are also not detected and therefore the dynamic range of the detector is lowered (see figure 1). The second approach is to introduce a deadtime after each pulse. This approach can work fairly well at low count rates but at high count rates this approach becomes inefficient. Closely separated pulses get rejected, which manifests with a lower detection efficiency.

We propose a new method with an aim to improve the signal processing. The proposed method is the adaptive triggering for scintillation signals (see figure 2). In our metod the trigger is raised by a finite value and for a finite amount of time after each pulse. Namely, after the detected pulse, the threshold is set to a low level. Immediately after the pulse is detected the threshold is raised by a finite value and for a finite value of time. After a finite value of time the threshold is lowered to the previous value. In this way the afterpulses which occur immediately after the detected pulse are not detected (It should also be noted that the probability of the ocurrence of this pulses decreases with a time from the real pulse due the their stohastic nature). At the same time the small pulses that precede or follow the real pulse can be detected.

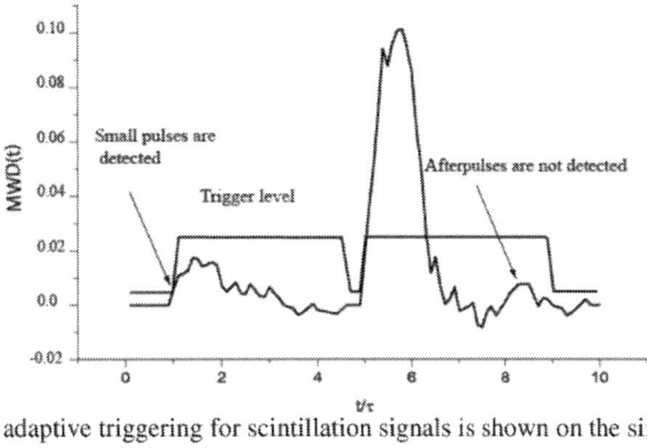

Figure 2. The method of adaptive triggering for scintillation signals is shown on the simulated signal. The small pulses are detected, while the afterpulses are not detected.

In order to determine the adaptive triggering parameters (the value and the time for which the trigger is raised) and in order to test our method, we firstly simulated the simple signals. The energy spectrum of the simulated signal is given as:

$$\frac{1}{N_0}\frac{dN(E)}{dE} = \frac{1}{2}\left(c\delta(E-E_1) + (1-c)\delta(E-E_2)\right), c = \frac{N_1}{N_o}, \quad (5)$$

where N is a total number of incident particles, and N_1 and N_2 represent the number of particles with energies E_1 and E_2, respectively.

As about the simulation parameters, for the beam frequency we chose the value of 10^5 Hz. For the decay time the value of 250 ns (which is fairly close to the decay time of the NaI(Tl) used in our experiment) was used. Our reference energy excited 1000 states, and for the bandwidth of energy histograms we choose the value of 200 MHz. The latter value corresponds to the bandwidth of the real measuring systems. The ideal energy spectrum should consists of two infinitely narrow peaks. Due to the finite precision of the simulated algorithm we expect the energy peaks to broaden. Also due to the afterpulses, we expect to have an increased low energy background. Due to the real pulses that appear immediately after the afterpulses (and therefore the concluded information about their amplitude is wrong) we also expect to have some background counts between the peaks.

As a criterion for determining the optimal triggering parameters we chose the relative number of low energy background counts, since this value strongly depends on the triggering parameters. For the optimal parameters we chose the values that after increasing do not decrease the relative number of background counts. Afterwards, the obtained optimal parameters for different energies were tested with the other simulated signal with simple energy spectrum given as:

$$h_{ideal}(\mathrm{E}) = \frac{1}{N_0}\frac{dN(E)}{dE} = \begin{cases} \frac{1}{E_1}, E < E_1 \\ 0, E > E_1 \end{cases}, \quad (6)$$

The same MWD parameters were used as in a previous case. The quality of parameters was deduced using a chi square method [4]:

$$\sigma_{spect} = \left(\frac{1}{n-1}\sum_{i=1}^{n}(h(E_i) - h_{ideal}(E_i))^2\right)^{1/2}, \quad (7)$$

where $h(E_i)$ denotes the measured value and $h_{ideal}(E_i)$ the correct value.

We employed the adaptive triggering by choosing the optimal parameters for the value and the time for which the trigger is raised, for every simulated pulse. As it can be seen from figure 3, the difference between the constant threshold and the adaptive triggering method is mostly pronounced at the lowest energies. With the constant threshold we basically lost the information about the low energy part of the spectrum, while with the adaptive triggering method this information is retained.

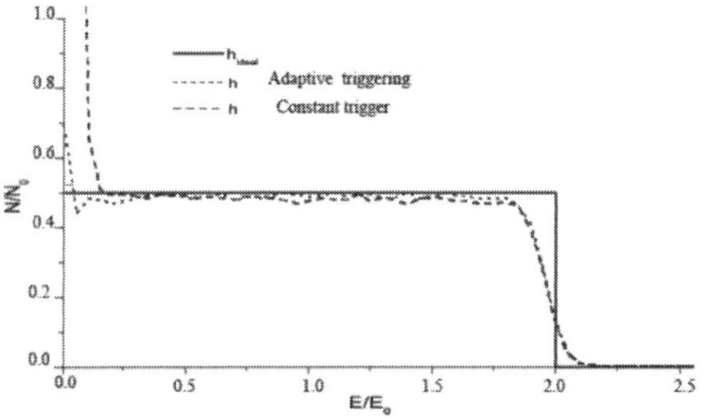

Figure 3. The energy spectrum given with eq. 7. The constant triggering (dashed line) and adaptive triggering method (dotted line) were employed.

In order to test the simulation parameters, we performed a simple experiment. As a radiation source ^{137}Cs was used. For the gamma ray detection NaI(Tl) scintillator coupled to a Photonis PMT was used. Raw anode output signal was sampled with a PicoScope digitizer with a sampling period of 200 ps and a resolution of 8 Bits. Two experimental setups were performed, by varying the distance of the source from the detection system. The one with a smaller pulse frequency was used as a reference measurement (i.e. for calibration, etc).

The obtained spectrum with employed constant threshold method is shown on figure 4. The K_α X-ray line from Ba [5] (decay product of ^{137}Cs) is not visible in the spectrum since the low energy part of the spectrum is noised up with the false pulses. With increasing the threshold the background lowers, however the dynamic range of the detectors lowers as well. Also, it can be noticed that the peak with energy approximately equal to the threshold energy is also detected. The peak is the consequence of the fact that the afterpulses are mainly detected at the energy near to the threshold energy.

The obtained spectrum with the adaptive threshold trigerring method is shown at figure 5. For the low values of the adaptive threshold, low energy peak is piled up with the afterpulses. For the higher values of the adaptive threshold, the low energy peak is clearly visible. However, also false peak with energy approximately equal to the constant threshold plus adaptive threshold can be noticed, as well. This peak is as in a previous case the consequence of the fact that the afterpulses are mainly detected at energy near to the threshold energy.

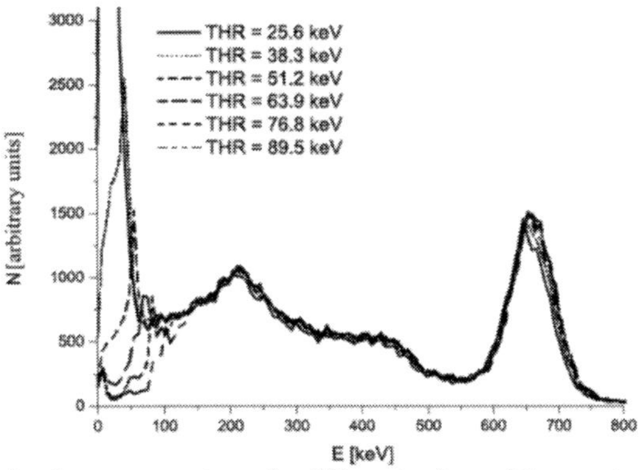

Figure 4.The obtained energy spectrum for different values of the constant threshold THR.

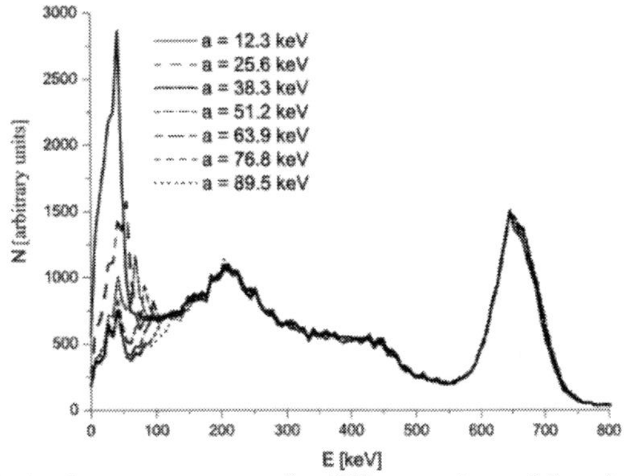

Figure 5 .The obtained energy spectrum for different values of the adaptive threshold a.

3. Conclusions

To conclude, we propose a new method for the signal processing. In our method, the trigger is raised by a finite value and for a finite amount of time after each pulse. The method was first tested on simple simulated signals. Consequently, based on the simulation results, the optimal triggering parameters were obtained. Using these parameters, adaptive triggering was then tested with a NaI(Tl) detector in an intense ^{137}Cs gamma-ray field. The results show that this method significantly extends the useful energy range on the low energy side of the spectrum when post-pulse inhibition is not applicable due to high count rates. This is crucial if the desired information is present there.
Last but not least, we have to state that our method is still in the development phase, more complicated synthetised signals should certainly be studied as well as the dependence of this method on the MWD parameters and initial pulse frequency.

4. References

[1] G.F. Knoll, Radiation Detection and Measurement, John Wiley & Sons, New York, (2000).
[2] H. Yang, D. K. Wehe, D. M. Bartels, Nucl. Instrum. Meth. A, **598**, 779 (2009).
[3] A. Georgiev, W. Gast R. M. Lieder, IEEE Trans. Nucl. Sci. 41, 1116 (1994).
[4] W. R. Leo, Techniques for Nuclear and Particle Physics Experiments, Springer, 1994.
[5] E. Browne, J. K. Tuli, Nuclear Data Sheets 108, 2173 (2007).

J. V. is grateful to European Social Fund and Republic of Slovenia, Ministry of Education, Science and Sport for financial support.

FAIRNESS 2014 IOP Publishing
Journal of Physics: Conference Series **599** (2015) 012036 doi:10.1088/1742-6596/599/1/012036

Radial modes of oscillations of slowly rotating magnetized compact hybrid stars

[1]**N R Panda,** [2]**K K Mohanta and** [1]**P K Sahu**

[1] Institute of Physics, Sachivalaya Marg, Bhubaneswar-751005, Odisha, India
[2] Rairangpur college, Mayurbhanj-757043, Odisha, India

E-mail: `pradip@iopb.res.in`

Abstract. Stars having a quark core surrounded by a mixed phase followed by hadronic matter are termed as hybrid stars(HS). For the equation of state(EOS) of hadronic matter, we have considered RMF(Relativistic Mean Field) theory, in the presence of strong magnetic fields leading to Landau quantization of the charged particles. For the EOS of quark phase we have used the simple MIT bag model also in presence of strong magnetic field. The bag pressure and magnetic field are parametrized to make density dependant. We have constructed the intermediate mixed phase by using Glendenning conjecture. The EOSs are constructed comprising the hadron, mixed and quark phases. Eigenfrequencies of radial pulsations of slowly rotating magnetised hybrid star are calculated using constructed EOSs in a general relativistic formalism given by Chandrasekhar and Friedman.

1. Introduction

If the central density of neutron stars exceed the nuclear saturation density ($n_0 \sim 0.15\,\mathrm{fm}^{-3}$), then the compact stars might contain deconfined quark matter in them. So along with the hadronic matter, HS have quark matter in their interior. In between the quark and hadronic phases, a quark-hadron mixed phase exists. The size of the core depends on the critical density for the quark-hadron phase transition and the EOS, describing the matter phases.

Recently[1] the mass measurement of millisecond pulsar PSR J1614-2230 and pulsar J1903+0327 have set a new mass limit for compact stars to be $M = 1.97 \pm 0.04\,\mathrm{M}_{\odot}$ and $M = 1.667 \pm 0.021\,\mathrm{M}_{\odot}$, respectively[2]. This measurement for the first time has set a very strong limit on parameters of the EOS of the matter under extreme conditions [3, 4]. New observations suggest that in some pulsars, the surface magnetic field can be as high as $10^{14} - 10^{15}$G. It has also been noticed that the observed giant flares, SGR 0526-66, SGR 1900+14 and SGR 1806-20 [5], are the manifestation of such strong surface magnetic field in those stars. If we assume flux conservation from a progenitor star, we can expect the central magnetic field as high as $10^{17} - 10^{18}$G. Such strong fields are bound to effect the properties of HS. It can modify the EOS of matter of the star. The effect of strong magnetic field, both for nuclear matter [6, 7, 8, 9, 10] and quark matter [11, 12, 13] have been studied earlier in detail.

Cameron [14] had suggested that vibration of neutron stars could excite motions that can have interesting astrophysical applications, there have been several investigations of the vibrational properties of neutron stars. Simple dimensional analysis suggests that the fundamental mode period of a vibrating neutron star would be of the order of milliseconds[15]. Typical neutron star model with mass about one solar mass and radius about 10 km give periods (3–5)ms, and the period is relatively insensitive to the exact value of the central density [16]. Since the rotation is a general property of all stellar bodies, it is interesting to study the normal modes of rotating HS. The determination of normal mode frequencies of a rotating HS is non trivial.

The value of period of oscillation strictly depends on the EOS along with some constraints on the parameters in both hadron and quark phase. Here we have investigated the effect of magnetic field on EOSs of both the matter phases (quark phase and hadron phase). Then we have constructed the mixed phase EOS in presence of strong magnetic field. Eigenfrequencies of radial pulsations of slowly rotating magnetised hybrid star are calculated in a general relativistic formalism given by Chandrasekhar & Friedman [17]. We have studied the square of the frequencies of the slowly rotating HS in the presence of strong magnetic field for various central densities.

2. Formalism

We have considered charge neutral, beta equilibrated matter. For the magnetic field we choose the gauge to be, $A^{\mu} \equiv (0, -yB, 0, 0)$, B being the magnitude of magnetic field. The momentum of the charged particles in x-y plane is quantized and hence the energy of the nth Landau level is [18] given by

$$E_i = \sqrt{{p_i}^2 + {m_i}^2 + |q_i|B(2n + s + 1)}, \tag{1}$$

where n=0, 1, 2, ..., being the principal quantum numbers for allowed Landau levels, $s = \pm 1$ refers to spin up(+) and down(-) and p_i is the component of particle(species i) momentum along the field direction. Setting $2n + s + 1 = 2\tilde{\nu}$, where $\tilde{\nu} = 0, 1, 2...$, we can rewrite the single particle energy eigenvalue in the following form

$$E_i = \sqrt{{p_i}^2 + {m_i}^2 + 2\tilde{\nu}|q_i|B} = \sqrt{{p_i}^2 + \widetilde{m}_{i,\tilde{\nu}}^2}\ , \tag{2}$$

where the $\tilde{\nu} = 0$ state is singly degenerate. The total energy density and pressure of the hadronic matter in presence of strong magnetic field can be written as, [19]

$$\begin{aligned} \varepsilon_{\rm HP} &= \frac{1}{2}m_{\omega}^2\omega_0^2 + \frac{1}{2}m_{\rho}^2\rho_0^2 + \frac{1}{2}m_{\sigma}^2\sigma^2 + \frac{1}{2}m_{\sigma^*}^2\sigma^{*2} + \frac{1}{2}m_{\phi}^2\phi_0^2 + \frac{3}{4}d\omega_0^4 + U(\sigma) \\ &\quad + \sum_b \varepsilon_b + \sum_l \varepsilon_l + \frac{B^2}{8\pi^2}, \\ {\rm P_{HP}} &= \sum_i \mu_i n_i - \varepsilon_{\rm HP}, \end{aligned} \tag{3}$$

where the last term in $\varepsilon_{\rm HP}$ is the contribution from the magnetic field.

The total energy density and pressure of the quark matter is given by [19],

$$\varepsilon_{\rm QP} = \sum_i \Omega_i + B_G + \sum_i n_i\mu_i, \qquad {\rm P_{QP}} = -\sum_i \Omega_i - B_G, \tag{4}$$

where B_G is the bag constant. With the above given hadronic and quark EOS, we now perform the Glendenning construction [20] for the mixed phase, which determines the range of baryon density, where both phases coexist. Allowing both the hadron and quark phases to be separately charged, and still preserving the total charge neutrality as a whole in the mixed phase. Thus the matter can be treated as a two-component system, and can be parametrized by two chemical potentials, usually the pair (μ_e, μ_n), i.e., electron and baryon chemical potential. To maintain mechanical equilibrium, the pressure of the two phases are equal. Satisfying the chemical and beta equilibrium the chemical potential of different species are connected to each other. The

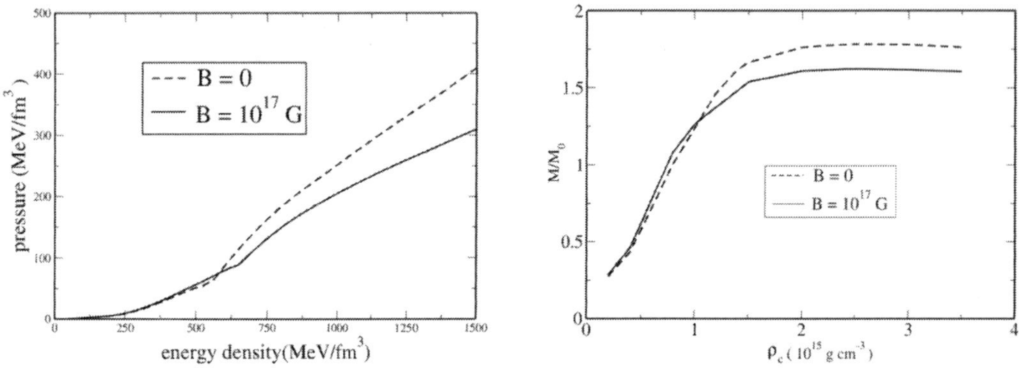

(a) (b)

Figure 1: (a)Pressure (MeV/fm^3) against energy density (MeV/fm^3) for zero and non zero value of magnetic field. B=0 (dotted line) and B=$10^{17}G$ (solid line). (b) Gravitational mass in solar mass unit against central density ($g\ cm^{-3}$) for zero and non zero value of magnetic field. B=0 (dotted line) and B=$10^{17}G$ (solid line).

Gibbs condition for mechanical and chemical equilibrium at zero temperature between both phases are given by

$$\mathrm{P_{HP}}(\mu_e, \mu_n) = \mathrm{P_{QP}}(\mu_e, \mu_n) = \mathrm{P_{MP}}. \tag{5}$$

This equation gives the equilibrium chemical potentials of the mixed phase corresponding to the intersection of the two phases. At lower densities below the mixed phase, the system is in the charge neutral hadronic phase, and for higher densities above the mixed phase, the system is in the charge neutral quark phase. As the two surfaces intersect, one can calculate the charge densities ρ_c^{HP} and ρ_c^{QP} separately in the mixed phase. If χ is the volume fraction occupied by quark matter in the mixed phase, we have

$$\chi\rho_c^{\mathrm{QP}} + (1-\chi)\rho_c^{\mathrm{HP}} = 0. \tag{6}$$

Therefore the energy density ϵ_{MP} and the baryon density n_{MP} of the mixed phase can be obtained as

$$\epsilon_{\mathrm{MP}} = \chi\epsilon_{\mathrm{QP}} + (1-\chi)\epsilon_{\mathrm{HP}}, \qquad n_{\mathrm{MP}} = \chi n_{\mathrm{QP}} + (1-\chi)n_{\mathrm{HP}}. \tag{7}$$

We parametrized the bag constant and magnetic field strength as given in Ref.[19].

The formalism given by Chandrasekhar & Friedman [17] is a fully general relativistic one to calculate the effect of rotation (to order Ω^2, where Ω is the angular velocity of rotation) on the eigenfrequencies of radial pulsations of stars. It gives an exact formula to calculate the frequency (σ') of oscillations of a 'slowly' rotating stellar configuration. The Chandrasekhar-Friedman formula is of the form:

$$\sigma'^2 I_1 = I_2 + I_3 + I_4. \tag{8}$$

The detail mathematical expression of I_1, I_2, I_3, I_4 are given in Ref.[21].

The equations governing infinitesimal radial pulsations of a non-rotating star in general relativity was given by Chandrasekhar [22], and it has the following form :

$$F\frac{d^2\xi}{dr^2} + G\frac{d\xi}{dr} + H\xi = \tilde{\sigma}^2\xi. \tag{9}$$

Here $\xi(r)$ is the Lagrangian fluid displacement and $c\tilde{\sigma}$ is the characteristic eigenfrequency (c is the speed of light). The quantities F,G,H depend on the equilibrium profiles of the pressure p and density ρ of the stars. The details are given in Ref.[15].

Journal of Physics: Conference Series **599** (2015) 012036 doi:10.1088/1742-6596/599/1/012036

Table 1: Pulsation frequency for non-rotating and rotating HS as a function of central density ρ_c for n=0, 1 and 2 radial modes in absence and presence of magnetic field.

ρ_c	n	(Without magnetic field)			(With magnetic field)		
10^{15} g cm^{-3}		M/M_0	$\tilde{\sigma}^2(10^8 s^{-2})$	$\sigma'^2(10^8 s^{-2})$	M/M_0	$\tilde{\sigma}^2(10^8 s^{-2})$	$\sigma'^2(10^8 s^{-2})$
0.25	0	0.32	5.11	6.31	0.32	5.10	5.78
	1		13.30	7.66		13.92	8.20
	2		36.38	2.48		38.12	2.62
0.5	0	0.56	3.38	1.13	0.60	3.92	1.33
	1		25.05	1.90		26.26	2.22
	2		62.04	5.09		64.54	5.92
1	0	1.16	9.53	-2.65	1.26	10.25	1.75
	1		41.53	2.68		47.43	3.08
	2		104.71	8.56		111.98	9.03
1.5	0	1.67	10.18	2.03	1.54	7.61	-5.83
	1		49.58	5.25		49.72	2.75
	2		127.01	1.03		102.64	8.83
2	0	1.77	8.40	-1.52	1.61	6.49	-1.69
	1		34.34	3.42		34.37	2.02
	2		85.88	9.84		83.39	5.49
2.5	0	1.78	6.50	-3.11	1.62	5.40	-2.66
	1		29.04	2.29		26.79	1.58
	2		71.62	8.52		71.39	4.43

For a given EOS $P(\rho)$ and given central density, we solved TolmanOppenheimerVolkoff (TOV)[23] equations to obtain the radius R and mass M = m(R) of the star. Therefore the basic input to solve the structure and pulsation equations is the EOS, $P = P(\rho)$. It has been seen [24] that structure parameters of neutron stars are mainly dominated by the EOS at high densities, specifically around the core. Since the oscillation features are governed by structure profiles of neutron stars, it is expected to possess marked sensitivity on the high density EOS as well.

3. Results and Discussion

The results of our calculation are presented in Fig.1a, Fig.1b and Table 1. In Fig.1a, we have plotted pressure against energy density having density dependent bag pressure 170MeV [19]. It is clear that magnetic field softens the EOS as well as broadens the mixed phase region. In Fig.1b, we have plotted gravitational mass(in solar mass unit) against central density for zero and non zero value of magnetic field. The maximum gravitational mass of HS for non-magnetic case is 1.8 solar mass and for magnetic case is 1.6 solar mass, which are close to observed values [2]. In Table 1 we have presented a calculation of the square of frequencies ($\tilde{\sigma}^2$ and σ'^2) of magnetised and non-magnetised hybrid star as a function of central densities for fundamental mode and for the first two harmonics by using the radial pulsation equations of non-rotating HS, as given by Chandrasekhar [22] as well as for rotating HS as given by Chandrasekhar and Friedman [17] in the general relativity formalism. Due to mixed phase region which is transition from quark to hadron phase, a kink is observed in the frequency with increase in central density. These kinks can be considered as a distinct signature of the mixed and quark matter in neutron stars. In both magnetic and non-magnetic cases for non-rotating HS there is a kink at $\rho_c = 0.5 \times 10^{15} g\ cm^{-3}$ in fundamental mode, then frequency increases and decreases slowly with increasing central density and for other two higher harmonics there is no remarkable kink but frequency increases and then slowly decreases with central density. In absence of magnetic field for rotating HS there is an oscillating behaviour of square of frequency starting at central density $\rho_c = 0.5 \times 10^{15} g\ cm^{-3}$ for all modes. In presence of magnetic field for the rotating HS there is an oscillating behaviour in square of frequency for first two modes and then slowly decreases but for next mode square of frequency increases and decreases with central density. These behaviour are observed due to presence of mixed and quark matter in HS.

4. Acknowledgments

N R Panda would like to thank the DST (Department of Science and Technology), India for the support to attend conference and K K Mohanta acknowledge the Institute of Physics, India for the support during this work.

References

[1] Demorest P, Pennucci T, Ransom S, Roberts M, Hessels J 2010 A two-solar-mass neutron star measured using Shapiro delay *Nature* **467** 1081
[2] Freire P C C, Bassa C, Wex N et al. 2010 On the nature and evolution of the unique binary pulsar J1903+0327 *Monthly Notices of the Royal Astronomical Society* **412** 2763
[3] Glendenning N K 2000 *Compact Stars: Nuclear Physics, Particle Physics and General Relativity* (Springer-Verlag, New York)
[4] Weber F 1999 *Pulsars as Astrophysical Laboratories for Nuclear and Particle Physics* (IOP Publishing Bristol)
[5] Palmer D M et al. 2005 A giant big gamma-ray flare from the magnetar SGR 180620 *Nature* **434** 1107
[6] Broderick A, Prakash M and Lattimer J 2000 The Equation of State of Neutron Star Matter in Strong Magnetic Fields *AstroPhys.* J **537** 351
[7] Mallick R, Sahu P K 2014 Phase transitions in neutron star and magnetars and their connection with high energetic bursts in astrophysics *Nulear Phys.* A **921** 96
[8] Chen W, Zhang P Q and Liu L G 2005 The influence of the magnetic field on the properties of neutron star matter *Mod. Phys. Lett.* A **22** 623
[9] Wei F X, Mao G J, Ko C M, Kissinger L S, Stoecker H and Greiner W 2006 Effect of isovector-scalar meson on neutron-star matter in strong magnetic fields *J. Phys.* G **32** 47
[10] Yuan Y F and Zhang J L 1999 The Effects of Interior Magnetic Fields on the Properties of Neutron Stars in the Relativistic Mean-Field Theory *AstroPhys. J.* **525** 2 950
[11] Chakrabarty S and Sahu P K 1996 Effect of magnetic field on strange star *Phys. Rev.* D **53** 8 4687
[12] Felipe R G, Martinez A P, Rojas H P and Orsaria M 2008 Magnetized strange quark matter and magnetized strange quark stars *Phys. Rev.* C **77** 015807
[13] Ghosh T and Chakrabarty S 2001 Can there be a quark matter core in a magnetar? *Phys. Rev.* D **63** 043006.
[14] Cameron A G W 1965 Possible Magnetospheric Phenomena associated with Neutron Stars *Nature* **205** 787.
[15] Sahu P K, Burgio G F and Baldo M 2002 Radial modes of neutron stars with a quark core, *ApJ* **566** L89.
[16] Culter C, Lindblom L and Splinter J 1990 Damping times for neutron star oscillations *Astrophys. J.* **363** 603
[17] Chandrasekhar S, Friedman J 1972 On the stability axisymmetric system to axisymmetric perturbation in general relativity.III. Vacuum Metrics and Carter's Theorem *Astrophys. J.* **176** 745
[18] Landau L D and Lifshitz E M 1965 *Quantum Mechanics* (Pergamon Press Oxford)
[19] Mohanta K K, Mallick R, Panda N R, Singh L P and Sahu P K 2014 Neutron star in the presence of strong magnetic field *Pramana-J Phys.* **82** 5 797
[20] Glendenning N 1992 First order phase transitions with more than one conserved charge: Consequences for neutron stars *Phys. Rev.* D **46** 1274
[21] Datta B, Hasan S S, Sahu P K and Prasanna A R 1998 Radial Modes of rotating neutron star in the Chandrasekhar-Friedman formalism *Int. Jour. Mod. Phys* D **7** 49.
[22] Chandrasekhar S 1964 The dynamical instability of gaseous masses approaching the Schwarzschild limit in general relativity *ApJ* **140** 417
[23] Shapiro S L and Teukolsky S A 1983 *Black Holes, White Dwarfs and Neutron Stars* (John Wiley & Sons, New York)
[24] Burgio G F, Baldo M, Sahu P K, Schulze H J 2002 Hadron-quark phase transition in dense matter and neutron stars *Physical Review C* **66** 2 025802; Burgio G F, Baldo M, Sahu P K, Santra A B and Schulze H J 2002 Maximum mass of neutron stars with a quark core *Phys.Lett.* B **526** 1 19

Non-perturbative relativistic calculations of the K-shell-vacancy production in Xe–Xe^{54+} collisions

Y S Kozhedub[1,2], **I I Tupitsyn**[1], **V M Shabaev**[1], **G Plunien**[3] **and Th Stöhlker**[4,5,6]

[1] Department of Physics, St. Petersburg State University, Ulianovskaya 1, Petrodvorets, 198504 St. Petersburg, Russia
[2] SSC RF ITEP of NRC "Kurchatov Institute", Bolshaya Cheremushkinskaya 25, Moscow, 117218, Russia
[3] Institut für Theoretische Physik, Technische Universität Dresden, Mommsenstraße 13, D-01062 Dresden, Germany
[4] GSI Helmholtzzentrum für Schwerionenforschung, 64291 Darmstadt, Germany
[5] Helmholtz-Institut Jena, D-07743 Jena, Germany
[6] Institut für Optik und Quantenelektronik, Friedrich-Schiller-Universität Jena, D-07743 Jena, Germany

E-mail: kozhedub@pcqnt1.phys.spbu.ru

Abstract. Non-perturbative calculations of the relativistic quantum dynamics of electrons in the Xe-Xe^{54+} collisions at 30 AMeV are performed. The calculations are based on an independent-electron model and use the coupled-channel approach with atomic-like Dirac-Fock-Sturm orbitals. Special attention is paid to the inner-shell processes. The role of the relativistic and many-particle effects is studied.

1. Introduction

Heavy-ion collisions play a very important role in investigations of the relativistic quantum dynamics of electrons in presence of strong electromagnetic fields [1]. Moreover, if the total charge of the colliding nuclei is larger than the critical one, $Z_1 + Z_2 > 173$, such collisions can provide a unique tool for tests of quantum electrodynamics at the supercritical regime [2]. The CRYRING [5] at the present GSI SIS18/ESR facility and, especially, the realization of the FAIR [3, 4] project will provide unique opportunities for studying new physics with low-energy heavy-ion-atom collisions. These studies require corresponding theoretical calculations which have to describe in details the relativistic quantum dynamics of electrons.

In the present paper, we perform the relativistic non-perturbative quantum-mechanical calculations of the Xe-Xe^{54+} collision at 30 AMeV within an independent electron model using the coupled-channel approach with atomic-like Dirac-Fock-Sturm orbitals [6, 7]. We evaluate the K-shell-vacancy production in this collision and investigate the role of the many-electron and relativistic effects.

Atomic units ($\hbar = e = m_e = 1$) are used throughout the paper.

2. Method

In the following, we briefly present the formalism used, for a complete description see Refs. [6, 8, 9]. Using the semiclassical approximation, where the atomic nuclei move along the classical trajectories and are considered as sources of a time-dependent external potential, we have to solve the time-dependent many-particle Dirac equation for electrons involved in the process. The method we employ is based on an independent particle model, where the many-electronic Hamiltonian $\hat{H}$ is approximated by a sum of effective single-electron Hamiltonians, $\hat{H}^{\text{eff}} = \sum \hat{h}^{\text{eff}}$, reducing the electronic many-particle problem to a set of single-particle Dirac equations for all electrons of the colliding system:

$$i\frac{\partial \psi_i(t)}{\partial t} = \hat{h}_i^{\text{eff}}(t)\,\psi_i(t) \quad \text{with} \quad i = 1, \ldots, N, \tag{1}$$

where the wave functions $\psi_i(t)$ have to satisfy the initial conditions for the N electrons:

$$\lim_{t \to -\infty} (\psi_i(t) - \psi_i^0(t)) = 0 \quad \text{with} \quad i = 1, \ldots, N. \tag{2}$$

As the effective single-electron Hamiltonian $\hat{h}^{\text{eff}}$ we use the two-center Dirac-Kohn-Sham Hamiltonian:

$$\hat{h}^{\text{eff}} = c(\vec{\alpha} \cdot \vec{p}) + \beta c^2 + V_{\text{nucl}}^A(\vec{r}_A) + V_{\text{nucl}}^B(\vec{r}_B) + V_C[\rho] + V_{xc}[\rho]\,, \tag{3}$$

where c is the speed of light and $\vec{\alpha}$, β are the Dirac matrices. Here $V_{\text{nucl}}^{\alpha}(\vec{r}_\alpha)$ and $V_C[\rho] = \int d^3\vec{r}^{\,\prime} \frac{\rho(\vec{r}^{\,\prime})}{|\vec{r} - \vec{r}^{\,\prime}|}$ are the electron-nucleus interaction and the electron-electron Coulomb potentials, respectively, and $\rho(\vec{r})$ is the electron density of the system. The exchange-correlation potential $V_{xc}[\rho]$ is taken in the Perdew-Zunger parametrization [10].

Solving the effective single-particle equations (1) is based on the coupled-channel approach with atomic-like Dirac-Sturm-Fock orbitals, localized at the ions (atoms) [9]. The many-particle probabilities are calculated in terms of the single-particle amplitudes employing the formalism of inclusive probabilities [11, 12], which allows one to describe the many-electron collision dynamics.

3. Results and discussion

In the present paper, we applied the methods described in Sec. 2 to calculate the inner-shell processes in low-energy collisions of neutral atoms with bare ions. The calculations were performed for the Xe-Xe^{54+} collision, which experimental investigation was recently carried out at GSI (Darmstadt) [13]. Since we were mainly interested in the dynamics of electrons in presence of the strongest electric field, we focused on the study of the K-shell electron population probabilities of the colliding ions.

Figure 1 shows the probabilities of the q-vacancy creation in the K shell of the target atom (neutral xenon) as functions of the impact parameter. In order to investigate the role of many-electron effects we performed the calculations using both the many-electron approach and the active-electron approximation. The results are indicated in Fig. 1 by bold and thin lines, correspondingly. In the active-electron approximation only the K-shell electrons of the target are considered as the active electrons and participate in excitation and charge-transfer processes, while the others provide a screening potential. In both cases the probabilities have a rather similar oscillatory behaviour, but the magnitudes of the peaks and bottoms are changed while their positions are preserved.

In order to investigate the role of the relativistic effects we performed the corresponding calculations in the nonrelativistic limit by multiplying the standard value of the speed of light by the factor 1000 (in atomic units). The comparison of the obtained relativistic and nonrelativistic results is presented in Fig. 2. As one can see from this figure, the relativistic (bold lines) and nonrelativistic (thin lines) curves have the same oscillatory behavior but the nonrelativistic curves are shifted toward larger impact parameters.

To summarize, the many-particle and relativistic effects on the probabilities of the q-K-shell vacancy production in the Xe-Xe^{54+} collision at 30 AMeV have been investigated. The contributions of these effects are of the same order of magnitude but of different type. The many-electron effects decrease the magnitudes of the oscillatory behavior, while the relativistic ones shift the oscillatory curves toward smaller impact parameters. These corrections are rather substantial and have to be taken into account.

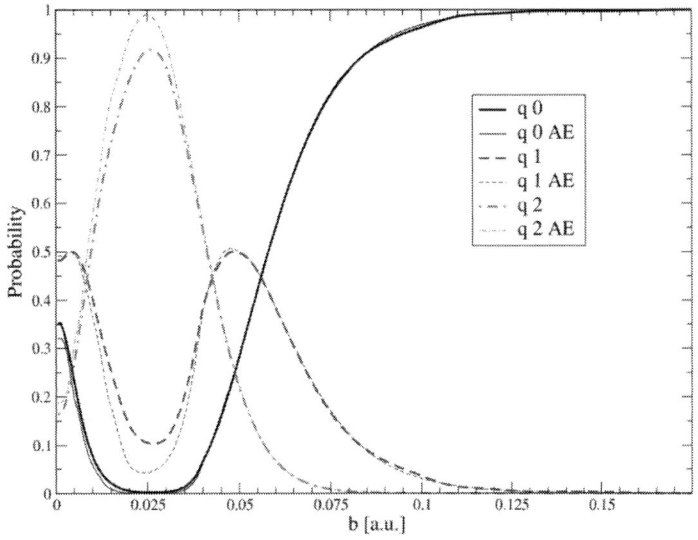

Figure 1. The probabilities of the target Xe q-K-shell-vacancy production as functions of the impact parameter b for the Xe-Xe^{54+} collision at the projectile (Xe^{54+}) energy of 30 MeV/u. The bold and thin lines indicate the results of the calculation using the many-electron approach and the active-electron (AE) approximation, correspondingly.

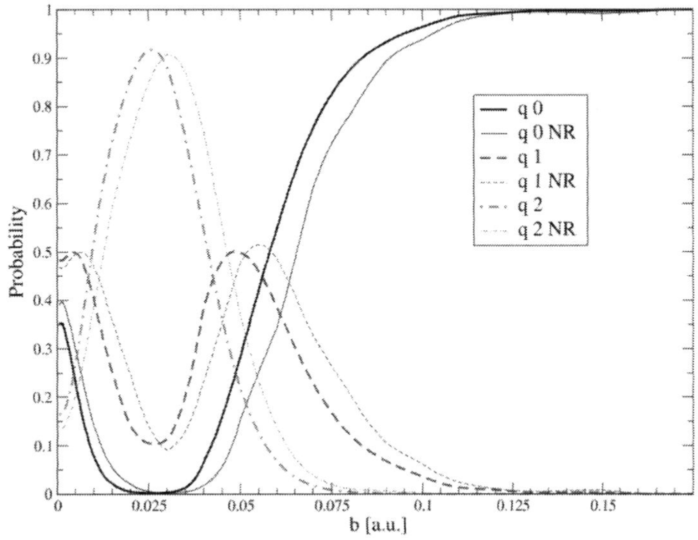

Figure 2. The probabilities of the target Xe q-K-shell-vacancy production as functions of the impact parameter b for the Xe-Xe^{54+} collision at the projectile (Xe^{54+}) energy of 30 MeV/u. The bold and thin lines indicate the relativistic and non-relativistic (NR) results of the calculation, correspondingly.

Acknowledgments
This work was supported by RFBR (Grants No. 14-02-31418, No. 13-02-00630, No. 12-03-01140-a, and No. 14-02-00241), GSI, and SPbSU (Grants No. 11.50.1607.2013, No. 11.38.269.2014, and No. 11.38.261.2014). Y.S.K. acknowledges the financial support from the Helmholtz Association and SAEC .

References
[1] Eichler J and Meyerhof W E 1995 *Relativistic Atomic Collisions* (New York: Academic Press)
[2] Greiner W, Müller B, and Rafelski J 1985 *Quantum Electrodynamics of Strong Fields* (Berlin: Springer-Verlag)
[3] http://www.gsi.de/en/research/fair.htm
[4] http://www.gsi.de/sparc
[5] Lestinnsky M *et al. CRYRING@ESR: A study group report* (unpublished)
[6] Kozhedub Y S, Shabaev V M, Tupitsyn I I, Gumberidze A, Hagmann S, Plunien G, and Stöhlker Th 2014 Phys. Rev. A **90** 014053
[7] Tupitsyn I I, Kozhedub Y S, Shabaev V M, Bondarev A I, Deyneka G B, Maltsev I A, Hagmann S, Plunien G, and Stöhlker Th 2012 Phys. Rev. A **85** 032712
[8] Kozhedub Y S, Tupitsyn I I, Shabaev V M, Hagmann S, Plunien G, and Stöhlker Th 2013 Phys. Scr. **T156** 014053
[9] Tupitsyn I I, Kozhedub Y S, Shabaev V M, Deyneka G B, Hagmann S, Kozhuharov C, Plunien G, and Stöhlker Th 2010 Phys. Rev. A **82** 042701
[10] Perdew J P and Zunger A 1981 Phys. Rev. B **23** 5048
[11] Lüdde H J and Dreizler R M 1985 J. Phys. B **18** 107
[12] Kürpick P and Lüdde H J 1993 Comput. Phys. Commun. **75** 127
[13] Hagmann S *et al. Electron Emission following* 1*s Adiabatic Ionization and Quasi-resonant* 1*s*-1*s Charge Transfer in Symmetric Ion-Atom Collisions* GSI proposal (unpublished)

FAIRNESS 2014 IOP Publishing
Journal of Physics: Conference Series **599** (2015) 012038 doi:10.1088/1742-6596/599/1/012038

^{8}B structure in Fermionic Molecular Dynamics

KR Henninger, T Neff, H Feldmeier

GSI Helmholtzzentrum für Schwerionenforschung GmbH, Planckstrasse 1, 64291 Darmstadt, Germany

E-mail: k.henninger@gsi.de

Abstract. The structure of the light exotic nucleus ^{8}B is investigated in the Fermionic Molecular Dynamics (FMD) model. The decay of ^{8}B is responsible for almost the entire high-energy solar-neutrino flux, making structure calculations of ^{8}B important for determining the solar core temperature. ^{8}B is a proton halo candidate thought to exhibit clustering. FMD uses a wave-packet basis and is well-suited for modelling clustering and halos. For a multiconfiguration treatment we construct the many-body Hilbert space from antisymmetrised angular-momentum projected 8-particle states. First results show formation of a proton halo.

Introduction

The physics of weakly-bound nuclear systems is a fruitful area of study, which investigates the location of the proton- and neutron-driplines, as well as the structures occuring at such "exotic" proton-to-neutron ratios, such as the formation of neutron- and proton-halos. While there are several well-known neutron-halo nuclei, there are very few proton-halo candidates: the first excited state of ^{17}F, and the ground state of ^{8}B are the best-known.

The nucleus ^{8}B also plays a role in astrophysics, occuring in the third branch of the proton:proton (pp) chain. In fact, the decay of ^{8}B to ^{8}Be is responsible for over 75% of the high-energy solar neutrino flux [1]. Since this decay rate depends on the rate of production of ^{8}B via ^{7}Be(p,γ)^{8}B, (which is itself dependent on ^{8}B structure), knowledge of the ^{8}B ground state wavefunction is important input into such reaction calculations.

A proton halo in ^{8}B is inferred from the low proton binding energy of 0.137 MeV [2], narrow longitudinal momentum distribution of ^{7}Be following breakup [3] and a matter radius of 2.38 $\pm$ 0.04 fm [4], but ^{8}B proton radius is not known. An ^{8}B proton halo would affect the capture cross-section for a proton on ^{7}Be quite strongly, since the radiative capture process is dominated by contributions from regions far outside the nuclear radius [5].

^{7}Be is in itself loosely-bound [6], which means that the ^{8}B core may have the structure α+^{3}He. What then, is the correct picture for ^{8}B? A shell-model type picture, a ^{7}Be core plus proton, or even the cluster structure $\alpha +^3\text{He} + \text{p}$? In order to access all these possibilities, one may choose to do such calculations in the Fermionic Molecular Dynamics (FMD) model, since it offers the possibility to access both shell-model type states and clustering (see [7] and refs. therein). FMD has been used for such diverse structures as the Hoyle state [8] and the neutron halo of ^{10}Be [9]. In this paper, I outline the model and present preliminary results for ^{8}B.

Fermionic Molecular Dynamics
Fermionic Molecular Dynamics (FMD) [10] is a microscopic model for light- to medium-mass nuclei. Overviews of the model are in [7, 10–12] and a summary is provided here.

Protons and neutrons are the relevant degrees of freedom in nuclear systems, so we choose intrinsic many-body configurations $|Q\rangle$, which are Slater determinants of single-particle states $|q\rangle$, or [10]:

$$|Q\rangle = \hat{\mathcal{A}}\left\{|q_1\rangle \otimes ... \otimes |q_A\rangle\right\} \tag{1}$$

where $\hat{\mathcal{A}}$ is the antisymmetrisation operator. The single-particle states are:

$$|q\rangle = \sum_i |a_i, \vec{b}_i\rangle \otimes |\chi_i^{\uparrow}, \chi_i^{\downarrow}\rangle \otimes |\xi_i\rangle c_i, \tag{2}$$

where a superposition of wave-packets $|a_i, \vec{b}_i\rangle$ can be used to describe extended distributions (*e.g.* halos). Parameters $\vec{b}$ relate to the mean position and mean momentum of the wave-packets and parameters a give the width of the wave-packets [11]. The ket $|\chi^{\uparrow}, \chi^{\downarrow}\rangle$ is the most general spinor, allowing all possible orientations of the nucleon spin. The isospin part of the state is given as $|\xi\rangle$ (*i.e.* proton or neutron).

In co-ordinate space one has:

$$\langle\vec{x}|a, \vec{b}\rangle = \exp\left\{-\frac{(\vec{x}-\vec{b})^2}{2a}\right\}. \tag{3}$$

Slater determinants like that in Equation 1 are not states of good angular momentum or parity, so these need to be projected out. The projection operators are given by [11]:

$$\begin{aligned}\hat{P}^{\pi} &= \frac{1}{2}(1+\pi\hat{\Pi})\\ \hat{P}^{J}_{MK} &= \frac{2J+1}{8\pi^2}\int d\Omega\, D^{J}_{MK}(\Omega)^{*}\hat{R}(\Omega),\end{aligned} \tag{4}$$

where $D^J_{MK}(\alpha, \beta, \gamma)$ is the Wigner D-matrix, $\hat{R}(\alpha, \beta, \gamma)$ is the rotation operator and $\hat{\Pi}$ is the parity operator [7]. The action of the projection operators is to project states $|Q; J^{\pi}MK\rangle$ from intrinsic states $|Q\rangle$. There are potentially $2J+1$ different K states for every state of good J, depending on the symmetries of the intrinsic state. By diagonalising the Hamiltonian in the set of K-projected states, we obtain the basis states

$$|Q; J^{\pi}M\alpha\rangle = \sum_K |Q; J^{\pi}MK\rangle c_K^{J^{\pi}\alpha}. \tag{5}$$

The parameters $\{q_\nu\}$ describing the intrinsic basis states $|Q\rangle$ are obtained by requiring:

$$\min_{\{q_\nu\}} \quad \frac{\langle Q; J^{\pi}M\alpha|\hat{H} - \hat{T}_{cm}|Q; J^{\pi}M\alpha\rangle}{\langle Q; J^{\pi}M\alpha|Q; J^{\pi}M\alpha\rangle}, \tag{6}$$

where $\hat{T}_{cm}$ is the operator for the centre-of-mass kinetic energy. The minimisation of the energy of the projected state (Equation 6) is referred to as Variation after Projection (VAP).

The VAP states $|Q; J^{\pi}M\alpha\rangle$ provide a good first approximation for the description of the nucleus. To improve the description, additional basis states can be generated by performing minimisation (Eq. 6) subject to constraints. These constraints could be on *e.g.* radii, moments or deformation parameters [7]. The FMD basis is general-enough to describe both shell-model like many-body states as well as clustering states [7].

^{8}B

Projected configurations with angular momenta and parities of 0+, 1+, 2+ and 3+ were created for ^{8}B, since with a ^{7}Be core (in its 3/2- ground state or first excited 1/2- state) plus a p3/2 proton, one can form states of total angular momentum 0 to 3. As per Equation 2, one Gaussian wave-packet was used to describe each of the neutrons and four of the protons in phase-space, while the fifth proton was described spatially by a superposition of two Gaussians, to allow for a possible extended distribution. A UCOM-transformed AV18 interaction [13] was used, for two reasons. Firstly, because AV18 is a realistic interaction (it reproduces the Nijmegen phase-shifts [14]), and secondly because, for application in FMD, the interaction is required to be in an operator representation, making UCOM the appropriate method to soften the interaction [13]. The Hamiltonian was diagonalised in this space, and ground- and excited state energies, radii and transition probabilities were extracted for the resulting eigenstates. Plots of the proton- and neutron-density distributions for the intrinsic states obtained by the VAP procedure are shown in Figure 1. One can note in this figure the appearance of clustering and of an extended distribution for a proton.

The level-scheme is provided as Fig 2. The calculated levels represent the energies for the eigenstates of the UCOM Hamiltonian, calculated in the space of VAP states. The calculated energies are compared to the calculated ^{7}Be threshold (the ground-state energy for ^{7}Be calculated in an FMD calculation); and one can note a slight underbinding with respect to threshold. Including more basis states (which will access more configurations) will improve this binding. The discrepancy between the experimental and theoretical binding energies is mainly due to missing tensor correlations in the FMD wave functions.

The radii and transition-strengths are provided in Tables 1 and 2 respectively. When considering the radii, one may note that the calculated matter radius is smaller than the measured radius. This is again due to the small basis set used: by including more basis configurations one may better-approximate the long tail for the spatial distribution of the loosely-bound proton and so improve the radius. One may note in Table 2 that the 3+→2+ transition is well-reproduced.

Table 1: Calculated rms matter-, charge-, proton- and neutron-radii, magnetic dipole moment (μ) and electric quadrupole moment (Q) for ^{8}B, for the basis of 4 VAP states (VAP). The calculations are compared to experiment, where values are known.

	R_{matter} [fm]	R_{charge} [fm]	R_p [fm]	R_n [fm]	μ [μ_N]	Q [efm^2]
Exp.	2.38±0.04 [4]				1.0355±0.0003 [15]	6.83±0.21 [16]
VAP	2.262	2.523	2.374	2.062	1.228	4.863

Table 2: Transition strengths $B(\Lambda M)$ calculated for the known transitions in ^{8}B (shown in black) and for some as-yet-unmeasured transitions (shown in red). Calculations were in the VAP basis as were those in Table 1. Where the character of the transition is not known, all possibilities are considered.

Energy [MeV]	States	Type	Basis	$B(\Lambda M)$ (meas.)	$B(\Lambda M)$ (calc.)
0.77	1+ → 2+	$M1$	VAP	2.63(12) μ_N^2 [2]	3.992 μ_N^2
2.32	3+ → 2+	$M1$	VAP	0.38(19) μ_N^2 [2]	0.274 μ_N^2
2.55	2+ → 2+	($M1$)	VAP		0.060 μ_N^2
3.30	1+ → 2+	($E2$)	VAP		5.438 e^2fm^4
	1+ → 2+	($M1$)	VAP		0.467 μ_N^2

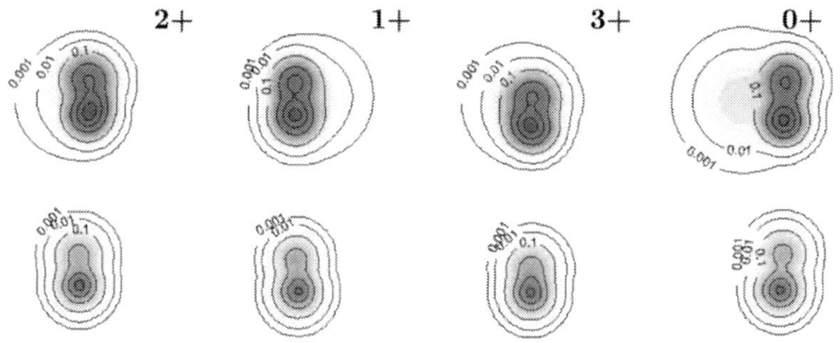

Figure 1: Proton- (red) and neutron-densities (blue) for the VAP basis configurations. The numbers on the contour lines give nucleon density in units of the saturation density $\rho_0 = 0.17\text{fm}^{-1}$. Numbers to the right show angular momentum and parity of the configuration.

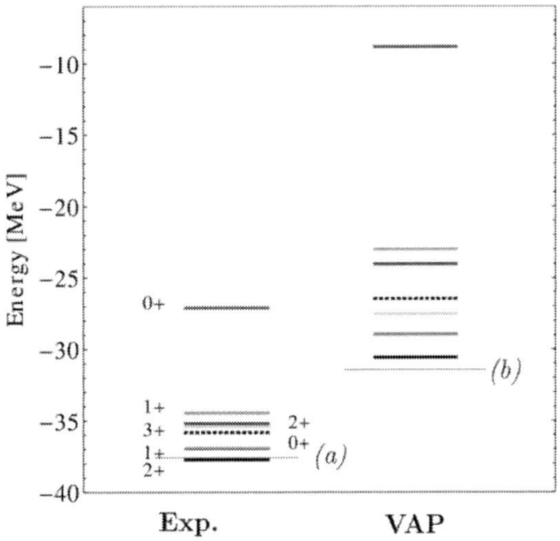

Figure 2: Level-scheme of ^{8}B comparing experimental and calculated ground and excited state energies. The thin lines indicate the ^{7}Be+p threshold in (a) the experimental and (b) the calculated case.

Conclusion

The large charge-, proton- and matter radii compared to neutron radius (Table 1) indicate proton halo formation in this model of ^{8}B. This is in line with previous studies (*e.g.* [3, 16, 17]). The extended proton distribution also shows up clearly in calculated proton densities (Fig. 1). Since all parameter values describing the nucleon states are dictated by the minimisation process (Eq. 6), including a second Gaussian wavepacket for the fifth proton *allows*, rather than *dictates*, its extended spatial distribution. If the second Gaussian were not needed to describe the proton's spatial distribution, its parameters would evaluate under minimisation such that its contribution would be negligible. Thus we may say that, under the interaction used, proton halo formation is indeed favoured in ^{8}B. Work on extending the basis by imposing constraints on matter-, proton- and neutron radii is underway.

References

[1] Xu H M *et al.* 1994. *Phys. Rev. Lett.*, **73**:2027.
[2] Tilley D R *et al.* 2004. *Nucl. Phys.* , **A745**:155.
[3] Smedberg M *et al.* 1999. *Phys. Lett.* B, **452**:1.
[4] Ozawa A *et al.* 2001. *Nucl. Phys.*, **A693**:32.
[5] Rolfs C S 1973. *Nucl. Phys.*, **A217**:27.
[6] Neff T 2011. *Phys. Rev. Lett.*, **106**:042502.
[7] Neff T and Feldmeier H 2010. *Cluster Structure in the FMD Approach.* (Research Signpost Publishing: Kerala). pp 67-94.
[8] Feldmeier H *et al.* 2007. *Phys. Rev. Lett.*, **98**:032501.
[9] Zakova A *et al.* 2010. *J. Phys.* G.: *Part. Nucl. Phys.*, **37**:055107.
[10] Feldmeier H and Schnack J 1997. *Prog. Part. Nucl. Phys.*, **39**:343.
[11] Neff T and Feldmeier H 2008. *Eur. Phys. J.* Special Topics, **156**:69.
[12] Bieler K Feldmeier H and Schnack J 1995. *Nucl. Phys.*, **A593**:493.
[13] Feldmeier H *et al.* 2005. *Phys. Rev.* C, **72**:034002.
[14] Wiringa R B Stoks V G J and Schiavilla R 1995. *Phys. Rev.* C, **51**:38.
[15] Shirley V S *et al.* (Eds.) 1996. *Table of Isotopes (Vol. 1).* (John Wiley and sons publishing: New York).
[16] Minamisono T *et al.* 1992. *Phys. Rev. Lett.*, **69**:2058.
[17] Werner R E *et al.* 1995. *Phys. Rev.* C, **52**:1106(R).

Material-related issues at high-power and high-energy ion beam facilities

M. Bender[1*], D. Severin[1], M. Tomut[1] and C. Trautmann[1,2]
[1]GSI Helmholtzzentrum, Planckstr. 1, 64291 Darmstadt, Germany
[2]Technische Universität Darmstadt, Alarich-Weiss-Str. 2, 64287 Darmstadt, Germany
E-mail: m.bender@gsi.de

Abstract. When solids are exposed to energetic ions (MeV-GeV), their physical and chemical structure can be severely modified. The change is governed by ultrafast dynamical processes starting from the deposition of large energy densities, electronic excitation and ionization processes, and finally damage creation in the atomic lattice system. In many materials, each projectile creates a cylindrical track with a few nanometers in diameter and up to many μm in length. To study and monitor the creation of damage, the GSI irradiation facility dedicated to materials science provides different in-situ and on-line techniques such as high resolution microscopy, X-ray diffraction, optical absorption spectroscopy, thermal imaging and residual gas analysis. The irradiation experiments can be performed under various gas atmospheres and under cryogenic or elevated temperature.

1. Introduction

When heavy ions above a specific energy of some 100 keV/u penetrate solids they mainly excite the electronic subsystem of the target [1]. On a time scale of 10^{-17} s, delta electrons are created that spread the energy radially around the trajectory of the projectile [2]. The following electron cascade distributes the energy inside of the electronic system within a time scale of up to 10^{-14} s. During this period, the atomic lattice is still cold. The redistributed electrons lead to a high space charge resulting in Coulomb repulsion of the ionized atoms, if the electrons do not compensate the space charge fast enough. After 10^{-13} - 10^{-12} s, the electrons transfer their energy to the atom subsystem due to electron-phonon-coupling leading to ultrafast heating of the lattice. Temperatures exceeding melting or vaporization of the target can be reached. Depending on the target material, this thermal spike may last up to 10^{-10} s. If the subsequent cooling of the liquefied cylinder along the ion path is rapid enough, quenching of atomic disorder occurs and a so-called ion track is formed. Track formation is pronounced in insulators such as polymers, ionic crystals, oxides as well as few III-V semiconductors and selected metallic alloys, but most pure metals and pure Si and Ge do not show track formation [3].
The track consists of a cylindrical zone of damaged material and has a typical diameter of several nanometers [4]. Tracks in particular in polymers can be selectively attacked by a suitable chemical etchant converting the damage trail into an open pore. The fact that each heavy ion produces an individual track which exhibits preferential etchability is the base of the ion-track nanotechnology. With this technique, small but extremely long (up to several tens of μm) channels can be fabricated. Their diameter is very uniform and can be adjusted by the etching process between about 10-20 nm and several μm. Track-etched membranes are available

Journal of Physics: Conference Series **599** (2015) 012039 doi:10.1088/1742-6596/599/1/012039

as commercial products and are applied in many different fields, e.g., as special filters, cell cultivation substrates, or as templates for the synthesis of micro- and nanowires [3, 5].
While individual ion tracks are utilized as nanostructures, the integral action of many overlapping tracks may lead to macroscopic changes and severe radiation damage. Damage effects are of particular concern for materials and devices exposed to high-intensity ion beams at the new high-power accelerators such as CERN and the future facility for antiproton and heavy ion research (FAIR). The existing GSI accelerator and the upcoming FAIR facility comprise several "hot spots" where materials face extreme radiation levels, e.g., stripper foils at the UNILAC, the future production target for the Super Fragment Separator (SFRS) and beam catchers [6]. Although they consist of carbon-based materials of rather high radiation hardness, the expected high beam intensities will result in structural and mechanical damage, requiring cost and time intensive maintenance. Of special concern are also organic insulating materials because under extensive beam exposure they embrittle and become conductive, a risk for magnet insulation and other electrical devices.
Another material-related issue is the deterioration of the vacuum of the accelerator system leading to enhanced beam losses. If the beam hits the inner surface of the beamline, the impacts of energetic ions initiate the release of gas [7, 8, 9]. Especially for high-intensity beams such as at the future synchrotron SIS100 at FAIR or at the S3 beamline of SPIRAL 2 (Caen, France), the amount of desorbed gas may be enormous, rendering beam-induced molecular desorption a critical issue for next generation accelerators. In the following, we provide a short description of the GSI irradiation facilities dedicated to materials research and ion beam induced effects. Selected examples of FAIR-relevant material issues including radiation hardness of specific components and desorption phenomena are presented.

2. Irradiation facilities for materials research

At the existing GSI facility, several beam lines are dedicated to materials research including target handling, beam monitoring and various online and in-situ analytical methods. The linear accelerator UNILAC provides all ion species up to uranium with energies up to 11.4 MeV/u where the typical range of the ions is about 100 μm. The X0-beamline is equipped with an automated sample exchange system for the efficient irradiation of up to 100 samples per hour. The beam is defocused in such a way that an area of 5 x 5 cm^2 is homogeneously illuminated. Applied fluences range from one single ion per sample to about 10^{13} ions/cm^2. Using the heavy-ion microprobe [10], single ions can be placed on a predefined position within a lateral accuracy of better than 1 μm. This microbeam allows radiation hardness tests of selected areas of electronic devices. For biophysics experiments, single ions are shot into the nuclei of living biologic cells allowing monitoring of protein processes responsible for repairing the radiation damage. The M-Branch comprises three different beamlines dedicated to high-fluence irradiations (up to $\sim 10^{14}$ ions/cm^2) combined with online and/or in-situ analysis of beam-induced materials modifications. At the M1-beamline [11] a scanning electron microscope (including an EDX system) and an atomic force microscope (under UHV condition) is installed. The M2-beamline houses an x-ray diffractometer suitable for monitoring beam-induced structural changes or amorphization of a sample. The M3-beamline is equipped with a large universal irradiation chamber to which various spectrometers [12] (e.g., Fourier transform infrared (FTIR) and Raman), a residual gas analyser (RGA), thermal camera, as well as resistivity or mechanical creep test equipment can be attached. The samples can be irradiated at temperatures between 15 K and almost 1000° C. A scheme of the M3 beamline is shown in figure 1.
The heavy-ion synchrotron SIS-18 delivers beams with much higher energies up to several 100 MeV/u into cave A. At such high energies the corresponding ion ranges are up to cm, thus the irradiation of bulk samples is possible. Besides radiation hardness tests of mm to cm thick samples or devices, materials research activities also concentrate on the irradiation of samples

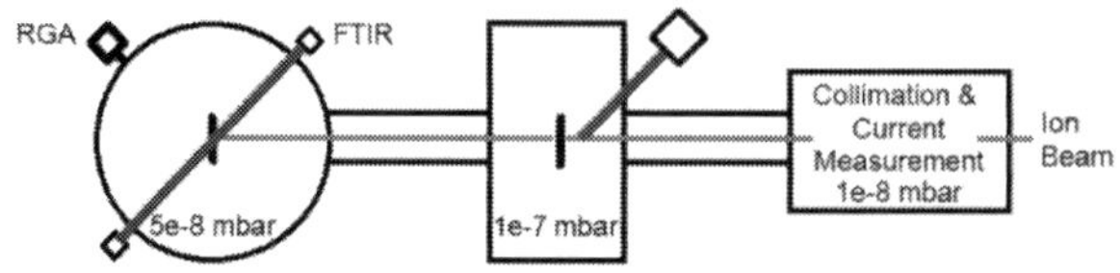

Figure 1. Schematic view of M3-beamline. The ion beam passes a diagnostic and collimation system (right) before entering the irradiation chamber. In the middle chamber, high temperature irradiation can be performed and the samples can be monitored with a thermal camera. In the left chamber, samples can be irradiated at cryogenic temperatures and various online spectroscopy techniques (e.g., FTIR or UV-vis) can be attached.

exposed to high pressures [16, 17, 18]. For this, miniaturized samples are mounted in diamond anvil cells which allow the generation of several tens of GPa. The cells are composed of two opposing 2-3 mm thick facetted diamonds which are squeezed together. Between the diamonds a gasket with a hole contains the sample in a pressure medium. The SIS-18 beam reaches the sample after penetrating the first diamond. The simultaneous exposure to pressure and radiation provides the possibility to simulate extreme conditions as present in the Earth mantle [4]. In the future also heating of the diamond anvil cell by a laser system is foreseen.

3. Ion-induced desorption

The release of gas into the vacuum system, triggered by impinging ions is a critical issue concerning beam losses and was investigated in the past at GSI and CERN [13]. For room temperature targets, the amount of desorbed gas per projectile ion (defined by the desorption yield η) scales with $(dE/dx)^2$ and can be described by a pure surface process, whereas the thermodynamic behavior of the substrate plays an important role for the release. In the above described thermal spike process, the ion impact leads to a region of elevated temperature resulting in enhanced thermal desorption. Several experiments with different beams and targets provided desorption yields which are in good quantitative agreement with thermal-spike model calculations [14].

Desorption effects are of concern when highest beam intensities are involved, e.g., at the SPIRAL-2 accelerator that will deliver some 10^{14} ions per second on a production target for rare isotopes. Even if the desorption yield is low, the high beam current will lead to enormous gas loads in the range of 10^{-3} mbar L s^{-1}. To the production target special rods are attached which stop the primary beam upon separating the rare isotopes. For thermal reasons the rods have to be made from a high melting material such as tungsten. The desorption yields of three different W samples were measured at the M1-beamline at GSI. For the irradiation 4.8 MeV/u Au ions with charge states 26+ and 53+ and respective currents of 3e-9 A and 3e-10 A were used. The pulse rate was 2 Hz and the pulse length 1.2 ms. The beam spot had a diameter of 6 mm. Two of the samples stem from the same batch of original rod material. They were brazed onto a Cu block as thermal substrate. The sample thickness was 3 mm which is much larger than the projected range of the Au projectiles ($\sim$ 17 μm). The third sample was cut from a pure W foil of 50 μm thickness. All samples were mounted on holders that could be inserted into the beam. The base pressure in the irradiation chamber was 1.8 e-8 mbar. During beam exposure, the dynamic vacuum, i.e., the total and partial pressures were recorded.

Figure 2 presents the results of the total desorption yield as a function of the accumulated fluence (left) and the partial pressure evolution for the thin W target irradiated with Au^{26+} ions (right). The yields were calculated from the pressure increase inside the vacuum chamber during irradiation. The overall uncertainty is estimated to be around 30% due to the large errors of

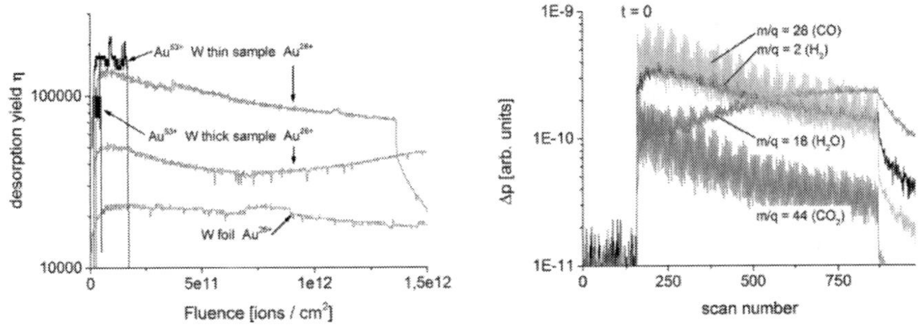

Figure 2. Left: desorption yields (released gas molecules per incident ion) for different W samples, measured with 4.8 MeV/u Au^{26+} and Au^{53+} ions. Right: partial pressure rise for the most relevant gas components. Irradiation starts at t = 0 and ends at scan number 875, corresponding to a fluence of 1.5e12 ions per cm^2.

the pressure and pumping speed measurements. Note that the beam current for the 53+ beam was one order of magnitude less than for the 26+ beam. Hence, within a fixed beamtime, much less fluence is accumulated. For a given sample, the higher charge state leads to higher desorption yields due to the higher energy loss of the 53+ beam [14]. The desorption yield tends to slightly decrease with increasing fluence due to beam scrubbing (cleaning of the surface with the ion beam). However, there are other effects which may increase the desorption yield (see m/q = 18 in Fig. 2 right). After the irradiation, the corresponding thick sample contained many macroscopic cracks where gas could be released under beam exposure. The composition of the desorbed gas is predominately H, CO and CO_2 (see Fig. 2, right) as observed so far for any sample and beam condition. The desorption yields measured for these tungsten samples are surprisingly high compared to earlier measurements using Au or Cu targets [15].

4. Radiation hardness of carbon-based materials

Graphite and other carbon based materials are the main choice for targets and beam protection parts facing intense primary beams. Their response to beam-induced thermal stress and stress waves is by far more favorable than standard materials such as stainless steel, due to their superior thermo-mechanical properties.

To test beam-induced damage processes and radiation hardness limits, thin carbon foils for electron stripping and bulk graphite as target for the production of secondary beams were irradiated with U and Au ions of 4.8 MeV/u kinetic energy. The beam was delivered from the high current source with a repetition rate of 1 - 2 Hz and pulse length of about 100 μs. This low duty cycle is leading to an enormous instantaneous power density deposition of 10 MW/g per pulse. For online investigation of beam-induced materials modifications, the infrared transmittance and radiance of 20 $\mu g/cm^2$ thin amorphous foils were monitored with the FTIR spectrometer and a fast thermal camera of type FLIR SC7500. This infrared camera is able to record fully radiometric images with kHz repetition rate and can be synchronized to the beam pulse. Therefore, the spatial and time decay of the heat load from the ion beam on the sample can be recorded with high precision and gives input for simulations of stress wave propagation. Figure 3 shows the thermography of a graphite foil exposed to a collimated U beam recorded at the maximum intensity of the beam pulse (left) and the evolution of the maximum temperature in the beam spot as a function of time (right).

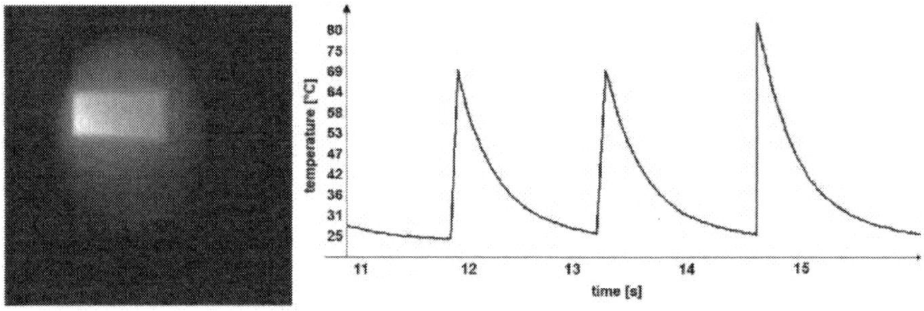

Figure 3. Left: Radiometric image of a graphite foil sample during irradiation with a low duty cycle 4.8 MeV/u U^{238} ion beam with a repetition rate of 0.6 Hz and a beam pulse duration of 150 μs. Image size is $\approx$ 5 x 5 cm^2. Temperature is coded in brightness. In this plot, bright colors within the rectangular beam spot are corresponding to temperatures higher than 40 °C, at the maximum intensity of the beam pulse. Right: maximum temperature within the beam spot versus time. The spikes correspond to individual ion beam pulses.

5. Conclusion and Outlook

Ion-induced molecular desorption from surfaces and radiation hardness of materials are important issues for FAIR and other next generation heavy-ion accelerators. At present, these processes are only partially understood and more detailed investigations are needed in particular for those accelerator components and devices that will be exposed to high-power and high-energy ion beams. Various analytical online systems allow the monitoring of radiation effects as a function of irradiation dose including processes such as material degradation, structural changes, outgassing, and temperature increase of the samples. The knowledge gain on damage creation is mandatory for limiting aging processes and providing life time estimates for safe device operation.

At the new FAIR facility, the BIOMAT beamline of the APPA cave will offer possibilities for irradiating samples at multiple extreme conditions including temperature, pressure and particularly high intensity irradiation not available at the existing SIS facility.

We acknowledge the support of FP7 EuCARD-2 under grant agreement no. 312453.

[1] Siegmund P. 2006 *Mat. Fys. Med.* **52-2** 557
[2] Toulemonde M., Assman W., Dufour C., et al. 2006 *Mat. Fys Med.* **52-2** 263
[3] Trautmann C. 2009 in *Ion Beams in Nanoscience and Technology* 369
[4] Lang M. Lian J., Zhang F. , et al. 2008 *Earth and Planetary Science Letters* **274** 355
[5] Toimil-Molares M. E. 2012 *Beilstein J. Nanotechnol* **3** 860
[6] Geissel H., Weick H., Winkler M., et al. 2003 *Nucl. Instr. Meth.* **B 204** 71
[7] Calder R. S. 1974 *Vacuum* **24** 437
[8] Mahner E., Hansen J., Küchler D., et al. 2002*Proc. EPAC* 2568
[9] Zhang S. Y. and Ahrens L. A. 1999 *Proc. PAC* 3294
[10] Fischer B. and Heiss M. 2007 *Nucl. Instr. Meth.* **B 260** 442
[11] Amirthapandian S., Schuchart F. and Bolse W. 2010 *Rev. Sci. Inst.* **81** 033702
[12] Baake O., Seidl T., Hossain, U. H., et al. 2011 *Rev. Sci. Inst.* **82** 045103
[13] Mahner E., Hansen J., Laurent, J.-M., et al. 2003 *Phys. Rev. ST-AB* **6** 013201
[14] Bender M., Kollmus H., Reich-Sprenger H., et al. 2009 *Nucl. Instr. Meth.* **B 267** 885
[15] Bender M., Kollmus H. and Assmann W., et al. 2007 *Nucl. Instr. Meth.* **B 256** 387
[16] Lang M., Zhang F., Lian J., et al. 2009 *J. Synch. Rad.* **16** 773
[17] Schuster B., Lang M., Klein R., et al. 2009 *Nucl. Instr. Meth..* **267** 964
[18] Glasmacher U. A., Lang M., Keppler H., et al. 2006 *Phys. Rev. Lett.* **96** 195701

Towards a fast calculator for the radiation characteristics of radiative recombination and radiative electron capture

G. Weber[1,2], H. Ding[1,3], M. O. Herdrich[1,3], A. Surzhykov[1,2]

[1] Helmholtz-Institut Jena, 07743 Jena, Germany
[2] GSI Helmholtzzentrum für Schwerionenforschung, 64291 Darmstadt, Germany
[3] Institut für Optik und Quantenelektronik, Friedrich-Schiller-Universität, 07743 Jena, Germany

E-mail: g.weber@gsi.de

Abstract. The radiative capture of free electrons (radiative recombination) and bound electrons (radiative electron capture) are among the most important charge changing processes for fast, highly-charged ions. While total cross sections can be obtained by an approximate formula with reasonable accuracy, the estimation of angular distributions and polarization properties of the emitted radiation requires a fully relativistic treatment that is numerical expensive. Therefore we recently started the development of a fast calculator for these radiation characteristics. The program is based on a grid of rigorously calculated data points for free-electron capture into bare ions, between which interpolation is performed to obtain radiation characteristics for specific collision systems. Also capture into few-electron systems is taken into account in an approximate way. We present first results from this development.

1. Introduction

The capture of electrons into bound states of ions is of significant importance for both experiment and theory in the fields of atomic and plasma physics as well as for astrophysics. The capture process is called radiative if it is accompanied by the emission of a photon that carries away the sum of the binding energy of the final state and the incident electron kinetic energy. If the initial electron is considered as free, the capture process is referred to as radiative recombination (RR) [1], being the time-reversal of the photoeffect [2], whereas the capture of a bound electron is called radiative electron capture (REC) [3, 4, 5]. The RR process is relevant for understanding the characteristics of hot plasmas [6, 7] being present in astronomical objects or in laboratory devices, for example electron beam ion traps and fusion devices, and it determines the charge change-induced beam losses in electron coolers [8, 9, 10]. On the other hand, REC is a prominent charge-changing process for fast, highly-charged ions interacting with dedicated target materials or with residual gas being present in the beam lines of accelerators and storage rings [11]. Moreover, when low- to medium-Z targets and heavy, highly-charged projectile ions are considered, the to-be-captured electrons can be treated as free particles having a momentum distribution equal to the one of the bound target states. This so-called impulse approximation reduces the REC description to the RR cross section folded with the incident electron momentum distribution. Consequently, both the REC and the RR processes, as well

Journal of Physics: Conference Series **599** (2015) 012040 doi:10.1088/1742-6596/599/1/012040

as the photoeffect, can be treated within the same theoretical framework. This fact gave rise to several REC measurements aiming for a deeper insight into the photoeffect while exploiting the advantageous experimental conditions present for the time-reversed process, see [5] and references therein.

For ongoing highly-charged ion studies at the experimental storage ring (ESR) and for future experimental campaigns planned at the CRYRING@ESR [12] and the high-energy storage ring (HESR) of the FAIR facility [13], a precise knowledge of REC characteristics in a broad range of collision energies from about 1 MeV/u up to 5 GeV/u is highly desirable. While total cross sections for RR and REC can be obtained by an approximate formula with reasonable accuracy [14] up to roughly 1 GeV/u, the estimation of angular distributions and polarization properties of the emitted radiation requires a fully relativistic treatment that is non-trivial and numerical expensive [15, 16]. Tabulated values for RR total and angular differential cross sections covering a broad range of projectile atomic number Z and collision energy E were published by Ichihara and Eichler [17, 18]. However, the grid points of this data set are too sparse to allow interpolation with a good accuracy. Moreover, the photon linear polarization, that attracted interest only recently due to the introduction of novel hard x-ray polarimeters [19, 20, 21, 22], is not addressed at all. Thus, we started the development of a fast and precise calculator for the differential cross section and linear polarization of radiation arising in the RR and REC processes. This calculator is based on a dense grid of rigorously calculated RR data points for the recombination into bare ions, between which interpolation is performed to obtain data for the collision systems of interest. Recombination into few-electron systems is taken into account in an approximative way. Finally, as described above, the capture of bound electrons is treated within the impulse approximation by folding the RR cross sections with the electron momentum distribution of the target electrons. In this report we present first results for the RR cross sections generated by our code, which is in the following referred to as the REC calculator (RECAL).

2. Radiative Recombination

Differential cross sections and linear polarization values of the RR photons are calculated using the algorithms provided by Fritzsche and Surzhykov which are part of the DIRAC toolkit, see [23, 24] and references therein for details. For first test calculations, we generated a RR data base of capture into bare projectiles for the $1s_{1/2}$ to $3d_{5/2}$ orbitals with atomic numbers $Z = 1, 7, 18, 36, 54, 66, 79$, and 92 and more than 100 kinetic energies between 2 MeV/u and 1 GeV/u (1.1 keV to 548.6 keV electron kinetic energy). The accuracy of the RR calculation is mainly determined by the accuracy of the continuum wavefunction representing the initial state of the incident electron, which is described by a series expansion in terms of partial waves. A careful choice of the number of partial waves, defined by $\nu = 2\kappa_{\rm max}$, with $\kappa_{\rm max}$ being the maximum number for the Dirac angular momentum to be taken into account, is of particular importance. More precisely, a too small choice of $\kappa_{\rm max}$ may lead to truncation errors, while, on the other hand, a too large value increases significantly computation time (approximately scaling with $\kappa^3_{\rm max}$) and eventually leads to an explosion of numerical errors due to the rapid oscillations of the radial part of the continuum wavefunction. With a proper choice of $\kappa_{\rm max}$ an uncertainty on the percent level can be reached for the parameter range concerned in this work.

The RR photon characteristics for a specific projectile and collision energy are obtained from an interpolation between the rigorously calculated data points using a method proposed by Akima [25]. To verify the quality of this procedure, in Fig. 1 the angular differential cross section for photons stemming from the recombination into the ground state and the $2p_{3/2}$ orbital of bare lead ions (Z=82) are compared to reference calculations published in [18]. For RR into the $1s_{1/2}$ orbital and electron kinetic energies lower than 300 keV (equals 547 MeV/u projectile energy) nearly perfect agreement is found, while an increasing deviation is seen for higher collision energies. This due to the fact that the proper $\kappa_{\rm max}$ number increases with higher energies and

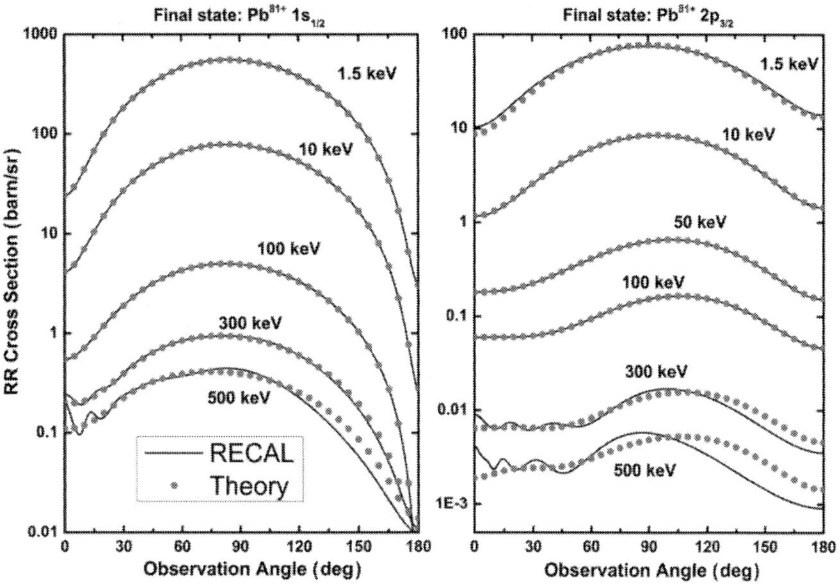

Figure 1. Angular differential radiative recombination cross section into the $1s_{1/2}$ and the $2p_{3/2}$ of bare lead ions for various collision energies. Fully relativistic calculations are compared to RECAL predictions. An increasing deviation at high collision energies is seen due to an improper choice of $\kappa_{\rm max}$ for the underlying RR data base of the RECAL calculator, see text for details.

in our first attempt we failed to chose large enough $\kappa_{\rm max}$ values for the relativistic regime of collision energies. Moreover, for RR into the $2p_{3/2}$ orbital, also at the lowest collision energy a small deviation between rigorous calculations and RECAL predictions is found in Fig. 1. We are still investigating if this feature is also due to a wrong maximum partial wave number or if it is caused by an improper interpolation along the Z axis. Here one has to stress that depending on $\kappa_{\rm max}$ the rigorous calculation for a single collision energy and a single final orbital may take up to a few hours of computation time on a standard desktop computer. We are currently performing systematic test calculations to obtain a map of optimum $\kappa_{\rm max}$ values as a function of the collision energy, final electron state and projectile Z. Therefore, we compare the obtained radiation characteristics to reliable reference calculations such as the data published in [17, 18]. As mentioned above, for the linear polarization of RR/REC photons no tables with references values are available, so here we rely on the convergence of the polarization characteristics with increasing $\kappa_{\rm max}$ as the only criterion.

So far only recombination into finally hydrogen-like ions was considered. However, there is the need to have RR/REC cross sections with a reasonable precision also for few-electron systems. It was previously shown that for few-electron, heavy ions and collision energies above a few MeV/u the RR process can be approximated with a single-electron model using an appropriate effective nuclear charge $Z_{\rm eff}$ as in that case the electron-electron interaction is very small compared to the interaction of each electron with the nuclear charge [26]. Effective Z values for the RECAL code were determined from a table of successive ionization potentials originally compiled for the SU program that is applied for electron beam ion trap simulations [27, 28].

In Fig. 2 we compare the RECAL results for RR into initially hydrogen-like and lithium-like uranium ions to rigorous multi-configuration Dirac Fock (MCDF) calculations presented

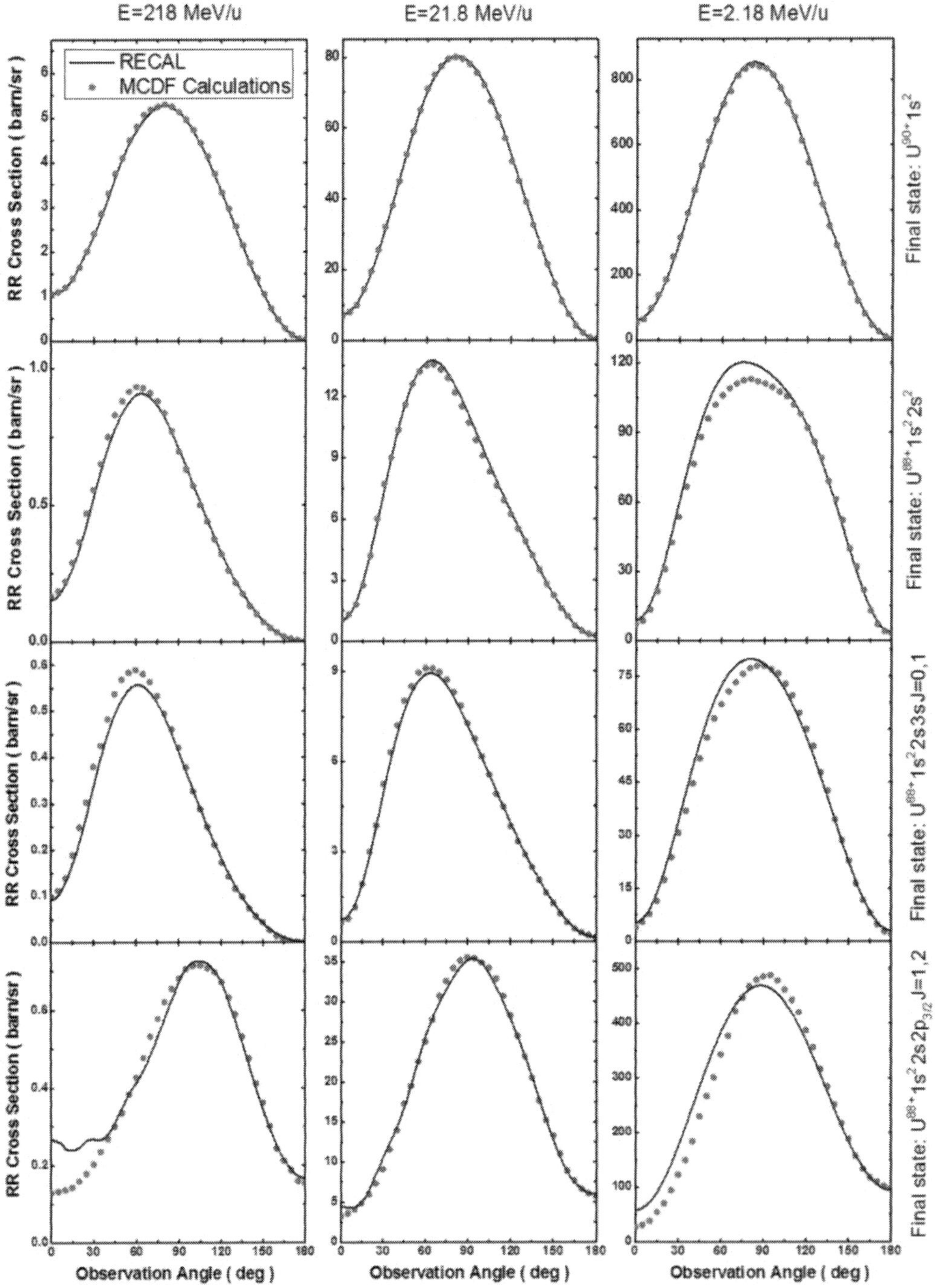

Figure 2. Angular differential radiative recombination cross section into hydrogen-like (final state: helium-like) and lithium-like (final state: beryllium-like) uranium ions for various collision energies. Rigorous MCDF calculations [26] are compared to RECAL predictions. Good agreement is found for the medium collision energy, while for the low and the high energy data significant deviations are seen for the finally beryllium-like system.

in [26] (results from Coulomb and Babushkin gauges were averaged). Note that the theory cross sections for recombination into neighboring J-levels of the $1s^22s3s$ and the $1s^22s2p_{3/2}$ configurations were summed up to allow a comparison to the RECAL predictions which rely on a single-electron data base. However, in many cases these levels differ only by a few eV and are hardly resolved by standard solid-state x-ray detectors. For recombination into the ground state of initially hydrogen-like uranium RECAL matches the theory values with good agreement. In this case one can expect that the huge binding energy of the final state suppresses any significant electron-electron interaction during the capture into the projectile. In contrast, for the initially lithium-like system only for a moderate collision energy of 21.8 MeV/u good agreement is found, while for the low and the high energy data significant deviations are seen. A possible explanation is that for low collision energies the electron-electron interaction can not be neglected. On the other hand at high energies the chosen $\kappa_{\max}$ might be too small as at least for the $1s^22s2p_{3/2}$ configuration and at observation angles below 30° an artefact due to a wrong $\kappa_{\max}$ value is visible in the RECAL data.

3. Summary

We presented first results from a fast calculator for the radiation characteristics of the RR and REC processes. Such a tool will be helpful for the planning and interpretation of energetic ion-atom collision experiments at the new FAIR facility as well as for ongoing heavy ion studies at GSI. A comparison to rigorously calculated data shows in general a reasonable agreement. However, significant deviations are observed for parameters where the underlying data base of the calculator was produced using improper values for the maximum number of partial waves $\kappa_{\max}$ for the series expansion representing the incident electron. A systematic study is needed to determine valid $\kappa_{\max}$ values throughout the full parameter range of interest.

4. References

[1] Oppenheimer J R 1928 *Phys. Rev.* **31** 349
[2] Scofield J H 1989 *Phys. Rev. A* **40** 3054
[3] Schnopper H W *et al.* 1972 *Phys. Rev. Lett.* **29** 898
[4] Ichihara A *et al.* 1994 *Phys. Rev. A* **49** 1875
[5] Eichler J and Stöhlker Th 2007 *Physics Reports* **439** 1
[6] Verner D A and Ferland G J 1996 *The Astrophysical Journal Supplement Series* **103** 467
[7] Badnell N R 2006 *The Astrophysical Journal Supplement Series* **167** 334
[8] Pajek M and Schuch R 1992 *Phys. Rev. A* **45** 7894
[9] Shi W *et al.* 2001 *The European Physical Journal D* **15** 145
[10] Banas D *et al.* 2014 *Physica Scripta* **2014** 014001
[11] Stöhlker Th *et al.* 1998 *Phys. Rev. A* **58** 2043
[12] Lestinsky M *et al.* 2012 CRYRING@ESR: A study group report *unpublished*
[13] Stöhlker Th *et al.* 2013 *Physica Scripta* **2013** 014085
[14] Stöhlker Th *et al.* 1995 *Phys. Rev. A* **51** 2098
[15] Eichler J *et al.* 1995 *Phys. Rev. A* **51** 3027
[16] Surzhykov A *et al.* 2003 *Phys. Rev. A* **68** 022710
[17] Ichihara A and Eichler J 2000 *Atomic Data and Nuclear Data Tables* **74** 1
[18] Ichihara A and Eichler J 2001 *Atomic Data and Nuclear Data Tables* **79** 187
[19] Tashenov S *et al.* 2006 *Phys. Rev. Lett.* **97** 223202
[20] Hess S *et al.* 2009 *Journal of Physics: Conference Series* **194** 012025
[21] Hess S *et al.* 2009 *Journal of Physics: Conference Series* **163** 012072
[22] Weber G *et al.* 2010 *Journal of Instrumentation* **5** C07010
[23] Surzhykov A *et al.* 2005 *Computer Physics Communications* **165** 139
[24] McConnell S *et al.* 2010 *Computer Physics Communications* **181** 711
[25] Akima H 1991 *ACM Trans. Math. Softw.* **17** 341
[26] Fritzsche S *et al.* 2005 *Phys. Rev. A* **72** 012704
[27] Becker R *et al.* 2007 *Journal of Physics: Conference Series* **58** 443
[28] Becker R 2014 *private communication*

Journal of Physics: Conference Series **599** (2015) 012041 doi:10.1088/1742-6596/599/1/012041

Towards Proton Therapy and Radiography at FAIR

M. Prall[1], P. M. Lang[1,2], C. LaTessa[1] F. Mariam[3] , F. Merrill[3], L. Shestov[1], P. Simoniello[1], D. Varentsov[1] and M. Durante[1]

[1]GSI Helmholtzzentrum für Schwerionenforschung GmbH, Planckstraße 1, 64291 Darmstadt, Germany

[2]TU Darmstadt, Schlossgartenstraße 9, 64289 Darmstadt, Germany

[3]Los Alamos National Laboratory, Los Alamos, NM 87545, USA

E-mail: m.prall@gsi.de

Abstract. Protons having energies in the GeV range have been proposed as an alternative to Bragg-peak hadron therapy. This strategy reduces lateral scattering and overcomes uncertainties of particle range and relative biological effectiveness. GeV protons could additionally be used for targeting in image guided stereotactic radiosurgery. We experimentally demonstrated the potential of GeV protons for imaging of biological samples using E=0.8 GeV protons and the pRad setup at Los Alamos National Laboratory (LANL). In this setup, a system of magnetic lenses creates a point-to-point mapping from object to detector. This mapping compensates image blur due to lateral scattering inside the imaged (biological) object. We produced 2-dim proton radiographs of biological samples, an anthropomorphic phantom and performed simple dosimetry. High resolution tomographic reconstructions were derived from the 2-dim proton radiographs. Our experiment was performed within the framework of the PANTERA (Proton Therapy and Radiography) project. In the future, the proton microscope PRIOR (Proton Microscope for FAIR) located in the FAIR facility (Darmstadt), will focus on optimizing the technique for imaging of lesions implanted in animals and couple the irradiation with standard radiotherapy.

1. Introduction

Proton therapy plays an important role in the treatment of cancer [1] and noncancer [2] diseases. More than 50% of patients with localized malignant tumors are nowadays treated with radiation [3]. The challenge in radiotherapy is to deliver ionizing irradiation such that the target region is ablated or sterilized, whereas the surrounding healthy tissue is spared. Protons having energies between about 60 and 250 MeV are well suited for this purpose due to their depth-dose profile (Bragg curve). The proton beam has a finite range in matter and deposits most of its energy at the end of its path (cf. Figure 1).

In treatment of cancer, typically several 10.000 Bragg-peaks are placed in a tumor volume. The depth (range) of each Bragg-Peak is adjusted via its initial energy. The lateral beam position can be adjusted by dipole magnets in the beam line. This so-called rasterscan technique was used to treat more than 450 patients in a pilot experiment at GSI [3] and is now implemented e.g. at Heidelberg Ion Beam Therapy Centre (HIT), where up to 1000 patients are treated each year. Yet, in Bragg-peak therapy the beam profile is broadened by multiple scattering in the patient's tissue. This produces a dose-halo around the target, which is typically a few mm wide. This can be problematic if life-critical structures e.g. in the brain are close (a few

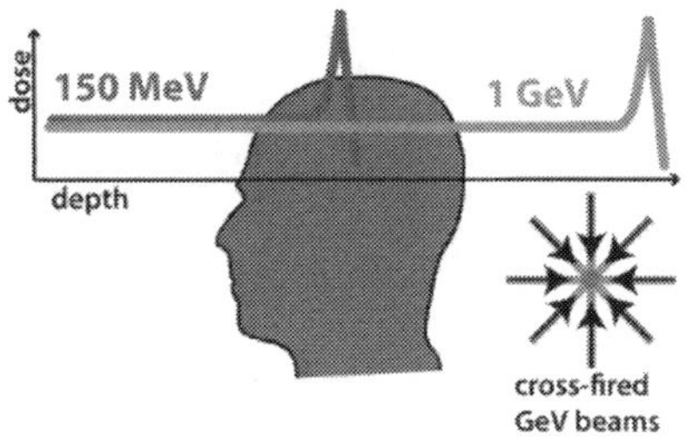

Figure 1. Comparison of Bragg peak therapy and GeV particle radiosurgery. The two graphs (150 MeV and 1 GeV) symbolize the depth-dose profiles in Bragg-peak and GeV therapy. Whereas the dose-maximum is placed inside the target in Bragg-peak therapy (here: the head), GeV beams have to be cross-fired to achieve a dose-maximum at their intersection (see bottom right).

mm or less). Lateral scattering of a proton is roughly proportional to $1/\beta^2$, where $\beta = v/c$ [4]. Therefore, lateral scattering can be significantly reduced by increasing the particle velocity β. Using protons having energies of about a GeV reduces lateral scattering to the sub-mm region [5]. However, a target cannot be treated from a single direction any more. Several GeV beams have to be cross-fired from different angles. This situation is sketched in the insert of Figure 1.

Clinical experience with cross-fired GeV proton beams has already been gained since 1975 at PNPI St. Petersburg with about 1300 patients [8]. Targets in the brain, e.g. cerebral artery-venous malformations, pituitary adenomas, but also outside the brain, like prostate cancer were treated successfully. Patients were aligned using X-rays.

Although GeV protons are used in the plateau region (cf. Figure 1), the depth-dose distribution is not flat [5]. Knock-out and evaporation hadrons, mostly secondary protons from nuclear interactions cause an increase of the overall dose deposition. These effects have already been studied for shielding of relativistic protons for space radiation protection [9]. Whereas the contribution of primary protons decreases by a few percent, the overall dose increases by $30 - 40$ % after about 20 cm of tissue [5]. These effects will be studied in more detail in order to include them into treatment planing for IGSpRS (Image Guided Stereotactic proton Radiosurgery). In IGSpRS, the beam, which has passed the patient is used for simultaneous radiography (imaging) of the treated region.

2. Image guided stereotactic radiosurgery with a proton microscope

Whereas a drawback of GeV radio surgery is the loss of the Bragg-peak, a valuable advantage is that IGSpRS becomes possible. IGSpRS allows to detect misalignments of the patient or a change of the target region online. This is not possible with the combination of computerized X-ray tomography, 2-dim X-ray imaging and immobilization devices used today in clinical practice. IGSpRS would be advantageous for patients with the above-mentioned diseases, especially close to life-critical structures.

The spatial resolution of conventional proton radiography based on energy loss is limited and thus its practical applicability [6]. A proton microscope (PM) compensates image blur using magnetic lenses to the sub-mm level. We are therefore aiming to realize IGSpRS with a PM at FAIR. Ignoring effects such an energy loss in the imaged object, a PM (cf. Figure 2) works as follows: After crossing a part of the patient with thickness d and radiation length L_{rad}, protons exit tissue with angles θ with respect to the initial beam axis z. The trajectory through the PM depends on the exit angle θ. A first set of magnetic lenses (A) transforms the angle θ into a distance from the beam axis at the position of the collimator (B). The larger θ, the larger the distance from the beam axis z. The collimator (B) suppresses protons having more than a certain distance from the beam axis and thus a scattering angle larger than a critical scattering angle θ_c. A second set of magnetic lenses (C) after the collimator reverses the effect of the lenses before the collimator (A) such that one obtains a point-to-point mapping between the exit coordinates at the patient (x_0, y_0) and at the detector (x_1, y_1). Magnetic fields can

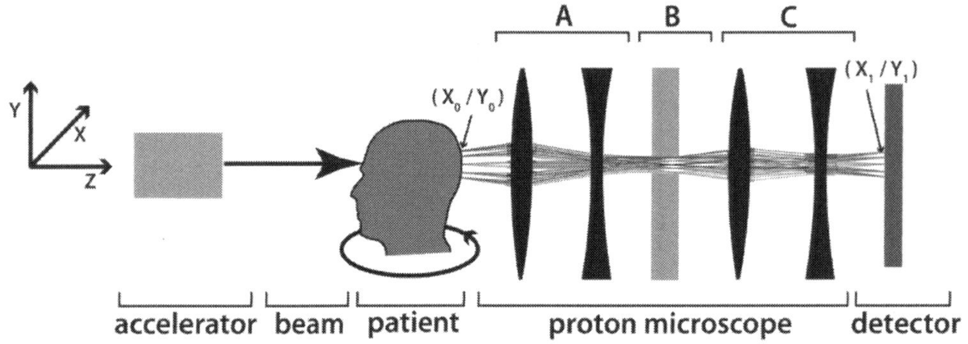

Figure 2. Image guided stereotactic proton radiosurgery using a proton microscope. The patient is rotated and beams are cross-fired through the target (e.g. a tumor). A proton microscope is used for simultaneous sub-mm resolution radiography of the treated region.

be chosen such that the image on the detector is larger than the radiographed object itself. Therefore the name proton *microscope*. In our case, fields were chosen such that a 1:1 mapping (without magnification) was achieved. The distribution of the exit angles θ can be described by a Gaussian having a standard deviation of $\sigma(\text{rad}) = \frac{14.1\,MeV}{\beta p}\sqrt{d/L_{rad}}$, where $\beta = v/c$ and p = proton momentum. The larger the path length d through tissue, the more protons leave the patients with larger angles θ, the more are suppressed by the collimator and the fewer reach the detector. Additionally, nuclear interactions lead to particle loss equal to $\exp(-d/\lambda)$, where λ is the nuclear interaction length. Putting together nuclear losses and suppression of scattering angles $\theta(d)$ above a critical angle θ_c by the collimator leads to a transmission probability $T(d)$:

$$T(d) = \exp(-d/\lambda) \cdot \left(1 - \exp\left(-\frac{1}{2}\frac{\theta_c \beta p}{14.1\ \text{MeV}}\frac{L_{rad}}{d}\right)\right) \quad (1)$$

The count rate on each point of the detector is proportional to $T(d)$. However, this formula cannot be inverted analytically to determine d from a count rate. In practice one uses careful calibration measurements instead. These also include the influence of any other material (e.g. beam windows) in the beam line, as well as the initial angular spread of the beam. At LANL, the detector is a scintillator screen (SC). Once a proton pulse, having a width comparable to the size of this screen crosses the patient, the instrument and eventually the SC, a camera makes a black and white image of the SC light output. This image shows the path length through the patient d with sub-mm resolution. Via (1) the grey values in this image correspond to the path length d through the patient.

In the above description, we ignored effects leading to image distortion and finite spatial resolution. In reality, lateral drift of protons inside the imaged object, energy inhomogeneity coming from both, the initial beam and inhomogeneous energy loss inside the imaged object, off-axis effects, the choice of the SC and the light optics through which the SC is filmed, determine the spatial resolution of the data. For our biological samples and an E=800 MeV beam, the overall resolution is in the order of tenths of millimeters. A thorough review of charged particle radiography is given in [10].

3. Proton radiography and tomography of biological samples

In order to experimentally demonstrate the potential of a PM for biomedical applications, we made measurements with the PM of the pRad facility of LANL which uses an 800 MeV proton beam [7]. The field of view of this instrument is about 10 cm x 10 cm. We produced 2-dim proton radiographs of biological samples, an anthropomorphic phantom and performed simple dosimetry. High resolution tomographic reconstructions using filtered backprojection [11] were derived from the 2-dim proton radiographs. Figure 3 gives an overview over our results, which will be presented more thoroughly in a forthcoming publication.

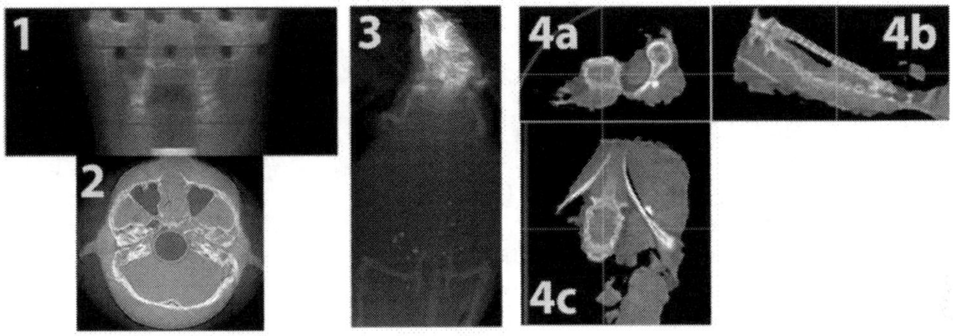

Figure 3. Some results: 1) Radiograph of a head phantom, 2) tomographic slice of a head phantom, 3) maximum intensity projection derived from a tomograph of a mouse, 4a) tomographic slice of a chicken sample, 4b) and 4c) orthogonal views of 4a).

4. Discussion

Measurements with thermoluminescent dosimeters inside the skull of the anthropomorphic phantom were made. These show that bony anatomy is clearly resolved with a dose of about $D_{image} \approx 10$ mGy. These structures could be used for position verification in IGSpRS. We expect that further measurements would allow to determine an even smaller value for D_{image}. At LANL, we use only two sets of thermoluminescent dosimeter (TLDs) for two independent dose measurements. For measurements, the TLDs were placed in slits inside the phantom and were later read out at DLR (Deutsches Zentrum fr Luft und Raumfahrt, German Aerospace Center) in Germany. This dose of $D_{image} = 10$ mGy is already negligible compared to target doses of up to 150 Gy used in proton radiosurgery [8]. The spatial resolution of the 2-dim proton radiographs is about 0.2 mm at high contrasts. 3-dim tomographic reconstructions derived from 2-dim radiographs show finest bony structures like ribs of the embedded mouse, substructure is visible in muscle-tissue of chicken samples, implanted steel and aluminum markers (not shown in Figure 3) are clearly resolved and do not produce artifacts in the 3-dim reconstructions. The world's first proton tomographic image of a skull was derived from our raw data. X-ray CTs of most samples were made using a state-of-the-art CT scanner (Biograph TruePoint by Siemens Healthcare, Forchheim, Germany). Reconstruction artifacts (streaks and ring artifacts) are visible in the proton tomographs. Yet the tomographic slices are comparable to the X-ray CTs (not shown). Artifact suppression is algorithmically challenging, a subject of current research [12] and beyond the scope of our study.

Recent studies on ion CT employ either complex or expensive detector designs [13, 14, 15, 16] or a time-consuming data acquisition strategy using a scanned beam [17]. In our method, the time for a proton CT is limited by the speed of the rotary table on which the imaged sample is placed and could be reduced to a few seconds for a tomographic scan. In the future, this feature might be important for clinical applications, where a proton CT scan should not take longer than a few seconds.

5. Outlook

Our study has clearly demonstrated the potential of a PM for IGSpRS and imaging in 2 and 3 dimensions. The encouraging clinical results obtained at PNPI with 1 GeV protons [8], support the rationale for IGSpRS for future clinical applications in radiosurgery at FAIR. Especially a subgroup of patients having life-threatening cancer- and noncancer diseases close (a few mm or less) to life-critical structures will benefit significantly from this bloodless treatment modality.

Within the PaNTERA (Proton Therapy and Radiography) project, we are mostly interested in imaging biological samples. Tomography is, of course not limited to these. Actually in many applications is becomes superior to x-ray CT, when the object is too dense or too thick for X-rays. At LANL, for example one has already studied damage detection in nuclear fuel pellets with this method [10]. A feature of a PM, that we did not use is the ability to produce short series of radiographs having microsecond spacing [10]. This allows to study dynamic processes like explosions or impacts in 2 dimensions.

At FAIR, the proton microscope PRIOR will not only be used for IGSpRS. A wide range of interdisciplinary experiments will be performed. These are e.g. experiments on fundamental properties of materials in extreme dynamic environments generated by external drivers (pulsed power drivers, high energy lasers, gas guns or HE generators) prominent for materials research and non-ideal plasma physics.

6. Acknowledgements

We thank Dr. Julia Bauer (HIT Heidelberg) for performing x-ray CT scans of our samples. Dr. Thomas Berger (DLR) and his team provided the human phantom and radiation detectors. They also took care of the detector readout. Last but not least, we thank the staff of pRAD for experimental support and fruitful discussions.

References

[1] M Durante and J Loeffler Nat. Rev. Clin. Oncol. 7 37
[2] C Bert, R Engenhart-Cabillic and M Durante Med. Phys. 39 1716
[3] Schardt D, Elsässer T, Schulz-Ertner D. Rev Mod Phys 2010;82:383
[4] H Paganetti Proton Therapy Physics, CRC Press, ISBN: 978-1-4398-3644-6
[5] M Durante and H. Stöcker Journal of Physics: Conference Series 373 (2012) 012016
[6] J M Schipper and A J Lomax 2011 Acta Oncol. 50 838
[7] N S P King, E Ables, K Adams, et al. NIMA, 1999, 424: 84-91.
[8] N K Abrosimov et al. Journal of Physics: Conference Series 41 (2006) 424-432
[9] M Durante and F A Cucinotta 2011 Rev. Mod. Phys. 83 1245
[10] C L Morris et al. Rep. Prog. Phys. 76 (2013) 046301 (26pp)
[11] A C Kak and M Slaney Principles of Computerized Tomographic Imaging (1988)
[12] T M Buzug, Computed Tomography From Photon Statistics to Moden Cone-Beam CT, Springer, ISBN: 978-3-540-39407-5
[13] Sadrozinski H-W et al 2003 Nucl. Instrum. Methods Phys. Res. A 511 27581
[14] Ohno Y, Kohno T, Matsufuji N and Kanai T 2004 Nucl. Instrum. Methods Phys. Res. A 525 27983
[15] Murashi H et al 2007 IEEE Nucl. Sci. Symp. Conf. Record pp 43548
[16] U Amaldi et al. Nucl. Instrum. Methods Phys. Res. A 629 33744
[17] J Telsemeyer, O Jaeckel and Maria Martisikova Phys. Med. Biol. 57 (2012) 79577971

FAIRNESS 2014 IOP Publishing
Journal of Physics: Conference Series **599** (2015) 012042 doi:10.1088/1742-6596/599/1/012042

Investigation of prompt γ ray emission for online monitoring in ion therapy

D Steinschaden[1], S E Brunner[1], H Dichtl[2], H Fuchs[3,4,5], D Georg[3,4,5], A Hirtl[6], J Marton[1] and A Pichler[1]

[1] Stefan Meyer Institute for Subatomic Physics, Austrian Academy of Sciences, Austria
[2] University of Applied Sciences Upper Austria, Austria
[3] Christian Doppler Laboratory for Medical Radiation Research for Radiation Oncology, Medical University of Vieanna, Austria
[4] Department of Radiation Onkology, Medical University of Vienna / AKH Vienna, Austria
[5] Comprehensive Cancer Center, Medical University of Vienna /AKH Vienna, Austria
[6] Division of Nuclear Medicine, Department of Biomedical Imaging and Image-guided Therapy, Medical University of Vienna / AKH Vienna, Austria

E-mail: dominik.steinschaden@oeaw.ac.at

Abstract.
The aim of this study was to investigate the principle of using prompt γ ray emissions for the determination of the position of the Bragg peak in radiotherapy with carbon ions. Developing a system for online monitoring in ion therapy with the required accuracy is an important step for the utilization of modern accelerator facilities for medical purposes. The investigations were carried out by Monte Carlo studies in the framework of the simulation environment GATE. In the course of these investigations, production parameters of prompt γ rays, like the emission rate and energy distribution were determined as a function of the primary carbon ion energy and the penetration depth in a water target. A possible connection between these parameters and the position of the Bragg peak and the dose delivery was investigated. Within the energy spectrum an energy range was identified in which the production rate of the γ rays shows a significant drop right after the Bragg peak.

1. Introduction

The number of medical centres with focus on cancer treatment based on ion therapy increased during the last years [1]. Near Vienna the treatment facility MedAustron will start the treatment of patients in 2015 with proton and carbon ion beams [2]. In order to perform a therapy the patients are examined by established imaging methods and a treatment plan is calculated. Up to now, no direct and precise method for online monitoring of the distributed dose exists, which fulfils the required high accuracy combined with fast data collection and processing. A few promising techniques have been proposed and some are being investigated in pre-clinical research. So far only Positron-Emission-Tomography (PET) offers a technically feasible method for monitoring the dose distribution shortly after the dose delivery [3]. A possible alternative to PET monitoring is the detection of prompt γ radiation during radiotherapy. In contrast to the method of PET monitoring which is based on the β^+ decay, the prompt γ ray emission which is based on nuclear reactions happens on a time scale of less than 1 ns after the energy deposition.

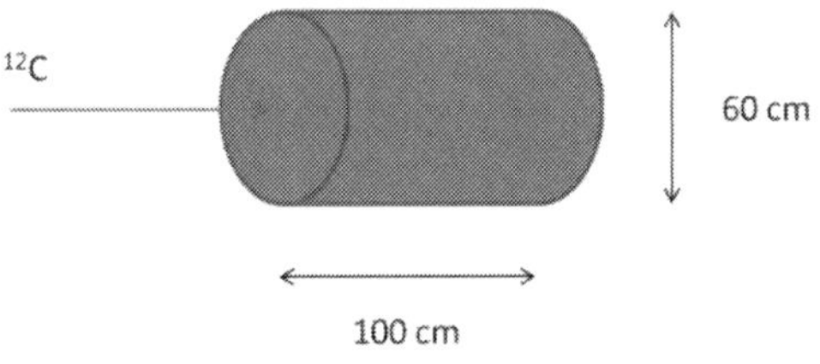

Figure 1. Simulation setup for investigations of the prompt γ ray production parameters. A carbon ion beam is aimed at a cylindrical water target along its longitudinal axis.

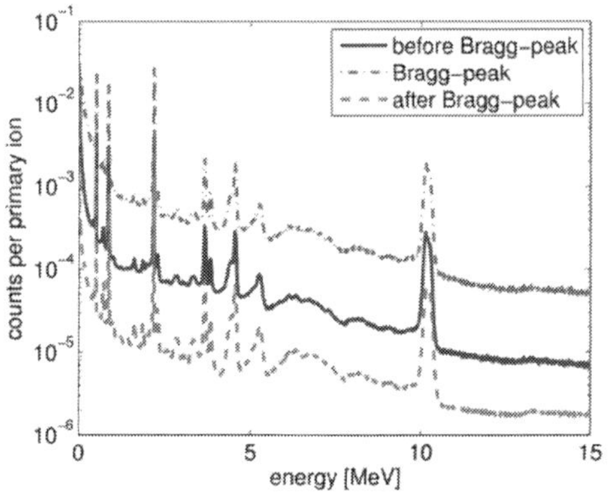

Figure 2. Spectra of γ rays produced by the irradiation of a water phantom with carbon ion at energies of 240 MeV/A. The solid line maps the energy spectrum before, the dashed dotted line at and the dashed line after the Bragg peak.

Therefore, prompt γ based monitoring does not suffer from physiological processes and high radiation background [4][5]. This approach was already experimentally demonstrated [6].

2. Material and Methods

The presented studies were realised within the simulation framework GATE (GEANT4 Application for Tomographic Emission) which is a modular and versatile toolkit for nuclear medicine [7], developed by the OpenGATE collaboration [8]. It is based on the simulation toolkit GEANT4 which contains modules to realise a complete range of functionalities and which can handle many different processes, particles and materials in an energy range from 250 eV up to a few TeV [9]. A verification of the simulated prompt γ spectra with experimental data has been performed with the results presented in reference [10]. Discrepancies of the simulated count rate up to the factor 2 has been observed in reference [11].

The simulation setup consisted of a cylindrical water target with a radius of 30 cm and a length of 100 cm which was assumed homogeneous. The phantom was irradiated with carbon ions along its longitudinal axis. The energy of the used ion beams varied from 120 MeV/A to 400 MeV/A corresponding to medical applications. The radial intensity distribution of the beam was Gaussian-shaped with an initial standard deviation of 3 mm and zero beam divergence (see Fig. 1).

In order to find a correlation between the Bragg peak, which was defined in this work as the position of the maximum dose deposit inside the irradiated target, ideal detectors were used to study the properties of prompt γ rays. The cylindrical target was divided into an inner section represented by a cylinder with a radius of 6 mm and a height of 100 cm centered along the longitudinal axis, which corresponds to 2 sigma of the used primary particle beam and an outer section which covered the rest of the water target. During the simulation, for each section the number of produced photons as a function of the penetration depth was accumulated in steps of 1 mm. Additionally, the number of produced photons was recorded up to 12 MeV in steps of 100 keV. Additional energy spectra of the produced γ rays were recorded for three subvolumes from the entrance face of the target until 5 mm prior to the Bragg peak, $\pm$ 5 mm around the Bragg peak and from 5 mm after the Bragg peak until the end of the target.

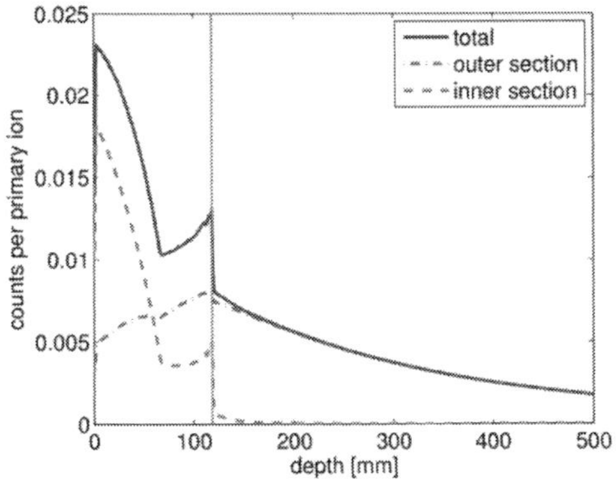

Figure 3. Production rate of prompt γ rays per primary carbon ion plotted for the different target regions as a function of the penetration depth. The red vertical line marks the position of the maximum dose deposit.

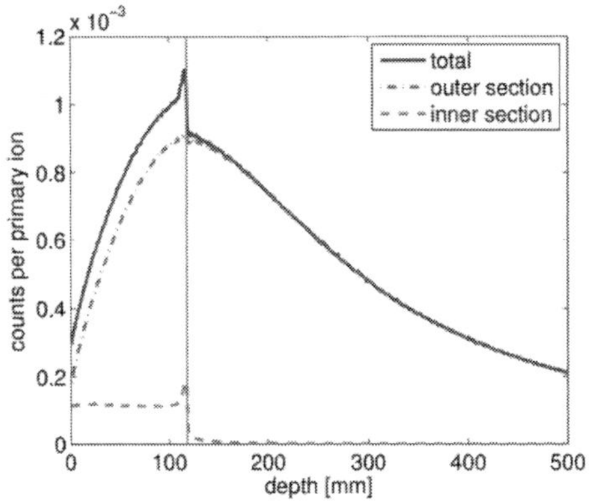

Figure 4. Photon production rates for photon energies of 2.2 ± 0.1 MeV plotted for different target regions as a function of the penetration depth. The position of the Bragg peak is marked by the red vertical line.

3. Results

Figure 2 shows the normalised energy spectra per cm for the 3 different target regions. Analysing the energy spectra did not result in finding a specific photon production process, which could be linked to the position of the Bragg peak. Apart from the count rate, the behaviour of the spectra remain the same for every penetration depth. A drop in the count rate could be identified after the Bragg peak.

An evaluation of the collected data with regard to the production rate for the different γ energies showed a significant difference. The total production rate is plotted in Figure 3 as a function of the penetration depth of the primary carbon ions with energies of 240 MeV/A.

Investigations showed that the high count rates in the area prior to the Bragg peak are mainly caused by photons with energies below 0.5 MeV. The rise in count rates at the Bragg peak and the drop afterwards is driven by photons with energies above 0.5 MeV. Figure 3 contains the production rate for photons separated into the inner and the outer section. While the production rate in the inner section drops almost immediately to zero after the Bragg peak, the drop in the count rate in the outer section is less significant. Consequently, the influence of this drop around the Bragg peak is less significant in the total photon production.

Similar results were observed for many of the prominent peaks of the energy spectra. As an example, in Figure 4 the production rates for 2.2 MeV γ rays are plotted as a function of depth for a primary energy of 240 MeV/A. These are mainly caused excited nuclei, which result from neutron capture of hydrogen and belong to the most prominent peak in the spectrum. Again, the production rate in the inner section shows a very significant drop after the Bragg peak, whereas the photon production of the outer section misses this effect. Due to the higher production rates in the outer section the significance of the drop is suppressed and thus, can not be used for the determination of the Bragg peak.

Although the effect is less drastic for other prominent peaks of the energy spectrum, the result remains comparable. Further analysis of the collected data showed that the significance of produced γ rays apart from the most prominent peaks rises. Also wider ranges of photon energies can be merged to increase both count rates and significance as Figure 5 shows for photon energies from 2.3 MeV to 6 MeV. In this energy range the production rate of γ rays inside the

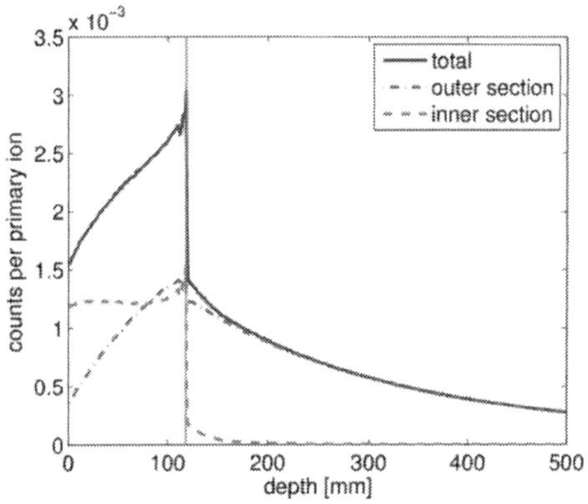

Figure 5. Production rates of γ rays with an energy from 2.3 to 6 MeV as a function of the penetration depth for the different target sections. The red vertical line marks the Bragg peak.

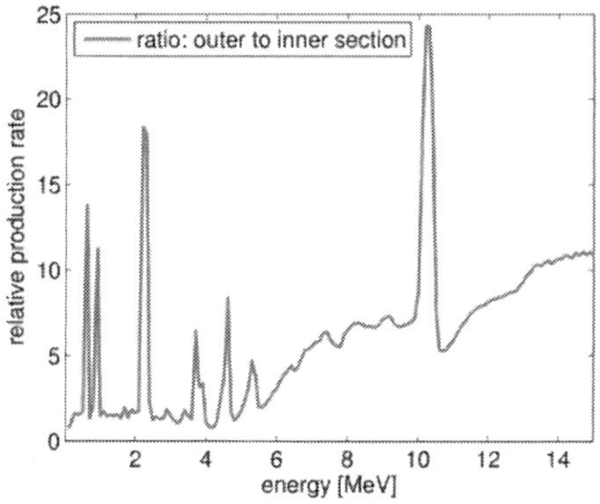

Figure 6. Relative γ ray production rate comparing the outer to the inner section of the target as a function of the γ ray energies. A lower value means less amount of outer photons, and therefore a more usable signal at the Bragg peak.

2 sigma range of the primary ion beam is even higher compared to the production in the outer section at the position of maximum dose deposition, which is marked by the red vertical line.

To get an overview of γ ray energies with a beneficial production rate, Figure 6 shows the relative photon production rate of the outer section to the inner section for 240 MeV/A primary carbon ions as a function of the photon energy. In confirmation of the studies above, the prominent peaks of the energy spectra show a significantly higher ratio of outer to inner γ rays. Also the values of the ratio for photons produced with energies higher then 6 MeV are above the average of the energy range of 2.3 - 6 MeV photons.

4. Discussion

Although the energy spectra do not reveal a significant photon energy which can be linked to the Bragg peak, the data show a strong correlation of the photon production rate and the position of maximum dose deposit visible by a drop in the production function. This signal is very sharp for photons produced in the inner section of the target most likely due to collisions of primary particles. γ rays produced outside of the 2 sigma range of the primary beam lower the visibility of the drop at the Bragg peak since they are produced with a high probability by secondary or scattered particles. An increase of the signal could be achieved by focusing on a energy range of produced γ rays with a beneficial ratio of outer to inner production rate. Such an energy range was found for γ rays between 2.3 MeV and 6 MeV which seems the most utilisable energy range for determining the Bragg peak.

5. Conclusion

The presented data provide a possibility to relate the production rate of prompt γ rays to the position of the Bragg peak, i.e., the point of the maximum dose deposition. Further investigations are necessary to develop a detector system to determine and monitor the spatial distribution of the deposited energy. Although optimization needs to be done, for typical medical beam rates online monitoring, based on prompt γ rays, seems to be feasible.

Acknowledgments
The main part of the work was performed on the computer cluster of the MOCCAMED group which enabled simulations with sufficient statistics in reasonable time. The project was financially supported by the Federal Ministry of Economy, Family and Youth and the National Foundation for Research, Technology and Development.

References
[1] Giap F N, Giap H B, Mazal A, Jermann M, Giap B, Levy R P and Blomquist E 2013 *Translational Cancer Research* **1**
[2] Medaustron `http://www.medaustron.at/` [Online; accessed 28-July-2014]
[3] Parodi K 2012 *Nuclear Medicine Review* **15** C37
[4] Testa E, Bajard M, Chevallier M, Dauvergne D, Foulher F L, Freud N, Letang J M, Poizat J C, Ray C and Testa M 2008 *Applied Physics Letters* **93** 093506
[5] Schardt D, Elsässer T and Schulz-Ertner D 2010 *Rev. Mod. Phys.* **82** 383–425
[6] Testa E, Bajard M, Chevallier M, Dauvergne D, Foulher F L, Freud N, Ltang J, Poizat J, Ray C and Testa M 2009 *Nuclear Instruments and Methods in Physics Research Section B: Beam Interactions with Materials and Atoms* **267** 993–6
[7] Jan S *et al.* 2011 *Physics in Medicine and Biology* **56** 881–901
[8] 2013 www.opengatecollaboration.org
[9] Agostinelli S *et al.* 2003 *Nuclear Instruments and Methods in Physics Research Section A: Accelerators, Spectrometers, Detectors and Associated Equipment* **506** 250–303
[10] Polf J C, Peterson S, Ciangaru G, Gillin M and Beddar S 2009 *Phys Med Biol* **54** 731–743
[11] Sarrut D, Bardis M, Boussion N, Freud N, Jan S, Ltang J M, Loudos G, Maigne L, Marcatili S, Mauxion T, Papadimitroulas P, Perrot Y, Pietrzyk U, Robert C, Schaart D R, Visvikis D and Buvat I 2014 *Medical Physics* **41**

FAIRNESS 2014 IOP Publishing
Journal of Physics: Conference Series **599** (2015) 012043 doi:10.1088/1742-6596/599/1/012043

The *g*-factor of highly charged ions

Sven Sturm[1], Florian Köhler[1,2] and Günter Werth[3]

[1] Max-Planck Institut für Kernphysik, Saupfercheckweg 1, 69117 Heidelberg, Germany
[2] GSI Helmholtzzentrum für Schwerionenforschung, Planckstraße 1, 64291 Darmstadt, Germany
[3] Institut für Physik, Johannes Gutenberg-Universität, Staudingerweg 7, 55128 Mainz, Germany

sven.sturm@mpi-hd.mpg.de

Abstract. Highly charged ions provide a unique opportunity to test our understanding of atomic properties under extreme conditions: The electric field strength seen by an electron bound to a nucleus at the distance of the Bohr radius ranges from 10^{10} V/cm in hydrogen to 10^{16} V/cm in hydrogenlike uranium. The theory of quantum electrodynamics (QED) allows for calculation *e.g.* of binding energies, transition probabilities or magnetic moments. While at low fields QED is tested to very high precision, new, hypothetical nonlinear effects like photon-photon interaction or a violation of Lorentz symmetry may occur in strong fields which then would lead to an extension of the Standard Model. The ultra-high precision determination of the magnetic moment of a bound electron in a highly charged ion provides a unique possibility to probe the validity of the current Standard Model in extreme conditions.

1. Introduction

Calculation of atomic properties reaches the smallest uncertainties for systems containing few or even only one electron. In this contribution we consider the magnetic moment of the electron bound in a hydrogenlike ion as test case for QED calculation. The magnetic moment μ is in general expressed by the dimensionless *g*-factor

$$\vec{\mu} = -g\mu_B \frac{\vec{s}}{\hbar} \qquad (1)$$

with μ_B the Bohr magnetron and *s* the electron´s spin. Calculations of the electron´s *g*-factor in the ground state of hydrogenlike ions as function of the nuclear charge *Z* have been performed by different groups [1 and references therein]. They include the relativistic effects of binding (Breit term), of virtual photon exchange in different orders (1- and 2- loop QED), and of nuclear properties and can be written as

$$g = 2\,(1 + a_{Breit} + a_{1loop} + a_{2loop} + a_{NuclearSize} + a_{Recoil} + \ldots)\,. \qquad (2)$$

Fig. 1 shows the result of the calculations for low and medium values of *Z*. In order to match the theoretical precision and to perform a significant test of the calculated terms the experiment requires

Journal of Physics: Conference Series **599** (2015) 012043 doi:10.1088/1742-6596/599/1/012043

an uncertainty of 10^{-10} or below. In this contribution we describe our experimental approach to reach this goal and our most recent results.

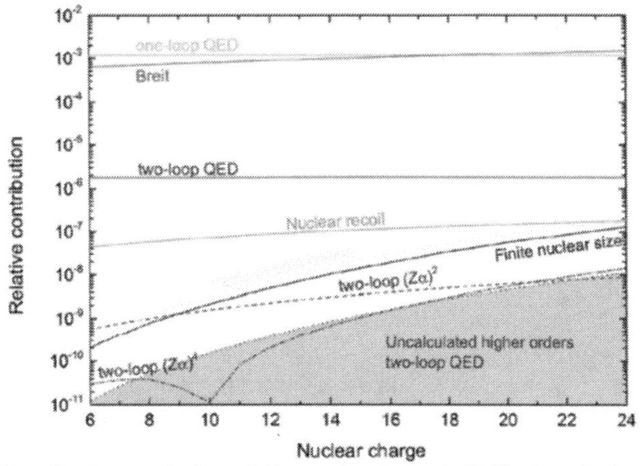

Figure 1: Contributions to the bound electron´s g factor in H-like ions for low and medium nuclear charges (courtesy Z. Harman)

2. Experiment

We use a cylindrical Penning trap [2] to confine a single hydrogenlike ion. Measuring the cyclotron frequency of the ion of mass M and charge q in the magnetic field B of the trap as well as the spin precession (Larmor) frequency with m the electrons mass, we obtain the g-factor of the electron through the ratio of these frequencies

$$g = 2\frac{\nu_L}{\nu_c}\frac{m}{M}\frac{q}{e} = 2\Gamma\frac{m}{M}\frac{q}{e} \quad . \qquad (3)$$

The masses of the electron and of the ion need to be known sufficiently well.

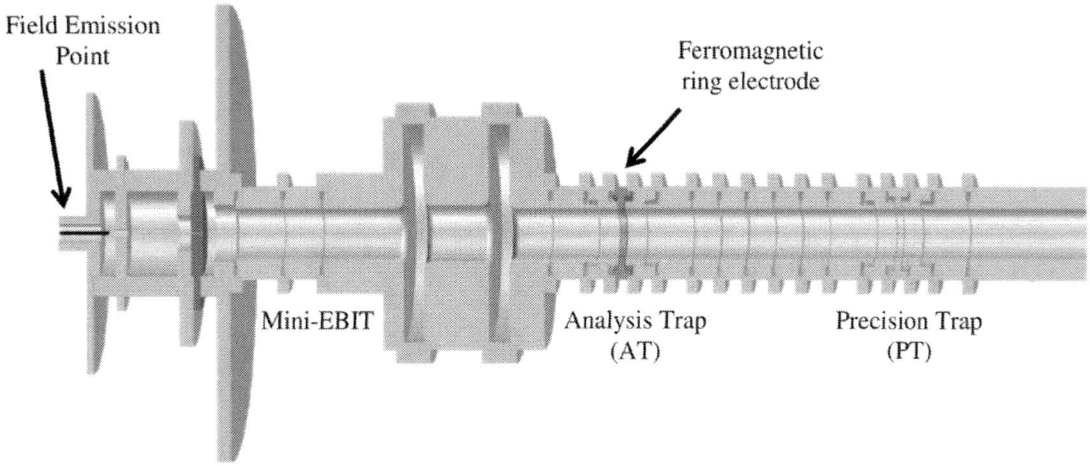

Figure 2: Penning trap arrangement including the Mini-EBIT for the in-situ creation of highly charged ions, the Analysis Trap (AT) for spin-state detection and the Precision Trap (PT) for the determination of the eigenfrequencies.

3. Apparatus

The core of our apparatus is a double Penning trap with an attached Mini-EBIT for highly charged ion creation (Fig. 2). Each trap consists of 5 electrodes of 3.5 mm inner radius made from gold plated OFHC copper: a central ring electrode, two endcaps, and two correction electrodes placed between endcaps and ring. A negative voltage applied to the ring electrode (nominally ~7V in our experiment) serves for ion confinement in the axial direction. Radial confinement is provided by a magnetic field of 3.7 T directed along the trap axis. Proper voltages on the correction electrodes make the axial potential harmonic. While one of the traps (precision trap, PT) is made as perfect as possible, in the second one (analysis trap, AT) the central copper ring is replaced by a ferromagnetic material which makes the magnetic field inhomogeneous. This is required, as shown below, for detection of the electron´s spin direction.

The trap tower is placed in a hermetically sealed vacuum vessel which is held at liquid helium temperature. Cryopumping serves for ultrahigh vacuum to avoid ion-neutral collisions. We achieved ion storage times of many months, from which an upper limit for the rest-gas pressure of $p<10^{-16}$ mbar can be deduced.

4. Ion creation and detection

Ions are released from a surface by electron ablation. They are stored in a simple 3-electrode trap arrangement and further ionized in an EBIT-like manner. The ion cloud is transferred to the PT, where unwanted species are removed by selective excitation of their oscillation and the ion number is reduced to the single particle by a careful reduction of the trapping potential. The ion motion is a superposition of three harmonic oscillations with frequencies $\omega_z/2\pi \approx 670$ kHz (axial oscillation), $\omega_+/2\pi \approx 27$ MHz (reduced cyclotron oscillation), and $\omega_-/2\pi \approx 9$ kHz (magnetron oscillation)

For single ion detection a high-*Q* resonator is attached to a correction electrode. The small (a few fA) image current which the ion induces by its axial oscillation in the trap is amplified by cryogenic electronics and appears as a peak signal in the Fourier transformed noise spectrum of the amplifier. When the ion´s axial oscillation is kept continuously in resonance with the detection circuit it will be damped by energy dissipation into the environment until thermal equilibrium at 4.2 K is reached. The ion signal is then a minimum in the noise spectrum as shown in Fig. 3

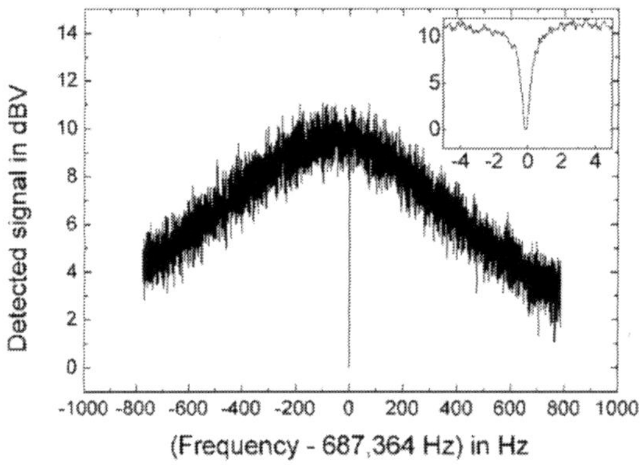

Figure 3: Fourier transform of the amplifier noise with a cooled single ion present in the trap, manifested by a minimum (dip) at the ion´s axial oscillation frequency. The insert shows the dip with high resolution.

5. Measurement of the ion's oscillation frequencies

Figure 3 indicates that the axial oscillation frequencies can be measured with very high resolution. The modified cyclotron and the magnetron oscillations can be measured by coupling them to the axial mode by an additional radio-frequency field at the frequencies $\omega_+ - \omega_z$ and $\omega_- + \omega_z$, respectively. In order to reach the required high precision of about 10^{-10} in the frequency measurement, an averaging time of several minutes is required. Magnetic field variations during this time may limit the precision. In order to reduce this time we developed a novel method measuring the phase of the radial motion, called PnA (Phase and Amplify) [3]: After ion cooling of the reduced cyclotron motion a fixed phase is imprinted on the motion by a short pulse at $\omega_+/2\pi$. Then a period of free oscillation follows. Finally the reduced cyclotron mode is coupled to the axial one by a strong pulse at $(\omega_+ + \omega_z)/2\pi$ which allows detection of the phase through the axial motion. Figure 3 shows the timing sequence. The PnA method reduced the time for the determination of the cyclotron frequency to a few seconds, which is essential to reach the required precision. The cyclotron frequency as required for the g-factor determination (see eqn (3)) is obtained by the so-called invariance relation[4] $\omega_c^2 = \omega_+^2 + \omega_z^2 + \omega_-^2$. Possible shifts of the eigenfrequencies caused by trap imperfections cancel in this relation to first order.

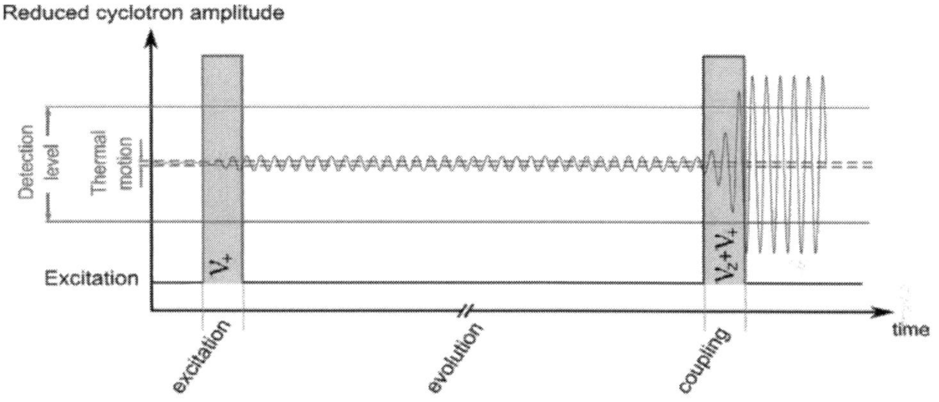

Figure 4: Timing sequence of the PnA method, depicting the low oscillation amplitude during the phase evolution, below the detection threshold of the tank circuit and the large oscillation following the parametric phase-sensitive amplification.

6. Measurement of the spin precession frequency

The axial oscillation frequency is the only access to the ion in our experiment. Therefore we have to couple the spin motion to the axial one. This is performed by the inhomogeneity of the magnetic field in the AT [5]. The force of the magnetic field on the magnetic moment associated with the electron's spin adds or subtracts to the electric trapping force, depending on the spin direction, thus changing the axial frequency. The size of this change is given by

$$\Delta\nu_z = \frac{g\,\mu_B\,B_2}{4\pi^2 M\,\nu_z} \quad . \qquad (4)$$

where B_2 is the quadratic coefficient in a series expansion of the inhomogeneous B-field. It amounts for our experimental conditions, depending on the ion's mass, to values between 140 and 600 mHz. Figure 5 shows that a spin flip, induced by a microwave field, can be unambiguously detected.

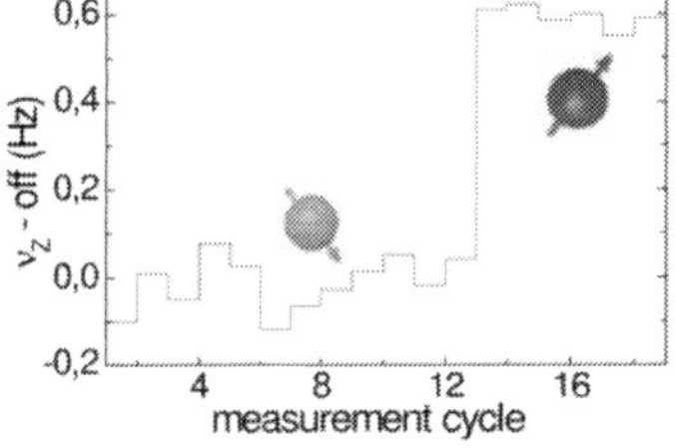

Figure 5: Change of the axial frequency upon an induced spin flip in a single $^{12}C^{5+}$ ion.

7. *g*-factor determination

The sequence to perform a *g*-factor determination starts by determination of the spin direction in the AT. The resistively cooled ion is adiabatically transferred to the PT, approximately 5 cm apart, where the eigenfrequencies are measured with high precision. Simultaneously, microwaves close to the spin precession frequency are injected into the trap, attempting to flip the spin. The ion is then transferred back to the PT, its spin direction is measured again in order to see whether or not it has changed in the PT. We plot the spin transition probability versus the scanned microwave frequency, divided by the simultaneously measured cyclotron frequency to obtain a resonance curve, from which the *g*-factor is obtained. Experiments have been performed on $^{12}C^{5+}$ [6], $^{16}O^{7+}$ [7], and $^{28}Si^{13+}$ [8]. The most precise value has been obtained for $^{28}Si^{13+}$: g = 1.995 348 958 7(5)(3)(8). The quoted uncertainties relate to the electron mass, statistical and systematical errors, respectively. The value agrees well with the theoretical prediction g = 1.995 348 958 0(17) [8], and represents to date the most precise test of QED in strong fields.

8. The electron's atomic mass

The experimental accuracy obtained in our experiments is independent of the ion under investigation. Therefore, we repeated the earlier experiment on $^{12}C^{5+}$ with now improved experimental precision. Moreover, the various contributions to the *g*-factor are for $^{12}C^{5+}$ significantly smaller than for $^{28}Si^{13+}$ and have been calculated to the 10^{-11} level. The mass of $^{12}C^{5+}$ shows only a small uncertainty due to the binding energies of the missing electrons, since the neutral carbon atom is the basis of the definition of the atomic mass scale. The agreement between theory and experiment in the case of $^{28}Si^{13+}$ lead to the conclusion that theory is correct at the level of our experiment. The theoretical uncertainty is dominated by the yet uncalculated QED terms of order $(Z\alpha)^5$ and higher, which in turn allows to even cancel the leading-order contribution by scaling the difference of experimentally measured and calculated *g*-factors for hydrogenlike $^{28}Si^{13+}$. Therefore, we can use experimental and theoretical results to obtain a new value of the atomic mass of the electron by rewriting eqn. (3):

$$m = \frac{g_{theo}}{2}\frac{\omega_c}{\omega_L}\frac{e}{q}M \qquad (5)$$

Our result of m = 0.000 548 579 909 067(16) u [9] improves the presently listed value in the CODATA tables of fundamental constants [10] by more than an order of magnitude.

9. *g*-factor of high-*Z* ions

For technical reasons the accessible hydrogenlike ions in our experiment are limited to Ca^{19+}. Since the QED contributions to the *g*-factor scale approximately with Z^2, higher-*Z* ions would allow more

Journal of Physics: Conference Series **599** (2015) 012043 doi:10.1088/1742-6596/599/1/012043

stringent tests of QED. To this end, a new setup at the Max-Planck Institute for Nuclear Physics in Heidelberg is currently in the commissioning phase which will allow injecting heavy highly-charged ions from an external source. While in the first phase these ions will be provided by the MPIK Super-EBIT, the system is also capable of accepting ions and possibly antiprotons from the HITRAP cooler trap at the future FLAIR facility.

10. The proton's atomic mass

Our method to use phase sensitive detection of the ions oscillation frequencies will be used to improve the present value of the proton´s atomic mass. Presently a significant contribution to our error budget is given by the fluctuations of the magnetic field strength during the time of the measurement. Its influence can be reduced by a measurement of the cyclotron frequency of ions simultaneously stored in different potential minima of a multi-electrode trap. We aim for an uncertainty in the proton´s mass below 10^{-11}.

11. Acknowledgements

We thank our colleagues from theory for useful discussions and our experimental collaborators, A. Kracke (MPIK Heidelberg), W. Quint (GSI Darmstadt), and G. Werth (Univ. Mainz). Financial support was provided by an Adv. Grant MEFUCO (#290870) and by the Helmholtz Alliance (HA216)

References

[1] D.A. Glazov et al. in: Springer Tracts in Modern Physics 256, 137 (2014)
[2] F.G. Major, V. Gheorghe, G. Werth, Charged Particle Traps, Springer, Heidelberg (2006)
[3] S. Sturm et al., Phys. Rev. Lett. 107, 143003 (2011)
[4] L.S. Brown, G. Gabrielse, Phys. Rev. A 25, 2423 (1982)
[5] N. Hermannspahn et al., Phys. Rev. Lett. 84, 427 (2000)
[6] H. Häffner et al., Phys. Rev. Lett. 85, 5308 (2000)
[7] J. Verdu et al., Phys. Rev. Lett. 92, 093002-1 (2004)
[8] S. Sturm et al., , Phys. Rev. Lett. 107, 023002 (2011)
[9] S. Sturm et al., Nature 506, 467 (2014)
[10] P.J. Mohr, B.N.Taylor, &D.B. Newell, CODATA recommended values of the fundamental physical constants: 2010. Rev. Mod. Phys. 84, 1527–1605 (2012)

AUTHOR INDEX

AUTHOR INDEX

Institute of Physics
Dirac House, Temple Back
Bristol BS1 6BE UK

CURRAN ASSOCIATES INC
proceedings
.com